T0319363

INTRODUCTION TO GROUND PENETRATING RADAR

INTRODUCTION TO GROUND PENETRATING RADAR

Inverse Scattering and Data Processing

Raffaele Persico

IEEE PRESS

WILEY

Copyright © 2014 by The Institute of Electrical and Electronics Engineers, Inc.

Published by John Wiley & Sons, Inc., Hoboken, New Jersey. All rights reserved.

Published simultaneously in Canada.

For general information on our other products and services or for technical support, please contact our Customer Care Department within the United States at (800) 762–2974, outside the United States at (317) 572-3993 or fax (317) 572-4002.

Wiley also publishes its books in a variety of electronic formats. Some content that appears in print may not be available in electronic formats. For more information about Wiley products, visit our web site at www.wiley.com.

Library of Congress Cataloging-in-Publication Data:

Persico, Raffaele, 1969–
  Introduction to ground penetrating radar: inverse scattering and data processing/Raffaele Persico.
    pages cm
  Includes bibliographical references and index.
  ISBN 978-1-118-30500-3 (hardback)
1. Ground penetrating radar.   I. Title.
  TK6592.G7P47 2014
  621.3848′5–dc23
                                                    2013039737

Printed in the United States of America.

10 9 8 7 6 5 4 3 2 1

To my wife Grazia Maria
and my daughter Luisa Anna

CONTENTS

FOREWORD

Ground penetrating (or probing) radar (GPR) is a vital technique on which the day-to-day safety of literally millions of people depend. The technology allows a very wide range of verifications, the most common being the safe and accurate location of the position of buried pipes and utilities, investigating the reinforcement and condition of roads, bridges, and airport runways, and identifying the structural integrity of buildings. Other important applications include locating buried potential hazards such as mine shafts and voids, investigating environmental and geological conditions (both of natural and man-made origin), studying glaciology, locating, identifying, and investigating archaeological sites, and uncovering forensic evidence including buried human remains and weapons.

Although the technology is widely used, it is a highly specialized area that requires a good understanding of the underlying science if it is to be applied successfully. In addition to the technical journals that regularly carry scientific papers on both the theory and application of GPR, there are two major biennial international conferences, namely the International GPR Conference and, in the intervening years, the International Workshop on Advanced GPR. In terms of books, the fundamental cornerstone of GPR in all its applications has long been David J. Daniels' *Ground Penetrating Radar*. However, one book, regardless of how well it is researched and written and how comprehensively it addresses its subject matter, cannot cover all aspects of the science to the equal satisfaction of all users. This current volume is not intended for the general reader or for anyone for whom this is a completely new subject. Rather it is aimed primarily at doctoral and post-doctoral students who wish to develop their understanding of the technology and, in particular, how the results may be developed and interpreted.

Dr. Persico has been a practitioner and researcher in GPR for many years, primarily interested in the resolution of inverse problems, with particular application to archaeological investigations and the conservation of cultural heritage. From his background in Physics and Mathematics and the expertise he has built up, he applies processing and interpretative techniques to GPR data collected with the primary aim of conserving, preserving, and rehabilitating buildings of historical importance and also archaeological remains. In 2010 he chaired the 13th International GPR Conference, held in Lecce, Italy. He is also an active member of the European GPR Association within which he has worked to build a virtual library, accessible to the membership through the Association website (www.eurogpr.org). His considerable expertise makes him an ideal candidate to share his knowledge and understanding of GPR interpretative techniques with other researchers and users.

This book consists of 15 chapters and 7 appendices, the aim being to introduce the reader to the complexities of using and interpreting GPR data step by step. An important feature of this book is the inclusion of questions at the end of almost all of the chapters, allowing the reader to assess his or her progress in understanding the subject. The answers to questions are in Appendix G.

Beginning with a general definition and description of GPR technology, this book goes on to consider an important basic characteristic of GPR operation, namely the inter-dependence of the nature of the survey medium with the transmission of the electromag-netic pulses through that medium. Chapter 3 considers the time and frequency aspects of GPR transmission and their implications. The following two chapters concentrate on the mathematical aspects of GPR transmission including Maxwell's equations, the effects of incident fields, and the relevant scattering equations considered in two dimensions for dielectric and magnetic materials (Chapters 4 and 5).

Continuing into methods of making the interpretation of GPR data more accessible, the next section describes a number of mathematically based constructs that can be used for this purpose. Chapter 6 introduces the reader to the nature of inverse scattering pro-blems and the associated mathematical uncertainties to be addressed in resolution. In Chapter 7, data processing steps typically associated with improving raw GPR data are described in detail and their effects are illustrated.

The Born approximation has become a standard algorithm to apply to GPR data and has also formed the basis of much of Dr. Persico's own research work. The Born approx-imation and its application to magnetic targets and also to weak and strong signal reflec-tors is described in full in Chapter 8. Leading on from this, the theoretical basis and application of diffraction tomography is the next topic, including: the consideration of horizontal and vertical resolution of targets; related sampling issues in space, frequency, and time; and the radiation characteristics of radar antennas (Chapter 9). Chapter 10 deals with two-dimensional migration algorithms in the frequency domain and the time domain.

Chapter 11, contributed by Drs. Lo Monte and Solimene together with Dr. Persico, extends data treatment into the development of three-dimensional scattering equations based on Maxwell's equations and includes particular consideration of Green's function applied to three-dimensional space. This is followed by extending the analysis of diffrac-tion tomography into three dimensions with careful consideration of the sampling para-meters required for the application of the technique in order to reconstruct the target(s) reliably from the GPR data (Chapter 12). This is important because improving target res-olution from appropriate survey parameters is of primary importance in all GPR applications.

Building on the previous chapters, Chapter 13 considers the corresponding deriva-tion of three-dimensional migration algorithms, first in the frequency domain and sub-sequently in the time domain before comparing two-dimensional and three-dimensional migration formulas in the time domain by means of a worked and illustrated example.

Chapter 14 considers the alternative technique of singular value decomposition. The same careful mathematical logic is used to derive the singular value decomposition before considering its application to GPR data.

Finally, Chapter 15 provides a variety of worked examples and exercises to illustrate some of the concepts covered in the preceding chapters. This begins with an examination of measuring propagation velocities followed by two sets of exercises dealing with target resolution, namely the interrelation of spatial sampling and horizontal resolution and the effects of frequency sampling on vertical resolution. Both of these latter two concepts are integral to understanding the capability of any GPR. The next set of exercises is concerned with trialing the number and categories of unknowns treated in the equations in order to optimize the quality of the GPR data without undue processing simply for the sake of it. This is followed by exercises examining frequency and spatial content of the data and consideration of the effects of measurement from above the soil–air interface (instead of directly coupled). Extending the area of investigation provides the basis for the next set of exercises, an important consideration given that this varies extensively from one survey to another. This is followed by exercises using background removal, the single most extensively applied processing technique for all GPR surveys since, by definition, it is the anomalous material that forms the targets. The added complication of complementary data sets in different orientations is then considered, based on a real archaeological example and including contributions from Drs. Ciminale, Leucci, and Matera. Lastly, the results of further 2D and 3D inversion techniques are compared (with the collaboration of Dr. Catapano).

Detailed mathematical workings in support of the content of the chapters are provided in full in the appendices.

This is not a volume for a beginner, but it is a careful and comprehensive enumeration and explanation of the mathematical concepts inherent to GPR. It should provide a useful platform for those who wish to delve deeper into the subsurface of the technology and equip themselves with the mathematical tools for handling their own data sets.

ERICA CARRICK UTSI
November 2013

ACKNOWLEDGMENTS

I sincerely thank the six contributors—Dr. Ilaria Catapano, Professor Marcello Ciminale, Dr. Giovanni Leucci, Dr. Lorenzo Lo Monte, Dr. Loredana Matera, and Dr. Raffaele Solimene—who helped me write this book. Their contributions were of the utmost importance to me, and the interaction with them has enhanced and deepened my knowledge of the covered topic. Moreover, I am grateful for the fruitful and interesting scientific discussions with Professor Massimiliano Pieraccini of the University of Florence and with his research group, with a special mention to Dr. Filippo Parrini and Dr. Devis Dei. These discussions inspired me to think about the calculation of the Hermitian images in a closed form, detailed in Chapter 3. I also would like to thank Dr. Jacopo Sala of 3d-Radar for our scientific collaboration, which in particular inspired me to write Section 15.7. I also would like to thank Dr. Erica Utsi of Utsi Electronics Ltd, Chair of the European GPR Association. She wrote the Foreword for this book, and this is an honor for me. Finally, I am grateful to Dr. Francesco Gabellone for his valuable aid in conceiving the cover for this book.

RAFFAELE PERSICO

ABOUT THE AUTHOR

Raffaele Persico was born in Napoli, Italy in 1969. After humanistic secondary school studies, he achieved his degree in Electronic Engineering from the University of Napoli Federico II and then his Ph.D. in Information Engineering from the Second University of Napoli. He has been Fellow Student at the Second University of Napoli, Research Scientist at the Consortium CO.RI.S.T.A., and then Research Scientist at the Institute for the Electromagnetic Sensing of the Environment IREA-CNR. Since 2007, he has been affiliated with the Institute for Archaeological and Monumental Heritage IBAM-CNR.

Dr. Persico's research activity has been devoted to microwave imaging, inverse problems, GPR data processing, and GPR systems. He has co-authored about 150 papers in international journals and conference proceedings and holds an Italian patent on the reconfigurability of the GPR systems. He is reviewer for several international journals and is Associate Editor of *Near Surface Geophysics*. He chaired the 13th International Conference on Ground Penetrating Radar, held in Lecce, Italy in 2010, and has devised a prototypal reconfigurable stepped-frequency GPR system together with the IDS Corporation and the University of Florence. Since 2009 he has been a Member of the EuroGPR Association.

CONTRIBUTORS

Ilaria Catapano, Institute for Electromagnetic Sensing of the Environment IREA-CNR, Italy

Marcello Ciminale, University of Bari Aldo Moro, Italy

Giovanni Leucci, Institute of Archeological Heritage–Monuments and Sites IBAM-CNR, Italy

Lorenzo Lo Monte, University of Dayton, USA

Loredana Matera, University of Bari Aldo Moro, Italy

Raffaele Solimene, Second University of Napoli, Italy

1

INTRODUCTION TO GPR PROSPECTING

1.1 WHAT IS A GPR?

Ground-penetrating radar (GPR), also known as surface-penetrating radar (SPR) (Daniels, 2004), is literally meant as a radar to look underground. Actually, it is used to look into both soils and walls and, recently, even beyond walls.[1]

In principle, the GPR can be viewed as composed by a central unit, a transmitting antenna, a receiving antenna, and a computer. The central unit generates an electromagnetic pulse or, more generally, an electromagnetic signal that is radiated into the soil by the transmitting antenna. The signal is radiated in all directions, but most energy is radiated within a conic volume under the antenna, as shown in Figure 1.1. When the electromagnetic waves meet any buried discontinuity (a buried object but also the interface between two geological layers, a cavity, a zone with different humidity, etc.), they are scattered in all directions according to some pattern depending on the buried scenario. Consequently, they are partially reflected also toward the receiving

---

[1] Actually the instruments that perform the so-called "through wall imaging" are not customarily considered GPR systems. However, we can say that conceptually they are at least a hybrid between a radar and a GPR.

Introduction to Ground Penetrating Radar: Inverse Scattering and Data Processing,
First Edition. Raffaele Persico.
© 2014 The Institute of Electrical and Electronics Engineers, Inc. Published 2014 by John Wiley & Sons, Inc.

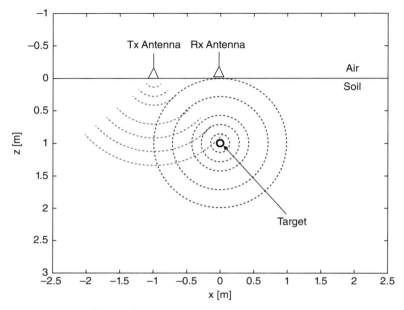

Figure 1.1. The working principle of a GPR.

antenna, again according to Figure 1.1. More precisely, Figure 1.1 is intended to be the central cut of a three-dimensional scenario.

Usually, both the transmitting and the receiving antennas are incorporated into a rigid structure[2] and move together. The gathered signal is customarily represented in real time on the screen of the computer[3] and is stored in the hard disk of the computer. It is implicit that the equipment of a GPR also includes suitable cables to connect the central unit, the antennas, and the computer, along with a device to provide energy in the field. The energy is usually supplied by rechargeable batteries in the form of a zero-frequency electrical voltage. The central unit transforms this energy into a signal in the microwave frequency range. Modern systems are often also equipped with a GPS, in order to geo-reference the probed areas.

In Figure 1.2, a photograph of a GPR is shown, and the main components are put into evidence. The trolley is facultative, but extremely useful for prospecting on the soil. Usually, the antennas are also equipped with an odometer that allows us to measure the covered distance.

---

[2] The couple of antennas is often enclosed in a unique box, and the whole box is improperly called the "GPR antenna."

[3] In the past, other recording systems were exploited as described in Daniels (2004) and in Jol (2009), because the GPR was invented much earlier than the laptop.

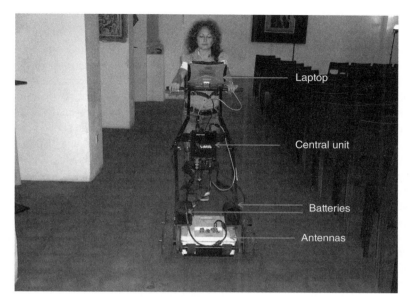

Figure 1.2. Photograph of a GPR (a Ris Hi Mode system) equipped with a double antenna at 200 and 600 MHz antenna.

The odometer is an important detail, because it allows us to compensate for the natural nonuniformity of the velocity of the human operator while driving the GPR: that is, it allows us to achieve a uniform sampling of the GPR signal along the observation line.

However, in some cases it is impossible to make use of the odometer—for example, because the prospecting is on a sandy area that hinders the rotation of the wheel. In these cases, periodical markers have to be recorded along the observation line, which is partitioned into segments of known length. The velocity of the instrument (and thus also the sampling) is considered constant within each segment but not along the entire observation line.

The working principle of a GPR is the same as that of a conventional radar. However, there are meaningful differences between the two instruments, in terms of technologies, exigencies, applications, and frequency bands (Daniels, 2004; Levanon, 1988). In particular, unlike the conventional radar, usually a GPR has to identify static targets, and in most cases the interpretation of the data is not requested in real time. On the other hand, in GPR prospecting the electromagnetic waves do not propagate in air but instead propagate in more complicated host media, customarily lossy and inhomogeneous, possibly dispersive, and in some cases anisotropic and/or magnetic (Daniels, 2004; Jol, 2009; Conyers, 2004). Last but not least, in GPR prospecting the characteristics of the host medium are usually not known a priori and have to be estimated from the data, as described in more depth in the next chapter.

1.2   GPR SYSTEMS AND GPR SIGNALS

There are essentially two kinds of GPR systems: the pulsed one and the stepped-frequency one. A pulsed GPR system radiates and receives the echoes to electromagnetic pulses. On the other hand, a stepped-frequency GPR decomposes the electromagnetic pulse into its spectral components and radiates them sequentially. Consequently, it radiates and receives trains of sinusoidal signals. The soil and the buried targets usually have a linear behavior with respect to the radiated GPR signal, in the sense that the signal scattered by the buried targets is a linear quantity (more details will be given in Chapter 6) with respect to the incident signal. Moreover, the soil can usually be considered a time-invariant medium within the time needed for the GPR measurement campaign. This makes the pulsed and the stepped-frequency GPRs theoretically equivalent. In practical terms, however, stepped-frequency systems are generally claimed more performing (Noon, 1996), even if the pulsed systems are quite more common and their technology has been assessed for a longer time. So, the debate about what kind of system is really the best one (or at least the most convenient one in dependence of the application) is still open. In this text we will not enter such a debate, which is mainly based on technological aspects, but will deal with both some analytical and practical aspects of the GPR prospecting in relationship with both systems. In particular, whatever the system, the GPR signal can be regarded as a function of the spatial point and of the time or the frequency indifferently, because of course we can Fourier transform pulsed GPR data in frequency domain and we can back-Fourier transform stepped-frequency GPR data in time domain.

Following a widely accepted terminology (Daniels, 2004), the GPR data relative to a single spatial point will be labeled as an A-scan or just a GPR trace, and the comprehensive set of GPR traces relative to an entire scanned line will be labeled as a B-scan. A B-scan corresponds to a matrix of numbers: N time (or frequency) samples times M spatial positions—that is, M traces each of which discretized into N samples. This is equivalent to assuming that the GPR "stops" in each A-scan position, gathers the data in that position, and goes on to the next position. Actually, in most cases the data are gathered in continuous mode—that is, the GPR gathers the data while moving—but the model "stop-gather-and-go-on" is in most cases acceptable because of the huge difference between the velocity of propagation of the electromagnetic signal and the velocity of the human operator, even if the time required to store an A-scan is actually quite longer than its formal time bottom scale. This happens because of several reasons such as sequential sampling (for pulsed systems), integration time of the harmonics (for stepped frequency systems), and stacking (for both). Here, we will not focus on these aspects, which are mainly technological and already explained elsewhere (Noon, 1996; Daniels, 2004; Jol, 2009). Let us just restrict ourselves to say that a nonexcessive (the quantification is case-dependent) and constant velocity during the data acquisition is always a good rule of thumb. The comprehensive set of GPR data relative to a series of parallel B-scans is usually labeled as a C-scan. In general, what is immediately visualized in the field is a B-scan in some color or gray tone scale. These data, usually called raw data, can allow us to identify targets of interest, but in general the image and its interpretation can improve meaningfully after a suitable processing.

1.3   GPR APPLICATION FIELDS

A meaningful overview about the GPR applications is beyond the purposes of this book and is not its goal. Notwithstanding, for sake of self-consistency, a brief outlook is provided.

Within the field of the archaeology (Conyers, 2004), GPR can allow us to identify the areas with alleged interesting buried remains, which in turn allows us to avoid an exhaustive and expensive (sometimes too expensive) excavation. Another issue of interest is the field of the preventive archaeology—that is, the preventive prospecting of areas where something is going to be built (a road, a building, an underground station, etc.). This mitigates the risk of destroying archaeological sites and also mitigates the economic risk that the works will be stopped by some Cultural Heritage Institution.

Monitoring of monuments as historical buildings, statues (Sambuelli et al., 2011), ancient fountains, historical bridges (Solla et al., 2011), and so on, is another subject of interest. In particular, GPR monitoring (possibly integrated with other geophysical investigations) can give information about the state of preservation of the monuments and can provide useful information in order to address a restoration project properly. In some cases, information of historical interest can also be achieved—for example, about the presence of walled rooms, crypts, hypogeum rooms, tombs, hidden frescoes, and so on (Pieraccini et al., 2006; Grasso et al., 2011).

GPR prospecting is also exploited in civil engineering (Grandjean et al., 2000; Utsi and Utsi, 2004). In particular, it can be used to identify structural damages and to investigate about hidden structures like sewers or water and gas pipes, whose presence is in many cases not precisely documented.

Demining is another important application. In particular, modern mines are customarily built with plastic materials with only little or even no metallic parts. Therefore, they are often hardly visible or completely invisible to a metal detector. Moreover, a metal detector is not able to provide all the details possibly available from a GPR system, namely the position (in particular the depth), the size, and (among certain limits) the shape of the buried target. Demining has been dealt with for years within the GPR community (Groenenboom and Yarovoy, 2002), and it has also been successfully performed many times (Sato and Takahashi, 2009).

GPR prospecting is also exploited for asphalt monitoring.[4] In particular, it is possible to identify subsidences or damaging before they become worse or even dangerous for drivers and pedestrians. These problems are even more pressing in areas where the roads frost in the winter and thaw in the spring (Hugenschmidt et al., 1998; Villain et al. 2010).

In several application fields, it can be particularly useful to make use of advanced GPR systems equipped with a large array of antennas (Sala and Linford, 2010; Böniger and Tronicke, 2010). These systems can gather simultaneously several (up to 14 and more) measurement lines with a unique *going through*. On the other hand, these systems need a quite flat scenario to provide good performances, because the arrays are rigid and possibly quite large (up to 2 m and larger).

---

[4] In this case, the antennas are usually mounted on a car.

GPR prospecting is also applied with regard to mines and pits. In particular, it can help to identify shallow veins of the mineral of interest (Ralston and Hainsworth, 2000; Francke, 2010) and can even help with regard to some safety issues. In fact, fractures, water infiltrations, or just obsolescence can badly affect the stability of the structure, in mines as well as in tunnels (Grodner, 2001; Cardarelli et al., 2003). In some cases, even explosive gases trapped in natural cruets can be met while digging, especially in coal mines (Cook, 1975).

Another application of interest is GPR prospecting on the ice (Arcone et al., 2005). In particular, polar ice contains information about the geological history of our planet and can also provide information about the occurring climatic and environmental changes. GPR prospecting can be successfully performed on both fresh and salty ice.

GPR prospecting on fresh water is a field of interest too—for example, for sedimentology applications in relationship with the bottom of lakes, ponds, or rivers[5] (Smith and Jol, 1997). Liquid seawater, instead, is customarily too lossy to allow a reliable GPR prospecting.

Industrial agriculture is another applicative field where it is of interest to devise an intelligent exploitation and distribution of the water (Friedman, 2005). In this framework, GPR can be a useful tool for the evaluation of the electromagnetic characteristic of the shallower layers of the soil, and in particular its dielectric permittivity. Some semiempirical relationships (Topp et al., 1980) can in some cases allow us to estimate the water content from the dielectric permittivity.

Let us also mention the subject of the GPR investigation on Mars,[6] where unmanned vehicles are gathering data, mainly looking for water and, consequently, the possible (current or past) presence of life (Picardi et al., 2005).

Finally, let us also prompt (a) forensic applications, where, for example, buried corpses or hidden weapons are looked for (Hammon et al., 2000), and (b) borehole prospecting (Ebihara et al., 2000), where antennas are lowered in one (reflection mode) or two (transmission mode) carrot-holes.

1.4 MEASUREMENT CONFIGURATIONS, BANDS, AND POLARIZATIONS

GPR data are mostly taken in reflection mode and, if possible, the antennas are preferably in contact with or at very short distance from the structure to be probed. However, in some cases the data are necessarily moved in a contactless configuration—as, for example, in demining (Sato and Takahashi, 2009) and asphalt monitoring (Hugenschmidt et al., 1998) or also for the monitoring of works of art that cannot be touched (Pieraccini et al., 2006).

In most cases, the antennas are rigidly placed in a unique box and move together, but in some cases the two antennas can be moved separately from each other. This is a

---

[5] In this case, the antennas are usually mounted on a boat.
[6] In the past also GPR data from the Moon have been analyzed.

valuable resource, especially in order to measure the electromagnetic characteristics of the soil by gathering common midpoint (CMP), wide angle reflection and refraction (WARR), or trans-illumination data (Davis and Annan, 1989; Daniels, 2004; Conyers, 2004; Jol, 2009). Actually, sometimes the WARR configuration (also called multistatic multiview) is exploited not only to measure the characteristics of the soil but also to improve the image of the buried scenario. However, this is unpractical on a large scale, and the improvement achievable on the image is usually quite marginal, because the information achievable from the spatial diversity is not independent from that achievable from the frequency diversity (Persico et al., 2005; Persico, 2006).

Customarily, the electromagnetic characteristics of the soil and/or of the buried targets also depend on the frequency. This is expressed by saying that the soil (more in general the propagation medium) and/or the targets are dispersive. Several dispersion laws are known (Lambot et al., 2004), but it might be not easy to establish in the field what is the most suitable dispersion law for the application at hand. Therefore, as a matter of fact, in most cases the dispersion is not accounted for in the data processing. More precisely, we can say that the dispersion phenomenon is more often considered if the purpose is to characterize the propagation medium in itself, and more rarely if the purpose is to focus the targets embedded in it.

In general, the needed band of frequencies depends on the particular application. Customarily, as is well known, lower frequencies penetrate the opaque structures better than higher frequencies but provide a worse image of the targets. This drives us to use low frequencies (below 200 MHz) if the required investigation depth overcomes 5–7 m or more (e.g., in some geological applications), radio frequencies (200–700 MHz) for applications where the depth to reach is of the order of 3 m (e.g., in most archaeological prospecting), higher radio frequencies (700–3000 MHz) for applications where the maximum required depth of investigation is of the order of 1 m (e.g., detections of fractures or asphalt monitoring), and sometime even higher microwave frequencies if the maximum investigated depth can be limited to the order of 50 cm (e.g., demining or determination of the water content in the shallower layers of the soil). This classification is sketchy: It just indicates an average distribution, and many exceptions might be found. In particular, it refers to GPR application in "temperate" soils: The ice constitutes an exception and can allow a much deeper penetration of the GPR signal. In general, the maximum penetration depth depends on the current case history and can be estimated in the field, on the basis of the data.

Several kinds of antennas are exploited in GPR prospecting. For the low-frequency cases, below 200 MHz, the antennas are customarily unshielded loaded dipoles, often quite long (depending on the central frequency, they can be up to 3 m long and even longer). The fact that the antennas are unshielded makes them gather reflections from targets in air too, and it makes the results more vulnerable to the electromagnetic interferences from external sources. On the other hand, a shield would make them quite weighty. Instead, beyond 200 MHz the GPR antennas are customarily shielded. In the range 200–1000 MHz, the most widely exploited antennas are probably the bow ties (Lestari et al., 2004), that are linearly polarized and customarily are fed with a coaxial cable. Sometimes a circular polarization can help for the discrimination of some targets. In these cases, large band spiral antennas (Daniels, 2004) can be used, even if their use is more rare. At higher frequencies, Vivaldi antennas and horns (Gentili and Spagnolini, 2000; Pieraccini et al., 2006) can be exploited too. They are linearly polarized and can be fed by a waveguide, which makes them more robust and suitable for high-frequency applications (Stutzman and Thiele, 1998). In some cases, the GPR system is equipped with an array rather than just a pair of antennas (Sala and Linford, 2010; Böniger and Tronicke, 2010). In these cases the single elements of the arrays are usually dipole-like antennas.

1.5   GPR DATA PROCESSING

The processing of GPR data is a large topic, and in particular the current application can require or at least can make more suitable some strategy with respect to some other. In particular, two fundamental categories of processing can be distinguished, namely deconvolution-based (Jol, 2009) processing and SAR effect-based processing. In deconvolution-based algorithms (also called 1D), one essentially processes the single GPR traces trying to retrieve the shape of the radiated pulse—that is, trying to equalize the distortion that the radiated pulse suffers because of the propagation in a dispersive inhomogeneous medium and because of the scattering from the target. This kind of processing is important especially in cases when the targets looked for have a foreseeable "signature" and tend to distort the impinging pulse in a known way. Examples of deconvolution-based processing in relationship with demining problems are probably the most common ones: In particular, usually the main trouble in this case is not to identify the mine (even if the difference between the dielectric characteristics of a plastic mine and those of the surrounding soil might be low), but rather to reduce the false alarm probability—this is, the probability to confuse the mine with any other target characterized by the same order of size and average depth (Timofei and Sato, 2004). In such a situation, a deconvolution can help in discriminating the nature of the reflecting target. The second category, which here will be labeled as SAR effect-based [the acronym stands for synthetic aperture radar, (Daniels, 2004)], is concerned with a processing that regards all of the traces within a B-scan or a C-scan and is aimed to focus within a vertical plane (2D models) or in a buried volume (3D models) the targets embedded in the host medium at hand. Within these SAR processing, it is then possible to distinguish a plethora of models and related algorithm, based on different kinds of approximation of the scattered field. In particular, there are linear algorithms based on the Born approximation (Chew, 1995), on the Rytov approximation (Devaney, 1981), on the extended Born approximation (Torres-Verdin and Habashy, 2001), on the Kirchhoff approximation (Liseno et al., 2004), and so on. Moreover, there are nonlinear approximations as the second-order Born approximation (Leone et al., 2003) or iterative algorithms that update up to convergence (according with some Cauchy-like criterion) the result of a single-step processing. Customarily, in these cases the single-step processing is linear (Moghaddam and Chew, 1992), but the comprehensive algorithm is nonlinear. There are also fully nonlinear approaches, based (for example) on the statistical minimization of some cost functional (Caorsi et al., 1991). Finally, let us list also the linear sampling method (Colton et al., 2003), a fast nonlinear inversion algorithm based on single-frequency multiview data that has been becoming popular in recent years.

Actually, most of these 2D and 3D models have been developed in a context wider than that of GPR data processing, which is the literature on microwave inverse scattering. Notwithstanding, they can be applied (in several cases they have been already applied) to the reconstruction of targets embedded in the soil or in masonry, which we can classify as a GPR application. Their utility and the trade-offs between them are related to the many possible specific applications.

In this text, we will focus only on GPR data processing based on the Born approximation. In particular, the core of the processing dealt with here will be the migration (Stolt, 1978; Schneider, 1978; Yilmaz, 1987) and the linear Born model-based inversion (Colton and Kress, 1992) algorithms, both of which considered either in a 2D or in a 3D framework. Let us also specify that, commonly, the 3D processing is meant as the suitable joining of several 2D results achieved from several B-scans. This is useful and practical, especially (but not only) in order to image horizontal buried layers where the plan of built structures can be identified (Conyers, 2004). However, rigorously this is not a 3D processing method. We will label it a pseudo-3D approach, in order to distinguish this method from a real 3D approach, that will be dealt within Chapters 11–13. A 3D approach is theoretically more refined than a pseudo-3D one, but it is also more difficult and computationally more demanding.

Let us outline, however, that, even if only these linear data processing will be dealt with, the complete scattering equations are derived, so that the intrinsic nonlinear nature of the scattering phenomenon is shown.

Several reasons underlie the choice to limit our discussion to linear Born model-based processing. First, an adequate discussion of all the listed techniques would require a book quite long (at least four times the size of the current book). Second, in the common GPR praxis the most exploited focusing algorithm is undoubtedly the migration, also because there are several commercial codes able to implement it. Third, we have preferred to give space to some extra topics that are not inverse scattering issues but, in the real world, are inseparable aspects of GPR data processing, namely the indirect measure of the characteristics of the embedding medium (that in general are not known a priori), the extraction of the scattered field data from the total field data, and some aspects specifically related to the gathering of the data either with a pulsed- or a stepped-frequency GPR system.

Let us outline that the GPR processing is not constituted by a mere focusing. For example, data filtering and gain variable versus the depth are often very important passages. It is also worth outlining that some practice in the field is unavoidably essential (better if the starting phase is performed with the assistance of a more skilled user): there is no book that can replace it.

The main aim of this text is to focus on some of the theoretical aspects that seem to the author particularly important for an "aware" execution of a GPR measurement campaign, followed by a proper processing and, when possible, a reasonable interpretation.

2

CHARACTERIZATION OF THE HOST MEDIUM

2.1 THE CHARACTERISTICS OF THE HOST MEDIUM

For a correct interpretation of the GPR signal, it is important to have some estimation of the electromagnetic characteristics of the background medium. A complete characterization theoretically means a measure of the dielectric permittivity and of the magnetic permeability, both meant as complex quantities to account for losses (in particular, in this way the dielectric permittivity can account for the electric conductivity too) and variables versus the frequency. These quantities are also functions of the buried point because in general the medium at hand is not homogeneous; and possibly they are tensor quantities instead of scalar ones, because the medium might be anisotropic (Slob et al., 2010). A complete characterization of the soil, therefore, is not an easy task. Fortunately, however, a complete characterization is usually not needed, and in many cases an average value of the propagation velocity of the electromagnetic waves in the soil is sufficient in order to focus and interpret satisfyingly the buried targets.

In general, the propagation medium is lossy too. However, in many cases it is a low-lossy medium; that is, the real part of the wavenumber is much larger (one order of magnitude) than the imaginary part. In these conditions, we can neglect the influence

Introduction to Ground Penetrating Radar: Inverse Scattering and Data Processing,
First Edition. Raffaele Persico.

of the losses on the propagation velocity and the dispersion that they arise (Daniels, 2004; Jol, 2009). This means that we can assume that (a) the GPR waves propagate at a velocity that can be evaluated neglecting the losses and (b) they don't get deformed while propagating, even if they attenuate exponentially versus the distance.

Most propagation media encountered in practical cases do not show magnetic properties. However, some exception can occur—for examples, in the case of a strongly polluted soil (Nabighian, 1987), in the case of some Martian soil (Stillman and Olhoeft, 2004), or in presence of some particular mineral, containing iron (Nabighian, 1987; Jol, 2009).

Under the hypothesis of a homogeneous, isotropic, nonmagnetic and low-loss medium, the propagation velocity of the electromagnetic waves, c, is related to the relative dielectric permittivity of the medium, ε_s (real), by means of the relationship $c = c_0/\sqrt{\varepsilon_{sr}}$ (Stutzman and Thiele, 1998), where c_0 is the propagation velocity of the electromagnetic waves in free space, about equal to 3×10^8 m/s. ε_{sr} is the relative permittivity of the soil, whereas the complex (absolute) dielectric permittivity of the soil is given by $\varepsilon_{eq} = \varepsilon_0 \varepsilon_{sr} - j\varepsilon_{sim} - j(\sigma/\omega)$, where ε_0 is the dielectric permittivity of the free space, equal to 8.854×10^{-12} Farad/m, ε_{sim} is the imaginary part of the dielectric permittivity (which accounts for dielectric losses), σ is the electric conductivity of the background medium (which accounts for conduction losses), and ω is the circular frequency. Actually, it is not easy to distinguish the nature of the losses experimentally, and so the two terms that compose the imaginary part of the dielectric permittivity can be also expressed by either an equivalent conductivity or an equivalent imaginary permittivity.

For many materials, expected values of the relative dielectric permittivity are presented in tabular form (Daniels, 2004; Jol, 2009; Conyers, 2004). However, it is customarily better to measure the propagation velocity of the waves in the background medium at hand from the data, because the current values of the soil characteristics depend on several environmental parameters, which usually are not known a priori (such as the water content, the compactness of the soil, the presence of mineral salts, and possibly even the temperature). Notwithstanding, some awareness of the average values found in tables can be useful in order to check the likelihood of the values retrieved in the field.

There are also tables regarding the exponential attenuation rate of the signal in several media (which is essentially related to the equivalent conductivity), but usually the range of possible values is much larger than that achievable for the relative permittivity.

2.2 THE MEASURE OF THE PROPAGATION VELOCITY IN A MASONRY

The electromagnetic characteristics of the propagation medium depends on its chemical composition, its water content, its porosity, its mineralogy, and possibly its temperature. These dependences are in general a complicated matter, as described (for example) in Daniels (2004); and Jol (2009). Consequently, in most cases one does not have at one's disposal all the instruments and competencies to retrieve the electromagnetic characteristics of the soil starting from a microscopic approach. Therefore, we will focus the attention on the subject of the measure of these characteristics from GPR (or at most TDR) data.

In order to measure the propagation velocity of the electromagnetic waves in the background medium, the first possibility is to make use of a cooperative target—that is, a known object placed on purpose.

This is easily performed in the case of a masonry or a pillar, or even a column.[1] In particular, in these cases a metallic sheet or a metallic bar[2] can be placed and removed at the opposite site of the structure, gathering a GPR scan for each of the two cases, as depicted in Figure 2.1. The average propagation velocity can be measured from the simple relationship

$$c = \frac{2d}{t} \tag{2.1}$$

where t is the return time of the echo corresponding to the opposite side of the structure and d is the depth of the reflector (i.e., the thickness of the structure). The factor 2 is due to the round-trip of the signal from the antenna to the reflector and vice versa. The time t is identified by comparing the two GPR traces gathered with and without the metal target behind the structure. Let us outline that, in many cases, the second interface of the masonry is visible as a flat anomaly from the GPR data, even without any metallic marker.

At any rate, the marker can be helpful because the GPR image might show several flat anomalies, possibly due to internal layers within the wall, ringing of the antennas (Conyers, 2004), multiple reflections (Daniels, 2004), or possibly some large obstacle beyond the probed masonry.

In such a case, the cooperative target allows us to identify which one among the visible layers is specifically due to the opposite side of the wall. In many cases, the thickness d can be measured directly with a tape or (more rarely) from a reliable scaled map of the building. The measure of the propagation velocity from the return time and the thickness of the structure can be also applied to pillars and columns (Masini et al., 2010).

However, there are cases where the direct measure of the thickness of the masonry is not immediate, because (for example) the masonry is long or it ends in contact with another orthogonal wall, or possibly its thickness is not uniform so that the two air–masonry interfaces are not parallel to each other (this can happen, especially in ancient buildings). In these cases, it might be more reliable to work out the propagation velocity of the signal in the masonry from the behavior of the gathered GPR signal, with the marker placed behind the wall. This can be done in the same way as described in the following section with regard to the case of a homogeneous soil.

There are also cases where the masonry is quite thick and/or lossy, so that that the signal from the far interface is hardly (or not at all) perceivable. A marker might be useful even in some of these cases. In particular, without the check of a marker we might misinterpret some interface internal to the masonry as being the opposite interface. In other cases, the metallic marker might make stronger the reflection from the far interface

---

[1] A masonry, if homogeneous, can be modeled as a homogeneous half-space for intra-wall imaging purposes. This is not mathematically rigorous but works for any layered medium if the targets of interest are embedded in the first (shallower) layer.

[2] The metallic sheet provides a stronger returned signal with respect to the bar, but a bar allows a slightly more refined analysis of the returned signal, as will be shown later.

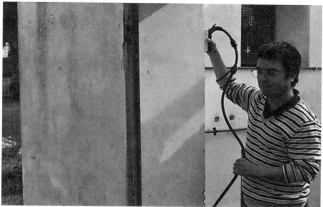

<u>Figure 2.1.</u> **Upper panel**: A metallic road is placed horizontally on the other side of the prospected wall. **Lower panel**: The metallic road is removed. The antennas have been moved in both cases downward along a vertical line.

so that it becomes more clearly perceivable as a result. In this last case, a metallic sheet is likely to work better than just a rod.

2.3 THE MEASURE OF THE PROPAGATION VELOCITY IN A HOMOGENEOUS SOIL

2.3.1 Interfacial Data in Common Offset Mode with a Null Offset: The Case of a Point-like Target

The measure of the propagation velocity of the electromagnetic waves in a homogeneous soil might be done, in principle, with a buried marker, similarly to what is described for the case of a wall. However, the excavation needed to put a metallic marker at a known depth would modify the compactness of the soil in that zone, and this would modify the electromagnetic characteristics of the soil in that point (Conyers, 2004; Soldovieri et al., 2009).

For the same reason, also laboratory analyses (Afsar et al., 1986) of soil or rocks samples require a particular attention: The samples should reach the laboratory with the same compactness and water content they had in the field; otherwise the result of the measure is not reliable. Sometimes a ground truthing (i.e., a limited set of localized excavations on some specific targets identified from the data) can be used to calibrate the propagation velocity (Conyers, 2004). These are useful when possible, but in some cases might be time-consuming, unauthorized, or not well-advised. So, in many cases the propagation velocity has to be estimated noninvasively from the same GPR data. The most common methods to do this is based on the diffraction curves, more commonly called diffraction hyperbolas.

To expose the method of the diffraction curves, let us start considering a point-like target—that is, a buried object small with respect to the minimum involved wavelength in the soil. In particular, let us refer to the scheme in Figure 2.2.

The transmitting antenna illuminates the target not only when it passes exactly over it, but also from a certain distance before it and up to a certain distance beyond it. So, the receiving antenna gathers an echo from the target not only when it crosses over

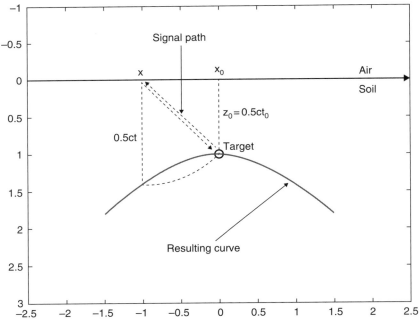

Figure 2.2. Measure of the permittivity of the soil from a diffraction hyperbola from a point-like target. The hyperbola is represented after time depth conversion (the units are arbitrary).

it, but usually within a segment centered on the target. The length of this segment depends on directivity of the antennas, the characteristics of the soil, and the depth of the target.

If the antennas are moved at the air–soil interface, and the offset between them is neglected, then the distance from the position of the source-observation point $(x,0)$ and the buried target in the position (x_0, z_0) is given by $\sqrt{(x-x_0)^2 + z_0^2}$. This quantity corresponds to the apparent depth at the abscissa x, which in terms of propagation velocity is given by $z(x) = ct/2$, where $t = t(x)$ the return time of the echo recorded at the position x. In the point $x = x_0$, the apparent depth of the target reaches its minimum value $z = z_0 = ct_0/2$, where t_0 is the minimum recorded return time. Putting together these equalities, we have

$$\frac{ct}{2} = \sqrt{(x-x_0)^2 + \left(\frac{ct_0}{2}\right)^2} \Leftrightarrow t = \frac{2}{c}\sqrt{(x-x_0)^2 + \left(\frac{ct_0}{2}\right)^2} \Rightarrow \frac{t^2}{t_0^2} - \frac{(x-x_0)^2}{0.25c^2t_0^2} = 1 \qquad (2.2)$$

Relationship (2.2) is the equation of a hyperbola (but of course only one of the two branches is considered) with the vertex at the target position (x_0, t_0).

Equation (2.2) is parametric with respect to the propagation velocity c, and so c can be estimated from Eq. (2.2). In principle, this might be done even just from two points, but of course a more extended fitting provides a more reliable result. There are commercial codes that allow us to do this fitting in an immediate graphical way [e.g., the GPRslices (Goodman and Piro, 2013) and the Reflexw (Sandmeier, 2003)] by depicting the model diffraction hyperbola at variance of a trial c on the data.

As is well known, by replacing 1 with 0 in the second term of Eq. (2.2) we achieve the equation of the asymptotes of the hyperbola, that is,

$$t = \pm\frac{2}{c}(x-x_0) \qquad (2.3)$$

Equation (2.3) shows that the asymptotes do not depend on the time depth of the target, and the two lines cross each other at the point $(x_0, 0)$—that is, over the point target and at the air–soil interface. Theoretically, the propagation velocity might be estimated just from the slant of the asymptotes, that is, $\pm 2/c$. This evaluation, however, is rarely possible, because the visible portion of the hyperbola is usually too short to allow a reliable identification of the asymptotes. However, the tangent line to the "tail" of the hyperbola—that is, the tangent to the hyperbola in a point far from the vertex—can also provide an estimation of the propagation velocity. In fact, from Eq. (2.2), the slant of the tangent to the diffraction hyperbola is given by

$$\frac{dt}{dx} = \frac{2}{c}\frac{(x-x_0)}{\sqrt{(x-x_0)^2 + \left(\frac{ct_0}{2}\right)^2}} \qquad (2.4)$$

The limit value of the expression (2.4) for $|x| \gg |x_0|$ is $\pm 2/c$, where sign depends on whether we consider positive or negative abscissas. Of course, this means that the tangent to the diffraction hyperbola becomes progressively more parallel to the asymptote. Equation (2.4) also allows us to quantify the closeness between the slants of these two lines. In particular, if $|x-x_0| \geq 3\,(ct_0/2)$, it is easy to recognize that to confound the tangent with the asymptote allows an evaluation of the propagation velocity with an error of about 5.5%.

The shape of the diffraction hyperbola depends also on the time depth of the target. In particular, for very shallow point targets ($t_0 \rightarrow 0$), the diffraction hyperbola tends to coincide with its asymptotes. This is the reason why shallow and small targets appear as little roofs in the raw GPR.

In some cases, it can be of interest to visualize the diffraction hyperbola in the (x,z) plane rather than in the (x,t) plane. This is trivial after evaluating the propagation velocity. In this case, the diffraction hyperbola becomes equilateral; that is, the asymptotes are orthogonal to each other. In fact, Eq. (2.2) in the plane (x,z) is just rewritten as

$$z = \sqrt{(x-x_0)^2 + z_0^2} \Rightarrow \frac{z^2}{z_0^2} - \frac{(x-x_0)^2}{z_0^2} = 1 \qquad (2.5)$$

and consequently the equation of the asymptotes is

$$z = \pm (x-x_0) \qquad (2.6)$$

whose slants with respect to the x-axis are 45 and -45 degrees, respectively. This means that the actual propagation velocity can be also viewed as that value that makes equilateral the diffraction hyperbola in the plane (x,z).

Let us note that because the real scenario is three-dimensional and the position of the buried target is not known a priori, the observation line may not pass exactly over the point-like object. In this case, the minimal distance z_0 between the target and the antennas is no longer equal to the depth of the target, and the scenario in Figure 2.2 should be tilted around the measurement line. However, the evaluation of the permittivity is not affected by this possibility, being essentially based on the distance–time correspondence and on Pythagoras' theorem.

In the performed evaluation, some approximations and/or assumptions are implicit. A first approximation is the use of an essentially optical model of the propagation. Actually, the propagation of the waves is not described just by means of rays: The physics of the phenomenon is more complicated and, as will be shown, can be fully described only by making use of Maxwell's equations.

The hypothesis that the target is point-like can be a strong assumption. Practically, the non-null size of the target tends to make larger the diffraction curve, thereby resulting in an underestimation of the propagation velocity. Consequently, in a real scenario, one should heuristically choose the narrower and "cleaner" hyperbolas among those visible at any given time depth: They are those most likely to be ascribable to electrically small targets.

2.3.2 Interfacial Data in Common Offset Mode with a Null Offset: The Case of a Circular Target

Circular targets of any size also provide a hyperbolic signature in the data (Li et al., 2012). In fact, with reference to Figure 2.3, let us consider a buried circular target with center C and radius R. In Figure 2.3, A is the source-observation point and E is the interfacial point at minimum distance from the target.

Thus, with reference to Figure 2.3 and making use of the same symbols exploited in the case of a point-like target (plus the radius R of the buried circular target), we have

$$AE^2 + EC^2 = AC^2 \Rightarrow \left(AE^2 + (ED + DC)^2 \right) = (AB + BC)^2$$

$$\Rightarrow (x-x_0)^2 + \left(\frac{ct_0}{2} + R \right)^2 = \left(\frac{ct}{2} + R \right)^2$$

$$\Rightarrow \frac{\left(t + \dfrac{2R}{c} \right)^2}{\left(t_0 + \dfrac{2R}{c} \right)^2} - \frac{(x-x_0)^2}{\left(\dfrac{ct_0}{2} + R \right)^2} = 1 \qquad (2.7)$$

Equation (2.7) describes the diffraction curve relative to a circular target. One can immediately recognize that this diffraction curve is still a hyperbola and reduces to

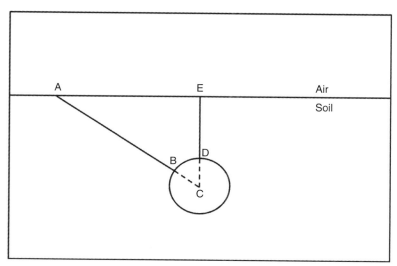

Figure 2.3. Measure of the permittivity of the soil from a diffraction hyperbola from a circular target.

the hyperbola of Eq. (2.2) for $R \rightarrow 0$. The vertex of the hyperbola is in the point (x_0, t_0), and the minimum time depth t_0 refers to the top of the target.

Replacing 1 with 0 in the last form of Eq. (2.7), we work out the equation of the asymptotes, which in this case is

$$t = -\frac{2R}{c} \pm \frac{2}{c}(x - x_0) \tag{2.8}$$

Equation (2.8) shows that the slant of the asymptotes of the hyperbola remains unchanged with respect to the case of point targets [see Eq. (2.3)], but they are rigidly translated so that they cross each other "in air" at the negative time $t = -2R/c$. This proves that it is not possible to find two different couples (R_1, c_1) and (R_2, c_2) that lead to the same diffraction hyperbola. In fact, in this case the slant of the asymptotes should be the same, so that necessarily $c_1 = c_2$, and then the intercept point between the asymptotes should be the same too, which leads to $R_1 = R_2$. Consequently, the problem of retrieving of the couple (R, c) from the data by means of a least square matching has a unique solution. Notwithstanding, usually the portion of visible hyperbola is not so long and noise-free to allow a clear identification of the asymptotes and thus, depending on the particular situation, the retrieving of both parameters (R, c) can be more or less difficult. In some cases (e.g., in the cases of buried pipes) the radius of the pipe can be known a priori, and it is possible to perform an immediate graphical matching between model and data that allows us to look for the propagation velocity c accounting for the correct value of the radius. This matching can be performed by means of commercial codes—as, for example, the Reflexw.

From Eq. (2.8), we can also appreciate that, if the circular target is not small but its real size is not accounted for, we overestimate the propagation velocity. In fact, for a given time depth t_0, for increasing values of the radius, the diffraction hyperbola keeps the same vertex but its asymptotes translate vertically toward the air half-space. Therefore, it becomes larger near the vertex. In the limit for $R \rightarrow \infty$ the cross point between the asymptotes, equal to $-2R/c$, tends to $-\infty$ and so the hyperbola degenerates into a flat interface.

2.3.3 Interfacial Data in Common Offset Mode with a Non-null Offset: The Case of a Point-like Target

In the presented evaluations we have considered data gathered at the air–soil interface and have neglected the offset between the antennas.[3] If one or both these hypotheses are removed, then the signature of the target is no longer a hyperbola. This is the reason why we prefer the more general term "diffraction curve" instead of the more specific one "diffraction hyperbola." In particular, in this section we consider again a point-like target but remove the hypothesis of null offset between the antennas, considering an

---

[3] Actually, an authentic monostatic (i.e., zero offset) measure occurs only when the same antenna is switched alternatively from transmitting to receiving mode. This solution is possible but rarely employed.

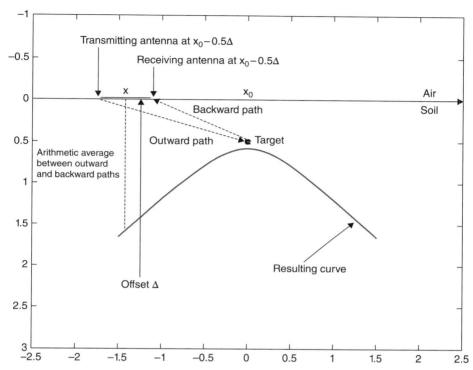

Figure 2.4. The diffraction curve with a non-null offset between the antennas. The curve is meant under time depth conversion with $z = ct/2$. The units are arbitrary.

offset $\Delta \neq 0$ between the transmitting and the receiving antennas. With reference to Figures 2.4 and 2.5, it can be recognized that Eq. (2.2) evolves into

$$ct = \sqrt{\left(x - x_0 - \frac{\Delta}{2}\right)^2 + \left(\frac{ct_{0\Delta}}{2}\right)^2 - \left(\frac{\Delta}{2}\right)^2} + \sqrt{\left(x - x_0 + \frac{\Delta}{2}\right)^2 + \left(\frac{ct_{0\Delta}}{2}\right)^2 - \left(\frac{\Delta}{2}\right)^2} \quad (2.9)$$

where x is the midpoint between the transmitting antenna and the receiving antenna, so that the abscissa of the transmitting antenna (source point) is $x - \Delta/2$ and that of the receiving antenna (observation point) is $x + \Delta/2$. It is not important which one between the two antennas comes geometrically first, because the offset can be considered either positive or negative. Moreover, x_0 is the abscissa of the target and $t_{0\Delta}$ is the return time recorded when $x = x_0$—that is, when the source point is equal to $x = x_0 - \Delta/2$ and the observation point is equal to $x = x_0 + \Delta/2$.

With reference to Figure 2.5, we see that, when $t = t_{0\Delta}$, the source point, the observation point, and the target compose an isosceles triangle, whose height is the depth of the target. This height, in terms of propagation time, is given by

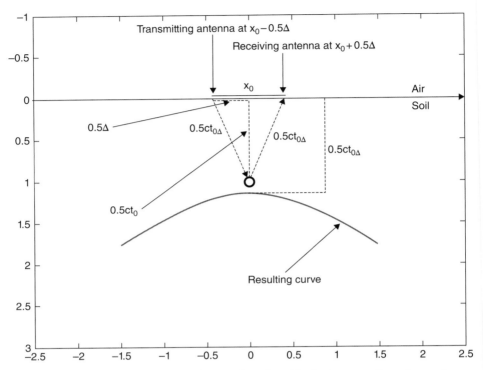

Figure 2.5. The scenario for $x = x_0$ —that is, when the midpoint between the antennas is over the target. The curve refers to the result after the time-apparent depth conversion $z = ct/2$.

$$z_0 = \sqrt{\left(\frac{ct_{0\Delta}}{2}\right)^2 - \left(\frac{\Delta}{2}\right)^2} \qquad (2.10)$$

Equation (2.10) makes it clear that Eq. (2.9) is just the sum of two distances, each of which is evaluated by means of Pythagoras' theorem.

By studying the function $t(x)$ provided by Eq. (2.9), $t_{0\Delta}$ is recognized to be minimum recorded time along the scan, so that it is provided by the data.

Squaring Eq. (2.9) twice, after some algebra we work out the following polynomial equation for the diffraction curve:

$$c^4 t^4 - 4c^2 t^2 \left((x - x_0)^2 + \frac{c^2 t_{0\Delta}^2}{4}\right) + \Delta^2 (x - x_0)^2 = 0 \qquad (2.11)$$

As can be seen, Eq. (2.11) is not a hyperbola but a fourth-degree curve. It reduces to the hyperbola of Eq. (2.2) if and only if the offset between the antennas is equal to zero. Note that the diffraction curve (2.11) admits the same two asymptotes of the diffraction hyperbola corresponding to a null offset between the antennas. In fact, considering high

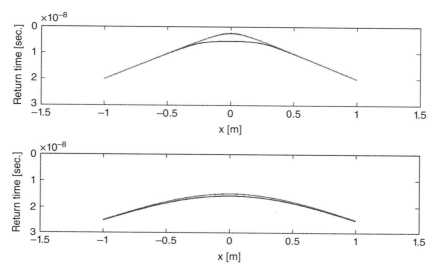

Figure 2.6. Quantitative comparison between the diffraction curve with offset equal to 0 cm (blue line), 10 cm (red line), and 50 cm (black line). The propagation velocity is 10^8 m/s. The depth of the target is 12.5 in the upper panel and 75 cm in the lower panel. For color detail, please see color plate section.

values of x (either positive and negative) and consequently of t [see Eq. (2.9)], we have that in Eq. (2.11) we can neglect the constant term $c^2 t_{0\Delta}^2/4$, so that Eq. (2.11) reduces to

$$c^4 t^4 - 4c^2 t^2 (x-x_0)^2 + \Delta^2 (x-x_0)^2 = c^4 t^4 - (x-x_0)^2 \left(4c^2 t^2 - \Delta^2\right) = 0 \qquad (2.12)$$

However, in Eq. (2.12) we can neglect the constant term Δ^2, so that it still reduces to

$$c^4 t^4 - 4c^2 t^2 (x-x_0)^2 = 0 \Rightarrow c^2 t^2 - 4(x-x_0)^2 = 0 \qquad (2.13)$$

which is equivalent to Eq. (2.3). This means that if the portion of the diffraction curve visible from the data is large enough, the slant of the asymptotes (and thus also the slant of the tangent to the diffraction curve far from the vertex) allows us to identify the propagation velocity independently of the value of the offset.

In order to provide a picture of the real "weight" of the offset, in Figure 2.6 three diffraction curves are shown, corresponding to three offsets respectively equal to 0 cm (blue line), 10 cm (red line), and 50 cm (black line).

The target is at depth 12.5 cm in the upper panel and at depth 75 cm in the lower panel. In these examples, the propagation velocity of the electromagnetic waves is 10^8 m/s. In the end, the effect of the offset is not dramatic in most cases. In particular, from the upper image of Figure 2.6, we can appreciate that even in a case when the offset is of the same order of the depth of the target, the two curves with null and non-null offset are quite similar. The physical reason of this similarity can be explained on the basis of

Figure 2.4. In fact, let us consider first the optical path with null offset, where the source and observation point collapse in the middle point of the segment between them. In this case, the optical path is the sum of a forward and a back half-path equal to each other. Instead, when considering a non-null offset, we have that the two half-paths are different from each other. However, if the forward half–path gets longer with respect to the case of null offset, then the back half-path becomes shorter and vice versa, and this partially counteracts (except just over the target) the variation of the comprehensive round-trip time of the signal.

2.3.4 Noninterfacial Data in Common Offset Mode with a Null Offset: The Case of a Point-like Target

Let us now consider the case with data gathered at fixed height $h > 0$ instead of at the air–soil interface. We restore for simplicity the hypothesis of null offset between the antennas. The target is point-like.

In this case, the path of the signal from the source–receiver point to the buried target and vice versa is not a linear segment but a bent line, as shown in Figure 2.7, because of the refraction of the wave at the air–soil interface. In particular, with reference to Figure 2.7, we have that the return time is given by

$$t = \frac{2}{c_0}\sqrt{(x-x_1)^2 + h^2} + \frac{2}{c}\sqrt{(x_1-x_0)^2 + \left(\frac{ct_{01}}{2}\right)^2} \tag{2.14}$$

where x is the abscissa of the source–receiver point, x_1 is the abscissa of the point at the air–soil interface where the optical path gets bent, x_0 is the abscissa of the target, c_0 and c are respectively the propagation velocity in free space and in the soil, and t_{01} is the minimum round-trip time from the air–soil interface to the target. This is a share of

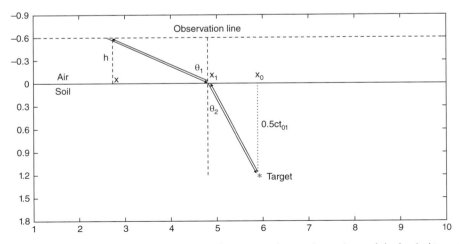

Figure 2.7. The bent optical path between the source–observation point and the buried target and vice versa (arbitrary length units).

the minimum recorded return time (from the source to the target and vice versa) corresponding to the path in the soil. In particular, the minimum recorded round-trip time is equal to

$$t_{min} = \frac{2}{c_0}h + t_{01} \Rightarrow t_{01} = t_{min} - \frac{2}{c_0}h \tag{2.15}$$

In particular, t_{01} is also experimentally distinguishable from t_{min} because the air–soil interface is usually very visible in raw contactless data.

In order to express the diffraction curve, we have to establish the value of x_1 versus x. This can be retrieved from the well-known Snell's law for the refraction (Franceschetti, 1997). In particular (see Figure 2.7), we have

$$\sin(\theta_1) = \frac{c_0}{c}\sin(\theta_2) \Leftrightarrow \frac{|x-x_1|}{\sqrt{(x-x_1)^2 + h^2}} = \frac{c_0}{c}\frac{|x_1-x_0|}{\sqrt{(x_1-x_0)^2 + \left(\frac{ct_{01}}{2}\right)^2}} \tag{2.16}$$

with the physical constraint that x_1 belongs to the interval $[x, x_o]$ if $x \le x_0$ and to the interval $[x_0, x]$ if $x \ge x_0$. A mathematical discussion of Eq. (2.16) might be done: in particular, considering the square of both members and multiplying both sides times the product of the denominators, after some algebra Eq. (2.16) becomes

$$x_1^4 - 2x_1^3(x + x_0) + x_1^2\left(x^2 + x_0^2 + 4xx_0 + \frac{\frac{c^4 t_{01}^2}{4} - c_0^2 h^2}{c^2 - c_0^2}\right)$$
$$- 2x_1\left(x^2 x_0 + xx_0^2 + \frac{\frac{c^4 t_{01}^2}{4}x - c_0^2 h^2 x_0}{c^2 - c_0^2}\right) + x^2 x_0^2 + \frac{\frac{c^4 t_{01}^2}{4}x^2 - c_0^2 h^2 x_0^2}{c^2 - c_0^2} = 0 \tag{2.17}$$

which is a fourth-degree equation in x_1 and is even resoluble in a closed form.[4] However, to deal with Eq. (2.17) analytically is not comfortable (Persico et al., 2013a), and in this text we will consider only a numerical solution of Eq. (2.16). Note that the monotonies of the involved functions allow us to easily infer that the solution of Eq. (2.16), under the said physical constraint, is unique. Therefore, Eq. (2.16) defines in a nonambiguous way a function $x_1(x,c)$. Thus, the formal expression of the diffraction curve in the case at hand is given by

$$t = \frac{2}{c_0}\sqrt{(x - x_1(x,c))^2 + h^2} + \frac{2}{c}\sqrt{(x_1(x,c) - x_0)^2 + \left(\frac{ct_{01}}{2}\right)^2} \tag{2.18}$$

---

[4] The solution of third- and fourth-degree equations was published for the first time in the treatise *Ars Magna*, written by Gerolamo Cardano in the year 1545.

Note that in the limit when the target is very deep—that is, for $t_{01} \to +\infty$, since $|x_1 - x_0|$ is a limited quantity (in particular, it is minorated by $|x - x_0|$) from Eq. (2.16)—we have that necessarily $x_1 \to x$. Substituting into Eq. (2.18), we obtain

$$t \to \frac{2}{c_0}h + \frac{2}{c}\sqrt{(x-x_0)^2 + \left(\frac{ct_{01}}{2}\right)^2} \tag{2.19}$$

which means that the limit path of the received GPR signal is composed by a vertical path in air from the source–receiver point to the interface followed the same path that the signal would show if the measurements were interfacial. Thus, this limit path is a hyperbola. On the other hand, in the limit for very shallow targets (i.e., for $t_{01} \to 0$), we have from Eq. (2.16)

$$|x-x_1|\sqrt{(x_1-x_0)^2 + \left(\frac{ct_{01}}{2}\right)^2} \to |x-x_1||x_1-x_0| = \frac{c_0}{c}|x_1-x_0|\sqrt{(x-x_1)^2 + h^2} \tag{2.20}$$

Therefore, x_1 should either tend to x_0 or solve asymptotically the equation

$$|x-x_1| = \frac{c_0}{c}\sqrt{(x-x_1)^2 + h^2} \Rightarrow (x-x_1)^2 = \left(\frac{c_0}{c}\right)^2\left((x-x_1)^2 + h^2\right)$$

$$\Rightarrow (x-x_1)^2\left(1 - \left(\frac{c_0}{c}\right)^2\right) = \left(\frac{c_0}{c}\right)^2 h^2 \tag{2.21}$$

However, since necessarily we have $c < c_0$, the only physical solution is that $x_1 \to x_0$. Substituting into Eq. (2.18), we obtain

$$t \to \frac{2}{c_0}\sqrt{(x-x_0)^2 + h^2} \tag{2.22}$$

Moreover, from Eq. (2.15) we have that $h \to c_0 t_{min}/2$, which substituted into Eq. (2.22) provides

$$t \to = \frac{2}{c_0}\sqrt{(x-x_0)^2 + \left(\frac{c_0 t_{min}}{2}\right)^2} \tag{2.23}$$

which is the diffraction hyperbola of a target "embedded in air," because the propagation of the signal actually occurs in air. Thus, in this opposite limit case the diffraction curve is again a hyperbola. However, in the intermediate cases between the two extremes, the diffraction curve is not a hyperbola in general. Incidentally, Eq. (2.23) also means that a very shallow target does not provide any information about the propagation velocity of the waves in the soil, because the propagation occurs entirely in air. In general, the measure of the propagation velocity from noninterfacial data is more critical than the measure from interfacial data: some more details on this aspect are provided in Persico et al. (2013b).

2.3.5 Interfacial Data in Common Midpoint (CMP) Mode

Let us now describe the measure of the propagation velocity of the electromagnetic waves from common midpoint (CMP) data (Daniels, 2004; Conyers, 2004), also reported as common depth point (CDP) data.

Let us assume that a buried point-like target has been identified at the abscissa x_0. In CMP mode, one records the return times achieved by placing the transmitting and the receiving antennas at the two positions x and $2x_0 - x$ symmetrical with respect to the target. According to Figure 2.8, the length of the comprehensive path from the transmitting antenna to the target and then from the target to the observation point is given by

$$ct = 2\sqrt{(x-x_0)^2 + \left(\frac{ct_0}{2}\right)^2} \Rightarrow \frac{t^2}{t_0^2} - \frac{(x-x_0)^2}{0.25c^2 t_0^2} = 1 \tag{2.24}$$

As can be seen, Eq. (2.24) is the same as Eq. (2.2). This is because (see Figures 2.2 and 2.8) the length of the path traveled by the signal is the same in the two cases even if, of course, only one-half of the diffraction curve is retrieved in CMP mode.

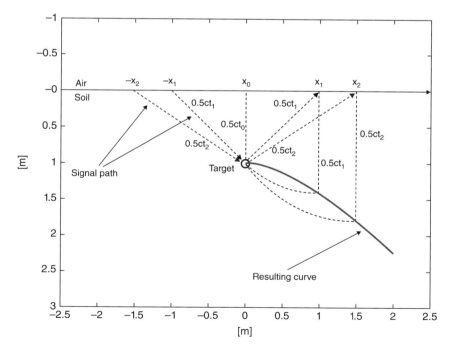

Figure 2.8. Scheme for common midpoint measurements.

Actually, in the case of CMP measurements, one doesn't need to assume that the target is point-like. In particular, a plane buried interface parallel to the air–soil interface provides similar results, as well as any roundish surface if its top is not tilted with respect to the air–soil interface. This is based on Snell's law on the reflected waves (Franceschetti, 1997), which assures (as is well known) the equality of the incident and reflected angles. Due to the unknown shape and size of the buried targets, this states the theoretical better performances of the CMP measure of the propagation velocity with respect to the common offset one.

Actually, CMP data can provide information about the electromagnetic characteristics of a stratified medium with horizontal layers even without any localized reflector [but in this case some care has to be taken with regard to possible waveguide effects (Yilmaz, 1987)].

Another possible advantage is that the increased distance between the antennas prevents the risk of saturation of the receiver. Theoretically, this allows us to increase the radiated power and can make visible a larger portion of the diffraction hyperbola, even if the radiated power has to respect some legal (further than technical) limits (Chignell, 2004).

Actually, in CMP mode we don't even need to identify a suitable buried target. In fact, the direct coupling signal between the antennas is somehow equivalent to a target at the air–soil interface put in the midpoint between source and receiver (see Figure 2.9). In particular, with reference to Figure 2.9, in CMP the receiving antenna gathers both a direct signal traveling in air and a direct signal traveling in the shallowest part of the soil. The signal traveling in air propagates at about the velocity of propagation of the waves in free space c_0.

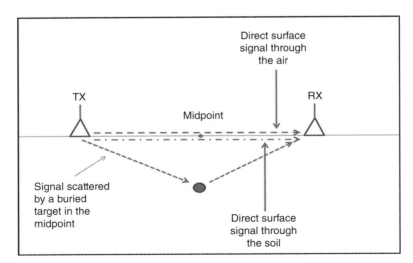

Figure 2.9. Geometric scheme for the received CMP signal. Beyond the signal scattered from the target (if any) buried under the midpoint, there is the direct signal between the antennas, propagating partly in air and partly in the shallowest part of the soil.

So, if the receiving antenna is at the generic (positive) abscissa x, the radiated pulse propagating in air is received at the time instant

$$t = \frac{2}{c_0}(x - x_0) \tag{2.25}$$

Instead, the pulse propagating in the soil at shallow depth propagates at the propagation velocity c of the waves in the soil and therefore is received at the time instant

$$t = \frac{2}{c}(x - x_0) \tag{2.26}$$

Equations (2.25)–(2.26) show that the same interface generates two diffraction curves $t = t(x)$, which are two straight lines with slants $2/c_0$ and $2/c$, respectively. Since $c < c_0$ the line relative to the propagation in the soil is the most tilted one. In particular, since both the distance between the antennas and the pulse arrival time are known, it is possible and well-advised to check whether the upper line is really relative to the propagation in air. Incidentally, the straight line relative to the propagation just below the surface corresponds to the degenerated diffraction hyperbola of a point-like target placed at the air–soil interface.

However, it is well-advised to make use not only of the surface reflections but also, if possible, of the reflection from some suitable buried target. In fact, the shallower layers of the soil might show a propagation velocity different from the average propagation velocity in the first 1–3 m, which in most cases is the depth range of interest. Moreover, the roughness of the soil might affect the superficial apparent propagation velocity.

An obvious drawback of CMP mode with respect to the common offset one is the fact that CMP requires two antennas that can be moved independently from each other. Moreover, the CMP procedure is intrinsically discrete in the sense that, customarily, the antennas are not moved in a continuous mode (keeping the two velocities equal and also keeping the directions constant can be a critical issue), which makes the measurement quite slower than that usually achievable in common offset mode.

In Section 15.1, the reader will find some exercises on the retrieving of the propagation velocity in the medium embedding the targets from the data.

2.4 LOSSY, MAGNETIC, AND DISPERSIVE MEDIA

Up to now we have devoted attention to the measure of the propagation velocity of the waves in the soil, which is in most cases the parameter of utmost interest. Let now devote some attention to other electromagnetic characteristics of the background medium.

With regard to the amount of the losses, in general they are more difficult to be measured with respect to the propagation velocity. Actually, since the attenuation due to the losses is exponential versus the length of propagation path, one can think of measuring it directly from the GPR data. However, in general we don't know the radar cross section (Levanon, 1988) of the buried targets and therefore in general we are not able to distinguish, from the mere amplitude of a received signal, the share of the transmitted energy that has been attenuated by the electromagnetic losses of the soil and the share that has been scattered by the target away from the receiving antenna. Thus, the simple comparison between the amplitude of the transmitted pulse and that of the received one might not be sufficient to establish the amount of the losses, and incidentally the amplitude of the transmitted pulse might be unknown too. This makes the measure of the losses from usual GPR data in common offset a hard task.

CMP data can offer some more possibilities, because we can create in a controlled way different propagation paths where the signal is reflected by the same target and so we can make a comparison between the amplitude of two or more different received pulses (Leucci, 2008). The radar cross section (even if unknown in general) is assumed to be the same for different position of the antennas. This is reasonable, even if not rigorous, especially if the reflector is a flat interface. Some care has to be taken with regard to the range of distances considered between the two antennas: If this range is electrically large, the amplitude of the received pulse is influenced also by the radiation pattern of the antennas, which should be therefore taken into account.

In the case of a homogeneous masonry, we can perform two measurements: the first one in transmission mode with the two antennas equal to each other on the opposite faces of the wall, and the second one in reflection mode with a metallic sheet put on one side of the masonry and the GPR antennas on the other side. Actually, in this case, it would be more correct to exploit the same antenna to transmit the pulse and to receive the echo. Then, we can compare the amplitude of a signal propagated in transmission mode with the amplitude of the same signal propagated in reflection mode along a path twice as long. No effect due to the antenna pattern is expected in this case, because both propagation paths develop in the broadside direction.

An alternative possibility is based on the use of a time domain reflectometry (TDR) probe (O'Connor and Dowding, 1999; Cataldo et al., 2011), briefly described in the following.

The TDR technique is in most cases a method in time domain. However, since now we are dealing with the measurement of losses in the soil, it is convenient for our purposes to propose an analysis of the TDR signal in frequency domain.

With reference to Figure 2.10, a TDR equipment is substantially a transmission line-based structure. Therefore, the basic theory is easily understood on the basis of the theory of the transmission lines (Franceschetti, 1997), which is (necessarily) taken for granted here.

With reference to Figure 2.10, suppose that an incident wave at fixed frequency propagates from the left-hand side along line 1. Line 1 is considered infinitely long in this scheme, which practically means that it is matched to the generator. Incidentally, this also means that the intrinsic impedance Z_c of line 1 is a given (non modifiable) parameter. After line 1, there is a piece of a different line, labeled 2, that has a fixed length L and is knocked in the soil. This second line is open at the end, which is schematized by means of a lumped load Z_{load} with infinite impedance (this amounts to neglect the energy radiated in the soil by the TDR probe). The intrinsic impedance of line 2 depends also on the soil where it is knocked in. In particular, if we label Z_{l0} the intrinsic impedance of line 2 in free space, then the intrinsic impedance of the same transmission line embedded in the soil

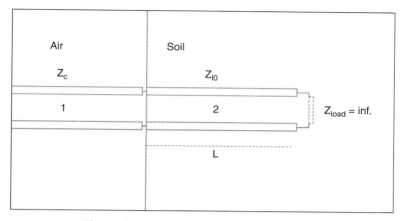

Figure 2.10. Circuital scheme for a TDR probe.

is given by $Z_l = \varsigma Z_{l0}$, where ς is the relative (dimensionless) impedance of the soil, given by

$$\varsigma = \sqrt{\frac{\mu_{sr}}{\varepsilon_{sr}}} \tag{2.27}$$

By means of the TDR probe, one can gather the reflection coefficient versus the frequency at the air–soil interface. To relate this quantity to the characteristics of the soil, we can first make use of the classical formula for the impedance transportation along the line (Franceschetti, 1997). Thus, the impedance seen at that air–soil interface, looking toward the soil, is given by

$$Z = Z_l \frac{Z_{\text{load}} + jZ_l \text{tg}(k_s L)}{Z_l + jZ_{\text{load}} \text{tg}(k_s L)} \tag{2.28}$$

where k_s is the wavenumber of the propagation medium that embeds line 2. The wavenumber is linked to the properties of the medium by means of the relationships

$$k_s = \frac{2\pi f \sqrt{\varepsilon_{sr} \mu_{sr}}}{c_0} \tag{2.29}$$

where f is the frequency and c_0 is, as usual, the propagation velocity of the electromagnetic waves in free space. In Eq. (2.29), ε_{sr} and μ_{sr} are meant as complex quantities, to account for losses.

At this point, the reflection coefficient at the air–soil interface, looking underground, is given by (Franceschetti, 1997)

$$\Gamma = \frac{Z - Z_c}{Z + Z_c} \tag{2.30}$$

where Z is given by Eq. (2.28).

After substituting Eq. (2.28) in Eq. (2.30) and after some algebraic manipulations, the following alternative expression for the reflection coefficient at the air–soil interface is achieved:

$$\Gamma = \frac{\Gamma_{12} + \Gamma_{load}\exp(-j2k_sL)}{1 + \Gamma_{12}\Gamma_{load}\exp(-j2k_sL)} \tag{2.31}$$

where Γ_{load} is the reflection coefficient at the load and Γ_{12} is given by

$$\Gamma_{12} = \frac{Z_l - Z_c}{Z_l + Z_c} = \frac{\varsigma Z_{l0} - Z_c}{\varsigma Z_{l0} + Z_c} \tag{2.32}$$

Γ_{12} represents the reflection coefficient between lines 1 and line 2—that is, the reflection coefficient that we would record at the air–soil interface if line 2 were infinite. As said, the load is an open circuit, and therefore we have

$$\Gamma_{load} = 1 \tag{2.33}$$

Substituting Eqs. (2.29), (2.32), and (2.33) in Eq. (2.31), eventually we retrieve the expression of the TDR datum versus the characteristics of the probed soil at fixed frequency, which is

$$\Gamma = \frac{\left(\sqrt{\frac{\mu_{sr}}{\varepsilon_{sr}}}Z_{l0} - Z_c\right) + \left(\sqrt{\frac{\mu_{sr}}{\varepsilon_{sr}}}Z_{l0} + Z_c\right)\exp\left(-j\frac{4\pi f}{c_0}\sqrt{\varepsilon_{sr}\mu_{sr}}L\right)}{\left(\sqrt{\frac{\mu_{sr}}{\varepsilon_{sr}}}Z_{l0} + Z_c\right) + \left(\sqrt{\frac{\mu_{sr}}{\varepsilon_{sr}}}Z_{l0} - Z_c\right)\exp\left(-j\frac{4\pi f}{c_0}\sqrt{\varepsilon_{sr}\mu_{sr}}L\right)} \tag{2.34}$$

Equation (2.34) can provide more information than that retrievable from the diffraction curves provided by GPR data.

In particular, all the presented diffraction curves, either in common offset or in CMP mode, are based on the propagation velocity of the waves. The propagation velocity, in low lossy soils (Jol, 2009), is related to the real part of the product $\mu_{sr}\varepsilon_{sr}$, and consequently the diffraction curves do not allow us to identify a contribution of the imaginary parts of the dielectric permittivity and/or of the magnetic permeability to the signal, and of course they do not allow us to distinguish the contribution of the dielectric permittivity from that of the magnetic permeability to the propagation velocity.

In Eq. (2.34), instead, the datum depends on both the impedance and the propagation velocity along the transmission line, which means that we have a decoupled dependence on the possibly complex dielectric permittivity and magnetic permeability. Consequently, a TDR probe can help in measuring both the dielectric permittivity and the magnetic permeability. In particular, for a magnetic nondispersive soil, the measure can be performed by matching Eq. (2.34) at several frequencies with the experimentally retrieved reflection coefficients.

If we can neglect the magnetic properties, Eq. (2.34) provides instead a relationship where the unknown is the complex equivalent dielectric permittivity, from which both the real dielectric permittivity and the equivalent conductivity can be retrieved (Daniels, 2004; Jol, 2009). Often, in these case, one matches a known dispersion law to the data at more frequencies, looking for some specific parameters involved in the predetermined dispersion law, such as the relaxation time (Cataldo et al., 2011).

In any lossy case, some attention has to be devoted to the determination of the square roots to be retained, and this depends on the convention adopted for expressing the waves propagating in the positive version of the chosen abscissas. In particular, in this section we have implicitly adopted the convention $\exp(j2\pi ft)$ to express temporal dependence. Consequently, since the signal has to attenuate far from the sources, we have that the imaginary part of k_s is nonpositive and the real part of the intrinsic impedance of the line is nonnegative.

In the more general case of a soil dispersive (with a meaningful but unknown dispersion law) and magnetic, even a TDR probe is theoretically insufficient, because Eq. (2.34) provides, at each frequency, one complex datum and two complex unknowns. In this "extreme" case, two or more TDR probes of different length and/or intrinsic free-space impedance can provide, in principle, the needed diversities in order to provide at least as many equations as unknowns or more equations than unknowns (but the reliability of the achievable results is another issue).

QUESTIONS

1. Can the electromagnetic characteristics of the soil be completely retrieved from GPR data?
2. Do conductivity losses influence the propagation velocity of the waves in the soil?
3. Can we retrieve the propagation velocity of the waves in the soil from two points of a diffraction curve from interfacial data?
4. Why can the evaluation of the propagation velocity in the soil from CMP data be superior to that achieved from common offset data?
5. In what aspect can the measure of the characteristics of the soil by means of a TDR probe be superior to that achieved from GPR data?
6. In what aspect can the measure of the characteristics of the soil by means of GPR data be superior to that achieved from a TDR probe?

3

GPR DATA SAMPLING: FREQUENCY AND TIME STEPS

3.1 STEPPED FREQUENCY GPR SYSTEMS: THE PROBLEM OF THE ALIASING AND THE FREQUENCY STEP

The working principle of a GPR system described in the previous chapter essentially refers to the "pulsed" GPR. Alternatively, it is possible to transmit a train of harmonic signals instead of a pulse. This procedure essentially involves transmitting the harmonic components of the pulse sequentially rather than at the same time. The systems that do that are, as well known, the stepped frequency GPRs (Robinson et al., 1974; Iizuka et al., 1984; Stickley et al., 1999; Alberti et al., 2002). The theoretical advantages of a stepped frequency system with respect to a pulsed one are essentially based on the possibility to have a trade-off between the duration of the harmonic signals (the so-called *integration time*) and the noise on the data. This allows us to improve the signal-to-noise ratio without being compelled to radiate a stronger signal (which might saturate the receiver). Of course, this is impossible for a pulsed system. Essentially, the "trick" of a stepped frequency GPR is that one can transmit more energy rather than more power. For a deeper discussion of these aspects, the interested reader is referred to Noon (1996). Furthermore, in Persico and Prisco (2008) and Persico et al. (2011), it is possible to find some recent

Introduction to Ground Penetrating Radar: Inverse Scattering and Data Processing,
First Edition. Raffaele Persico.
© 2014 The Institute of Electrical and Electronics Engineers, Inc. Published 2014 by John Wiley & Sons, Inc.

studies about further possibilities related to the flexibility of a stepped frequency system, and in particular to the possibility to reconfigure the system versus the frequency.

Here, with regard to data processing aspects, it is of interest to deal with the effects of the frequency sampling, in order to provide the calculation of the needed frequency step.

To introduce this subject, let us consider the Fourier transform, or spectrum, of a finite energy real signal $g(t)$, given by

$$\hat{g}(f) = \int_{-\infty}^{+\infty} g(t) \exp(-j2\pi ft)\, dt \tag{3.1}$$

where $g(t)$ a real function and $\hat{g}(f)$ is a Hermitian function, that is, it has the property that $\hat{g}(-f) = \hat{g}^*(f)$, where the asterisk stands for the conjugate value. Equation (3.1) is inverted as

$$g(t) = \int_{-\infty}^{+\infty} \hat{g}(f) \exp(j2\pi ft)\, df \tag{3.2}$$

Due to the Hermitianity of $\hat{g}(f)$, Eq. (3.2) can be also written as

$$g(t) = 2\mathrm{Re}\left\{ \int_{0}^{+\infty} \hat{g}(f) \exp(j2\pi ft)\, df \right\} \tag{3.3}$$

where Re indicates the real part of the quantity within accolades. Equation (3.3) expresses the signal as a sum of elementary harmonic functions.

At this point, let us suppose to have at our disposal a sampled version of the spectrum of the signal $\hat{g}(f)$—that is, a sequence of samples $\hat{g}(f_n)$ spaced from each other by a fixed frequency step Δf—and let us try to calculate the signal in time domain from these samples.

In a first moment, let us suppose that we have at our disposal an odd number of samples $2N+1$, starting from the initial frequency f_0. Thus, the considered frequency band ranges from f_0 up to $f_0 + 2N\Delta$, and the extension of this band is $B = 2N\Delta f$. The central frequency is given by $f_c = f_0 + N\Delta f$. The nth frequency sample is equal to $f_n = f_0 + (n-1)\Delta f$, with n ranging from 1 to $2N+1$, which is the same as $f_n = f_c + n\Delta f$ with n ranging from $-N$ to N.

From the discretization of Eq. (3.3), we have that the corresponding signal in time domain (customarily labeled as the *synthetic* time) is given by[1]

---

[1] It is well known that the Fourier direct and inverse transforms can be performed by means of fast numerical FFT and IFFT algorithms. However, for simplicity of exposition, we follow the method of a formal discretization of the Fourier integral.

$$g(t) \approx 2\text{Re}\left\{\sum_{n=1}^{2N+1} \hat{g}(f_n)\exp(j2\pi f_n t)\Delta f\right\} = 2\Delta f \text{Re}\left\{\sum_{n=-N}^{N} \hat{g}_n \exp(j2\pi[f_c + n\Delta f]t)\right\}$$

$$= 2\Delta f \text{Re}\left\{\exp(j2\pi f_c t)\sum_{n=-N}^{N} \hat{g}_n \exp(j2\pi n\Delta f t)\right\} \tag{3.4}$$

where $\hat{g}_n = \hat{g}(f_{n+N+1})$. Let us now write

$$g_1(t) = \sum_{n=-N}^{N} \hat{g}_n \exp(j2\pi n\Delta f t) = g_{1r}(t) + jg_{1i}(t) \tag{3.5}$$

where g_1 is a complex quantity, and thus in Eq. (3.5) the real and imaginary parts have been put into evidence. At this point, we can express our signal in synthetic time as follows:

$$g(t) \approx 2\Delta f \cos(2\pi f_c t)g_{1r}(t) - 2\Delta f \sin(2\pi f_c t)g_{1i}(t) \tag{3.6}$$

Let us now put

$$\begin{cases} g_{1r}(t) = \rho(t)\cos(\theta(t)), \\ g_{1i}(t) = \rho(t)\sin(\theta(t)) \end{cases} \tag{3.7}$$

After some straightforward calculations, we obtain

$$g(t) \approx 2\Delta f \sqrt{g_{1r}^2(t) + g_{1i}^2(t)} \cos\left(2\pi f_c t + \text{tg}^{-1}\left(\frac{g_{1i}(t)}{g_{1r}(t)}\right)\right) \tag{3.8}$$

Let us now consider the behavior of $g(t)$ at the time $t + 1/\Delta f$. Furthermore, let us preliminarily note that the function $g_1(t)$ is periodical with period $1/\Delta f$, as can be easily checked directly from Eq. (3.5). Consequently, we have

$$g\left(t + \frac{1}{\Delta f}\right) = 2\Delta f \sqrt{g_{1r}^2(t) + g_{1i}^2(t)} \cos\left(2\pi f_c t + \text{tg}^{-1}\left(\frac{g_{1i}(t)}{g_{1r}(t)}\right) + \Delta\phi\right) \tag{3.9}$$

where

$$\Delta\phi = 2\pi \frac{f_c}{\Delta f} \tag{3.10}$$

Equations (3.9) and (3.10) show that when transforming in time domain a stepped frequency signal, ideally "conceived" as the spectrum of a synthetic pulsed signal, we don't obtain a single pulse but rather a train of pulses spaced $1/\Delta f$ from each other. It is sometimes loosely said that the signal in time domain is a periodic train of pulses. Actually, from Eqs. (3.9)–(3.10), we see that actually the pulses of the train don't have the same shape, unless the central frequency f_c is an integer multiple of the frequency step Δf. At any rate, the physically relevant phenomenon is the repetition of the pulses, called as well-known *aliasing*. Due to the aliasing, the maximum synthetic time at which we can retrieve the signal reliably is $1/\Delta f$, and consequently the maximum reliably investigable depth [the so-called nonambiguous depth (Noon, 1996)] is given by the maximum depth from which we can receive a signal within the time interval $1/\Delta f$ starting from the beginning of the radiation of the synthetic time pulse. Beyond this time interval, the radar answer in synthetic time is ambiguous because any received echo might be due both to a deeper target that reflects a previously transmitted pulse or a shallower target that reflects a subsequently transmitted pulse. The nonambiguous depth D is therefore given by

$$D = \frac{c}{2\Delta f} \qquad (3.11)$$

The factor 2 is due to the round-trip of the GPR signal (the offset between the antennas has been neglected). Actually, one establishes a priori the depth up to which he/she thinks there are perceivable targets. So, Eq. (3.11) provides the needed frequency step (i.e., the maximum allowed value of the frequency step) that allows us to reach nonambiguously the chosen maximum depth. This frequency step is achieved simply inverting Eq. (3.11):

$$\Delta f = \frac{c}{2D} \qquad (3.12)$$

If the propagation medium is lossless or low lossy and does not show magnetic properties, Eq. (3.12) can still be rewritten as

$$\Delta f = \frac{c_0}{2\sqrt{\varepsilon_{sr}}D} \qquad (3.13)$$

where c_0 is the propagation velocity of the electromagnetic waves in free space and ε_{sr} is the relative dielectric permittivity of the host medium.

It might be thought that the maximum reachable depth can be chosen arbitrarily within the maximum depth "visible" by the system [which depends on the noise level and on the dynamic of the system (Noon, 1996)], and then the frequency step given by Eqs. (3.12) or (3.13) guarantees a nonambiguous result, but there is a subtlety to be considered:

In particular, for a given chosen maximum depth D, the frequency step given in Eqs. (3.12) or (3.13) guarantees that any object enclosed within the depth range up to D will not provide spurious replicas of its real echo within the probed depth. However, any actual target deeper than D is a potential source of aliasing, because its echo is in any case replicated with time periodicity $1/\Delta f$. The only way to mitigate this phenomenon is to choose D large enough so that the signal scattered by a target deeper than D reaches the surface very attenuated and is not perceived by the receiving antenna. This means to guarantee that the entire duration in the time domain of the perceivable signal is accounted for. In particular, it is an error to choose D on the mere basis of the expected depth range of the targets of interest; insteads, the choice should be based on the expected perceivable maximum depth.

Up to now, we have considered an odd number of frequency samples. Let us now consider the case with an even number of frequency samples $2N$. This time, the considered frequency band ranges from some f_0 up to $f_0 + (2N - 1)\Delta f$ and the extension of this band is $B = (2N - 1)\Delta f$. The central frequency is $f_c = f_0 + N\Delta f - \Delta f/2$. Consequently, Eq. (3.4) can be rewritten as

$$
\begin{aligned}
g(t) &\approx 2\mathrm{Re}\left\{ \sum_{n=1}^{2N} \hat{g}(f_n)\exp(j2\pi f_n t)\Delta f \right\} \\
&= 2\Delta f\,\mathrm{Re}\left\{ \sum_{n=-N+1}^{N} \hat{g}_n \exp\left(j2\pi \left[f_c + \left(n - \frac{1}{2} \right)\Delta f \right] t \right) \right\} \\
&= 2\Delta f\,\mathrm{Re}\left\{ \exp(j2\pi f_c t) \sum_{n=-N+1}^{N} \hat{g}_n \exp\left(j2\pi \left(n - \frac{1}{2} \right)\Delta f t \right) \right\}
\end{aligned}
\tag{3.14}
$$

where $\hat{g}_n = \hat{g}(f_{n+N})$. Let us now write

$$
g_1(t) = \sum_{n=-N+1}^{N} \hat{g}_n \exp\left(j2\pi \left(n - \frac{1}{2} \right)\Delta f t \right) = g_{1r}(t) + jg_{1i}(t)
\tag{3.15}
$$

At this point, the calculations steps are the same as those exposed in the case of an odd number of samples. Actually, this time the period of $g_1(t)$ is not $1/\Delta f$ but $2/\Delta f$. However, it is easily recognized from Eq. (3.15) that $g_1(t + k/\Delta f) = (-1)^k g_1(t)$, and thus expression (3.9) keeps unchanged with respect with the previous case. This shows that the considerations exposed or for an odd number of frequency samples hold also for an even number of frequency samples.

3.2   SHAPE AND THICKNESS OF THE GPR PULSES

The previous section suggests to insert a brief description of what kind of shape and above all what order of thickness the GPR pulses are expected to show. From Eq. (3.8), we see that the received pulse can be formally described as a sinusoidal quantity with amplitude and phase changing versus the time. In the case of GPR pulses, the central frequency f_c and the band are roughly of the same order (Jol, 2009). Starting from this basic consideration, let us provide an example considering a case where the frequency band of the pulse is found in the interval $\left[\frac{1}{2}f_c, \frac{3}{2}f_c \right]$; in this way, f_c represents both the central frequency and the extension of the band. In a first moment, let us suppose to deal with a pulsed GPR. The shape of the pulse depends of course on the function $\hat{g}(f)$. Let us investigate the case where $\hat{g}(f)$ is a constant function. This constant is generally complex, and thus we will indicate it as $\hat{g}(f) = K\exp(j\theta)$. Therefore, the pulse in the time domain is given by

$$g(t) = 2K\operatorname{Re}\left\{ \exp(j\theta) \int_{0.5f_c}^{1.5f_c} \exp(j2\pi ft)df \right\} = 2Kf_c \operatorname{sinc}(\pi f_c t)\cos(2\pi f_c t + \theta) \qquad (3.16)$$

where the sinc function is defined as $\operatorname{sinc}(x) = \sin(x)/x$.

In Figure 3.1 the graphs of the pulses relative to four different values of θ are provided. The graphs are expressed versus the dimensionless variable tf_c: As can be seen, the duration in this normalized time is about unitary; that is, the actual duration of the pulse is inversely proportional to its central frequency (and thus to its band). Moreover, the pulses have an oscillatory time behavior, due to the fact that their spectrum does not extend up to the zero frequency and thus their average value is substantially equal to zero. This is experimentally well known too.

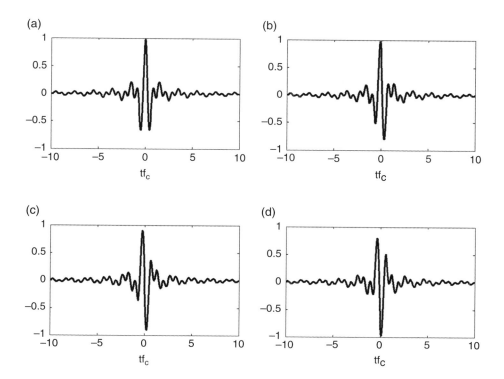

Figure 3.1. Graphs of expression (3.16), normalized to $2Kf_c$ and represented versus the dimensionless variable tf_c. **Panel a:** $\theta = 0$. **Panel b:** $\theta = \pi/4$. **Panel c:** $\theta = \pi/2$. **Panel d:** $\theta = 3\pi/4$.

Let us now discuss the case of a stepped frequency system. In this case, we are interested in achieving the shape of the synthetic pulse starting from a sampled version of its spectrum. We will limit ourselves to the case of an odd number of samples for sake of brevity, but the case with an even number of samples is analogous. Thus, we particularize Eq. (3.4) to the case $\hat{g}_n = K \exp(j\theta)$. The lower and upper frequencies of the band are equal to $f_c - N\Delta f = \left(\frac{1}{2}\right)f_c$ and $f_c + N\Delta f = \left(\frac{3}{2}\right)f_c$, which is to say that $\Delta f = f_c/2N$.

After this premix, the synthetic pulse in time domain is given by

$$g(t) \approx 2\Delta f K \operatorname{Re}\left\{ \exp(j2\pi f_c t + j\theta) \sum_{n=-N}^{N} \exp(j2\pi n\Delta ft) \right\} \tag{3.17}$$

and we have

$$\sum_{n=-N}^{N} \exp(j2\pi n\Delta ft) = \exp(-j2\pi N\Delta ft) \sum_{n=0}^{2N} \exp(j2\pi n\Delta ft)$$

$$= \exp(-j2\pi N\Delta ft)\frac{1 - \exp(j(2N+1)2\pi\Delta ft)}{1 - \exp(j2\pi\Delta ft)}$$

$$= \frac{\exp(-j2\pi N\Delta ft)\exp(j(2N+1)\pi\Delta ft)}{\exp(j\pi\Delta ft)}$$

$$\times \frac{\exp(-j(2N+1)\pi\Delta ft) - \exp(j(2N+1)\pi\Delta ft)}{\exp(-j\pi\Delta ft) - \exp(j\pi\Delta ft)}$$

$$= \frac{\sin\left((2N+1)\pi\Delta ft\right)}{\sin\left(\pi\Delta ft\right)} \tag{3.18}$$

Equation (3.18) shows that the result of the summation is a real function given by the ratio of two sines. This ratio is known as the *sine of Dirichlet* (Collin, 1985). The sine of the Dirichlet in Eq. (3.18) is a periodic function with period $1/\Delta f$ and with main lobes large $2/(2N+1)\Delta f = 4N/(2N+1)f_c$. Substituting Eq. (3.18) in Eq. (3.17), we obtain

$$g(t) \approx 2K\Delta f \frac{\sin\left(\dfrac{2N+1}{2N}\pi f_c t\right)}{\sin\left(\dfrac{\pi f_c t}{2N}\right)} \cos\left(2\pi f_c t + \theta\right)$$

$$= 2Kf_c \frac{\sin\left(\dfrac{2N+1}{2N}\pi f_c t\right)}{2N\sin\left(\dfrac{\pi f_c t}{2N}\right)} \cos\left(2\pi f_c t + \theta\right) \tag{3.19}$$

A graph of this function is provided in Figure 3.2, for several values of N, f_c, and θ. The result can be compared with the result in the "continuous case" only within the nonambiguous interval

$$I = \left[-\frac{1}{2\Delta f}, \frac{1}{2\Delta f} \right] = \left[-\frac{N}{f_c}, \frac{N}{f_c} \right]$$

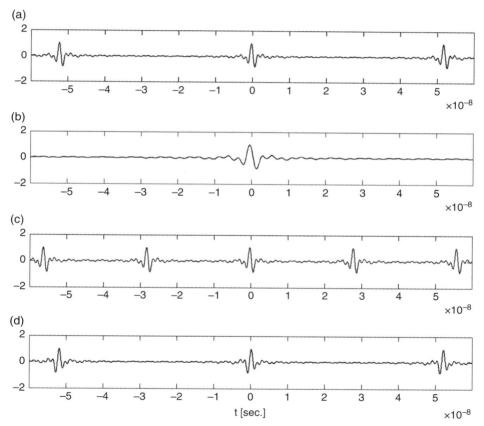

Figure 3.2. Graphs of the synthetic pulses of Eq. (3.19), normalized to $2Kf_c$, represented versus the synthetic time in seconds. **Panel a:** $f_c = 500$ MHz, $N = 13$, $\theta = \pi/4$. **Panel b:** $f_c = 250$ MHz, $N = 13$, $\theta = \pi/4$ (the replicas are 200% farther and the pulses are 200% larger with respect to panel a). **Panel c:** $f_c = 500$ MHz, $N = 7$, $\theta = \pi/4$ (the replicas are closer with respect to panel a, because of the larger frequency step). **Panel d:** $f_c = 500$ MHz, $N = 13$, $\theta = -\pi/4$ (the pulses have equal thickness but different shape with respect to panel a, because of the different constant phase term).

Even within the nonambiguous time interval, the two functions (3.16) and (3.19) are not equal to each other.[2] However, in the limit for high values of N (i.e., for $\Delta f \to 0$) the expression (3.19) tends uniformly[3] to the expression (3.16) within the nonambiguous interval, which in turn enlarges progressively toward the entire real axis.

In fact, if $t < < N/f_c$ and $N >> 1$, we can approximate

$$\sin\left(\frac{\pi f_c t}{2N}\right) \approx \frac{\pi f_c t}{2N}; \quad \frac{2N+1}{2N} \approx 1 \tag{3.20}$$

which, when substituted in Eq. (3.19), provides Eq. (3.16).

Of course, the first approximation in Eq. (3.20) is not licit any longer when t becomes comparable with N/f_c, but in this case both expressions (3.16) and (3.19) vanish and so the convergence of (3.19) to (3.16) is guaranteed all over the nonambiguous interval.

The nonambiguous interval, in turn, enlarges progressively versus N. Thus, in the end there is a punctual convergence[4] of the expression (3.19) to the expression (3.16) for any time instant.

3.3 STEPPED FREQUENCY GPR SYSTEMS: THE PROBLEM OF THE DEMODULATION AND THE FREQUENCY STEP

The digression on shape and thickness of the pulses facilitates the introduction of the problem of the Hermitian images, which arises from the possibility of an imprecise demodulation of the received signal. This is a technical problem regarding the hardware of a stepped frequency system. However, it can influence the choice of a suitable frequency step, and therefore it is relevant to our discussion.

That said let us consider a case when the harmonic component of the synthetic pulse of Eq. (3.17) are sequentially transmitted. Thus, the transmitted signal is a train of harmonics, whose time behavior (apart the temporal truncation) is given by

$$s_{tx}(t) = K \cos\left(2\pi f_{N+n+1} t + \theta\right) = K \cos\left(2\pi (f_c + n\Delta f)t + \theta\right), \ n = -N, \ldots, N \tag{3.21}$$

Let us now consider a target that produces an echo after a time $\bar{t}$ (smaller than the nonambiguous time interval $1/\Delta f$). Let us neglect the variation versus the frequency of the reflection behavior of this target. Thus, the received signal is a train of harmonic

---

[2] Actually, due to the hypothesis of a rigorously limited band, the theoretical support of the signal in time domain is infinite, and this makes it impossible to retrieve perfectly the function from its spectral samples in frequency domain.

[3] Which means that the maximum modulus of the difference between the two expressions vanishes for a large N.

[4] However, neither uniform convergence nor a convergence in the least square sense.

functions equally attenuated and shifted in time, with respect to the transmitted harmonics. The harmonics sequentially received are given by

$$s_{rx}(t) = K_1 \cos\left(2\pi f_{N+n+1}(t-\bar{t}) + \theta_1\right) = K_1 \cos\left(2\pi(f_c + n\Delta f)(t-\bar{t}) + \theta_1\right), \ n = -N, \ldots, N \tag{3.22}$$

where K_1 is the level of the received harmonic (smaller than the level K of the transmitted one) and a generic $\theta_1 \neq \theta$ accounts for a possible difference of phase between the incident and the reflected harmonics.

In order to work out the received synthetic pulse, we have to extract the amplitude and the phase of each received harmonic function.

This is accomplished by means of a demodulation. Here, for simplicity we will refer to a homodyne demodulation scheme even if an heterodyne scheme, making use of an intermediate frequency, is more common [the interested reader is referred to Noon (1996) for deeper technological details].

The block diagram of a homodyne system is provided in Figure 3.3. With reference to Figure 3.3, the received harmonic is multiplied times the carrier (i.e., a reference signal at the same frequency of the transmitted harmonic function) on one branch and multiplied times the carrier delayed by $\pi/2$ on the other branch of the flux diagram. Consequently, at point A the signal will be equal to

$$s_{pn}(t) = 2K_1 \cos\left(2\pi f_{N+n+1}(t-\bar{t}) + \theta_1\right)\cos\left(2\pi f_{N+n+1}t\right)$$
$$= K_1 \cos\left[2\pi(2f_{n+N+1})t - 2\pi f_{n+N+1}\bar{t} + \theta_1\right] + K_1 \cos\left[2\pi f_{n+N+1}\bar{t} - \theta_1\right] \tag{3.23}$$

where the subscript p stands for "in phase" and the subscript n refers to the current nth harmonic. Thus, the signal is the sum of a constant quantity plus a signal oscillating at frequency $2f_{n+N+1}$. The filter in cascade erases the oscillating term, and at point B the signal is equal to

$$p_n = K_1 \cos\left(2\pi f_{n+N+1}\bar{t} - \theta_1\right) \tag{3.24}$$

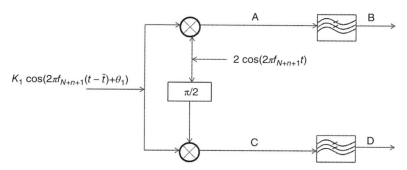

Figure 3.3. Homodyne demodulation scheme.

Similarly, at point C the signal is given by

$$s_{qn}(t) = -2K_1 \cos\left(2\pi f_{N+n+1}(t-\bar{t}) + \theta_1\right) \sin\left(2\pi f_{N+n+1}t\right)$$
$$= -K_1 \sin\left[2\pi(2f_{n+N+1})t - 2\pi f_{n+N+1}\bar{t} + \theta_1\right] - K_1 \sin\left[2\pi f_{n+N+1}\bar{t} - \theta_1\right] \tag{3.25}$$

where the subscript q stands for "in quadrature." Again, the filter in cascade erases the oscillating term, and at point D the signal is equal to

$$q_n = -K_1 \sin\left(2\pi f_{n+N+1}\bar{t} - \theta_1\right) \tag{3.26}$$

After storing the components in phase and in quadrature for all the received harmonics, the synthetic received signal is constructed making use of the complex quantities

$$p_n + jq_n = K_1 \cos\left(2\pi f_{n+N+1}\bar{t} - \theta_1\right) - jK_1 \sin\left(2\pi f_{n+N+1}\bar{t} - \theta_1\right)$$
$$= K_1 \exp(-j(2\pi f_{n+N+1}\bar{t} - \theta_1)) = K_1 \exp(-j(2\pi(f_c + n\Delta f)\bar{t} - \theta_1)), \quad n = -N, \ldots, N \tag{3.27}$$

The quantities (3.27) constitute the values of $\hat{g}_n$ to be substituted in Eq. (3.4). This substitution provides in fact

$$g_{rec}(t) = 2\Delta f \mathrm{Re}\left\{ \exp(j2\pi f_c t) \sum_{n=-N}^{N} \hat{g}_n \exp(j2\pi n\Delta f t)\right\}$$
$$= 2\Delta f K_1 \mathrm{Re}\left\{ \exp(j2\pi f_c(t-\bar{t}) + \theta_1) \sum_{n=-N}^{N} \exp(j2\pi n\Delta(t-\bar{t}))\right\} \tag{3.28}$$
$$= 2\Delta f K_1 \cos\left(2\pi f_c(t-\bar{t}) + \theta_1\right) \frac{\sin\left((2N+1)\pi\Delta f(t-\bar{t})\right)}{\sin\left(\pi\Delta f(t-\bar{t})\right)}$$

where the subscript rec stands for received. Now, the transmitted equivalent train of pulses is given by Eq. (3.17). Therefore, accounting also for Eq. (3.18), the transmitted synthetic train of pulses is expressed as

$$g_{tr}(t) = 2\Delta f K \cos\left(2\pi f_c t + \theta\right) \frac{\sin\left((2N+1)\pi\Delta f t\right)}{\sin\left(\pi\Delta f t\right)} \tag{3.29}$$

where the subscript tr stands for transmitted. Apart from an unimportant phase difference between θ and θ_1 (which means that in general the received pulses don't have the same shape as the transmitted ones), and apart from the obvious attenuation, the comparison between Eq. (3.28) and Eq. (3.29) shows that, within the nonambiguous interval, the GPR receives a synthetic pulse about as large as the transmitted one and centered at the time $\bar{t}$, as expected in a nondispersive medium.

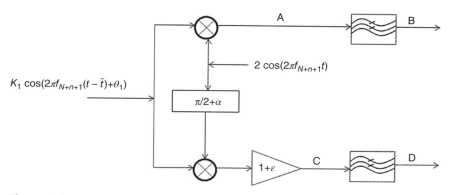

Figure 3.4. Homodyne demodulation scheme with a small phase and amplitude error.

At this point, let us introduce a demodulation error as shown in Figure 3.4. In particular, Figure 3.4 schematizes the fact that, for some technological imprecision, in general the phase shift between the two carriers sent to the p and q channels might be not exactly equal to $\pi/2$, and also some difference in the two amplitudes can occur. These phenomena are accounted for by means of an extra phase shifting and an "undesired" amplifier-attenuator on the q channel. Both the spurious phase factor α and the spurious amplitude factor ε are dimensionless quantities. Their modulus is in general much smaller than 1, and both can be either positive or negative.

In the same way exposed before, it can be worked out that in this case we have at point B

$$p_n = K_1 \cos\left(2\pi f_{n+N+1}\bar{t}-\theta_1\right) \tag{3.30}$$

whereas at point D we have

$$q_n = -K_1(1+\varepsilon)\sin\left(2\pi f_{n+N+1}\bar{t}-\theta_1+\alpha\right)$$

$$= -K_1(1+\varepsilon)\left[\sin\left(2\pi f_{n+N+1}\bar{t}-\theta_1\right)\cos\left(\alpha\right) + \cos\left(2\pi f_{n+N+1}\bar{t}-\theta_1\right)\sin\left(\alpha\right)\right] \tag{3.31}$$

At this point, we can substitute in Eq. (3.31) the first-order approximations

$$\sin\left(\alpha\right)\approx\alpha \quad \text{and} \quad \cos\left(\alpha\right)\approx1 \tag{3.32}$$

thereby obtaining

$$q_n \approx -K_1\sin\left(2\pi f_{n+N+1}\bar{t}-\theta_1\right)-\alpha K_1\cos\left(2\pi f_{n+N+1}\bar{t}-\theta_1\right)-\varepsilon K_1\sin\left(2\pi f_{n+N+1}\bar{t}-\theta_1\right) \tag{3.33}$$

where we have also neglected the higher-order term proportional to the product $\alpha\varepsilon$.

Consequently, the complex quantities retrieved from the demodulation become

$$
\hat{g}_n = p_n + jq_n \approx -K_1 \cos\left(2\pi f_{n+N+1}\bar{t}-\theta_1\right) - jK_1 \sin\left(2\pi f_{n+N+1}\bar{t}-\theta_1\right)
$$
$$
- jaK_1 \cos\left(2\pi f_{n+N+1}\bar{t}-\theta_1\right) - j\varepsilon K_1 \sin\left(2\pi f_{n+N+1}\bar{t}-\theta_1\right)
$$
$$
= K_1 \exp(-j(2\pi f_{n+N+1}\bar{t}-\theta_1)) - \frac{jaK_1}{2}\exp(j(2\pi f_{n+N+1}\bar{t}-\theta_1))
$$

$$
- \frac{jaK_1}{2}\exp(-j(2\pi f_{n+N+1}\bar{t}-\theta_1)) - \frac{\varepsilon K_1}{2}\exp(j(2\pi f_{n+N+1}\bar{t}-\theta_1))
$$

$$
+ \frac{\varepsilon K_1}{2}\exp(-j(2\pi f_{n+N+1}\bar{t}-\theta_1))
$$

$$
= K_1\left(1 - \frac{ja}{2} + \frac{\varepsilon}{2}\right)\exp(-j(2\pi f_{n+N+1}\bar{t}-\theta_1)) - K_1\left(\frac{ja}{2} + \frac{\varepsilon}{2}\right)\exp(j(2\pi f_{n+N+1}\bar{t}-\theta_1))
$$

$$
= K_1\left(1 - \frac{ja}{2} + \frac{\varepsilon}{2}\right)\exp(-j(2\pi(f_c + n\Delta f)\bar{t}-\theta_1))
$$

$$
- K_1\left(\frac{ja}{2} + \frac{\varepsilon}{2}\right)\exp(j(2\pi(f_c + n\Delta f)\bar{t}-\theta_1)) \tag{3.34}
$$

Substituting Eq. (3.34) into Eq. (3.4), we have

$$
g_{rec}(t)
$$

$$
\approx 2\Delta f K_1 \mathrm{Re}\left\{ \exp(j(2\pi f_c(t-\bar{t}) + \theta_1))\left(1 - \frac{ja}{2} + \frac{\varepsilon}{2}\right)\sum_{n=-N}^{N}\exp(j2\pi n\Delta f(t-\bar{t}))\right\}
$$

$$
- 2\Delta f K_1 \mathrm{Re}\left\{ \exp(j(2\pi f_c(t+\bar{t}) - \theta_1))\left(\frac{ja}{2} + \frac{\varepsilon}{2}\right)\sum_{n=-N}^{N}\exp(j2\pi n\Delta f(t+\bar{t}))\right\}
$$

$$
= 2\Delta f K_1\left\{\left(1 + \frac{\varepsilon}{2}\right)\cos\left[2\pi f_c(t-\bar{t}) + \theta_1\right] + \frac{a}{2}\sin\left[2\pi f_c(t-\bar{t}) + \theta_1\right]\right\}\frac{\sin\left((2N+1)\pi\Delta f(t-\bar{t})\right)}{\sin\left(\pi\Delta f(t-\bar{t})\right)}
$$

$$
- 2\Delta f K_1\left\{\frac{\varepsilon}{2}\cos\left[2\pi f_c(t+\bar{t}) - \theta_1\right] + \frac{a}{2}\sin\left[2\pi f_c(t+\bar{t}) - \theta_1\right]\right\}\frac{\sin\left((2N+1)\pi\Delta f(t+\bar{t})\right)}{\sin\left(\pi\Delta f(t+\bar{t})\right)}
$$

$$
= 2\Delta f K_1\sqrt{\left(1 + \frac{\varepsilon}{2}\right)^2 + \frac{a^2}{4}}\cos\left(2\pi f_c(t-\bar{t}) + \theta_1 - tg^{-1}\frac{a}{2+\varepsilon}\right)\frac{\sin\left((2N+1)\pi\Delta f(t-\bar{t})\right)}{\sin\left(\pi\Delta f(t-\bar{t})\right)}
$$

$$
- 2\Delta f K_1\sqrt{\frac{\varepsilon^2}{4} + \frac{a^2}{4}}\cos\left(2\pi f_c(t+\bar{t}) - \theta_1 - tg^{-1}\frac{a}{\varepsilon}\right)\frac{\sin\left((2N+1)\pi\Delta f(t+\bar{t})\right)}{\sin\left(\pi\Delta f(t+\bar{t})\right)} \tag{3.35}
$$

Equation (3.35) shows that the demodulation error essentially provides two Dirichlet sines, that is, two trains of pulses.

Apart from some unimportant amplitude and phase change in the carrier, the first train of pulses is the same achieved without any demodulation error; and, in particular, within the nonambiguous time interval $I = [0, 1/\Delta f]$, it provides the due echo at $t = \bar{t}$. Instead, the second term provides a train of pulses shifted in the opposite verse. Consequently, within the nonambiguous interval, it provides a spurious echo at the time instant $t = 1/\Delta f - \bar{t}$. This echo provides a false target called Hermitian image (Noon, 1996). The amplitude of the Hermitian image depends on the amounts of the amplitude and phase errors of the demodulation system.

Neglecting the effect of the relative shifts between the two carriers and the relative Dirichlet sines (that incidentally depend also on $\bar{t}$, i.e., depend on the position of the target), the average ratio between the level of the "true" pulse and that of its Hermitian image is given by

$$\frac{\sqrt{\left(1 + \frac{\varepsilon}{2}\right)^2 + \frac{\alpha^2}{4}}}{\sqrt{\frac{\varepsilon^2}{4} + \frac{\alpha^2}{4}}} \approx \frac{2}{\sqrt{\varepsilon^2 + \alpha^2}} \tag{3.36}$$

The inverse of Eq. (3.36) provides the average weight of the Hermitian image of a target with respect to the authentic image of the same target. Whether the Hermitian image is tolerable or not depends on the application and on the dynamic range of the system. In particular, if the Hermitian image is below the minimum signal perceivable by the system it does not cause any problem.

In the case where we need to avoid the possibility of any Hermitian image, instead, we have to halve the frequency step at a parity of nonambiguous depth. In this way, all the Hermitian images relative to the investigated depth range will occur beyond it.

Consequently, in order to account for demodulation errors, Eq. (3.12) evolves into

$$\Delta f = \frac{c}{4D} \tag{3.37}$$

Under the further hypothesis that the propagation medium does not show magnetic properties, Eq. (3.37) can be still rewritten as

$$\Delta f = \frac{c_0}{4\sqrt{\varepsilon_{sr}} D} \tag{3.38}$$

3.4 ALIASING AND TIME STEP FOR PULSED GPR SYSTEMS

The aliasing is a problem to be accounted also for pulsed GPR systems. In particular, in this case the received signal is sampled in time domain; thus, because the direct and the inverse Fourier integrals are formally similar, we will have replicas of the spectrum of the signal in the frequency domain. The replicas in frequency domains occurs with frequency step $1/\Delta t$ and therefore we have to guarantee that

$$\Delta t = \frac{1}{B} \approx \frac{1}{f_c} \tag{3.39}$$

where B is the total band of the signal, roughly equal to its central frequency f_c.[5] This time step is called the Nyquist rate, as is well known.

Actually, both the needed frequency step for a stepped frequency system and the time step for a pulsed system are driven by the Nyquist rate, of course with respect to the duration in time domain in the first case and to the band (i.e., the duration in frequency domain) in the second case. An important difference consists in the fact that the band of the antennas is in general apriori known (actually, the behavior of the antennas depends on the background scenario, but the band variation versus the background are in general quite marginal), whereas the maximum depth really reached by the signal is much more case-dependent. This makes the aliasing more difficult to be controlled apriori for a stepped frequency GPR system. On the other hand, the effects of the aliasing for a stepped frequency system are more easily recognizable from the achieved data because, after transformation in time domain and processing, aliased data would provide duplications of the actual targets along the depth and at a fixed distance from the real objects.

Usually, in pulsed systems we have the possibility to set the time bottom scale and the number of time samples, which implicitly defines the time step. Generally, several hundred time samples can be chosen without any computational problems, and this prevents us from aliasing in most situation if one chooses a reasonable time bottom scale with respect to the expected penetration of the signal. Thus, for pulsed systems, the traces are likely to be oversampled (an exercise that shows this is proposed in question 2). However, even if the case is rare from a practical point of view, it is well advised to have some awareness about the problem of the aliasing for a pulsed GPR system.

In stepped frequency systems, usually one can set directly the frequency step, and customarily the maximum option is of the order of 5 MHz (in most cases the available choices are very few, two, or even only one). In this way, the nonambiguous depth is kept very deep (depending on the soil at hand, it often goes beyond 10 m), and the targets deeper than this range should provide echoes too weak to be perceived.

Thus, in both cases, the parameter setting options are usually conceived to provide some oversampling with respect to the Nyquist rate. This is also useful in order to amortize possible uncertainties that we can have about the band of the antennas, the depth of the targets of interest, the characteristics of the soil, and so on.

However, it is also useful for practitioners to be aware of the fact that it is useless to exaggerate the oversampling, maybe thinking that more data provide an indefinitely better resolution. Actually, any band-limited function can be perfectly reconstructed from its samples in the time domain, and dually any time-limited function can be perfectly

---

[5] It is often reported that $\Delta t = 1/2B$, but the question is conventional depending on what we mean by B. Here we mean the range of positive frequencies of a modulated signal, which, after demodulation, extends from $-B/2$ to $B/2$.

reconstructed by its samples in the frequency domain if the samples are gathered at the Nyquist rate (Higgins, 1996). Therefore, it is substantially useless (further than time-consuming and computationally burdening) to oversample the GPR traces with a time (or a frequency) step an order of magnitude smaller than that prescribed by the Nyquist rate.

QUESTIONS

1. Let us suppose we have to perform a measurement campaign where the targets of interest are embedded in the first 50 cm. Let us also suppose that the soil is dry and sandy, with no magnetic properties and with relative permittivity equal to 4. Finally, let us suppose we have a stepped-frequency system. Is a frequency step of 75 MHz presumably adequate?

2. Suppose we handle the same scenario of question 1 and also suppose we have at our disposal a pulsed GPR system, equipped with antennas at 2 GHz. Which is the required time step? How many time samples are needed?

3. What is the effect of the aliasing for a stepped-frequency GPR system?

4. What is the effect of the aliasing for a pulsed GPR system?

5. Given a stepped-frequency GPR system, can we mitigate the effects of an under-sampling in the frequency domain by means of a truncation of the data in the time domain?

6. Given a pulsed GPR system, can we mitigate the effects of an undersampling in the time domain by means of a truncation of the data in the frequency domain?

THE 2D SCATTERING EQUATIONS FOR DIELECTRIC TARGETS

4.1 PRELIMINARY REMARKS

The behavior of the electromagnetic signal radiated by a GPR and scattered by buried targets is governed by Maxwell's equations. So, in order to provide a hopefully deep enough and self-consistent discussion of GPR data processing, we start here from the beginning and derive the whole formulation up to the migration and the linear inversion; this necessarily implies some nontrivial mathematics. In this chapter we introduce the subject within a two-dimensional framework, as often done in the literature (Colton and Kress, 1992; Chew, 1995; Pastorino, 2010). Therefore, with reference to Figure 4.1, we will suppose the reference scenario, the buried targets, and the sources invariant along the y-axis, which outgoes from the sheet. The problem of retrieving the shape, the position, and the electromagnetic nature of the buried targets (or more in general of the buried scenario) is mathematically expressed as the problem of reconstructing the absolute (possibly complex) dielectric permittivity $\varepsilon(x,z)$ and the absolute (possibly complex) magnetic permeability $\mu(x,z)$, both varying with the buried point. In particular, it is comfortable to recast the problem in terms of the dielectric and magnetic contrasts as follows:

Introduction to Ground Penetrating Radar: Inverse Scattering and Data Processing,
First Edition. Raffaele Persico.
© 2014 The Institute of Electrical and Electronics Engineers, Inc. Published 2014 by John Wiley & Sons, Inc.

$$\chi_e(x,z;f) = \frac{\varepsilon(x,z) - \varepsilon_s}{\varepsilon_s} \tag{4.1}$$

$$\chi_m(x,z;f) = \frac{\mu(x,z) - \mu_s}{\mu_s} \tag{4.2}$$

where ε_s is the absolute dielectric permittivity of the soil (in particular $\varepsilon_s = \varepsilon_{sr}\varepsilon_0$, where $\varepsilon_0 = 8.85 \times 10^{-12}$ Farad/m is the absolute permittivity of the free space), and μ_s is the absolute magnetic permeability of the soil (in particular $\mu_s = \mu_0\mu_{sr}$, where $\mu_0 = 1.26 \times 10^{-6}$ Henry/m is the absolute permeability of the free space). The contrasts are functions of the frequency, because of the dispersion. In many cases the dependence of the contrasts on the frequency is just neglected, also because it is not easy to have at one's disposal a reliable dispersion law for the case history at hand. So, we also will not consider this dependence.

At this point, with reference to the scenario depicted in Figure 4.1, let us consider Maxwell's equations in the frequency domain.

In particular, following the most common convention adopted within the electromagnetic literature, differently from Chapter 2, this time we will make use of the circular frequency $\omega = 2\pi f$ instead of the angular frequency f. Consequently, the Fourier transform is now meant as

$$\hat{f}(\omega) = \int_{-\infty}^{+\infty} f(t)\exp(-j\omega t)\,dt \tag{4.3}$$

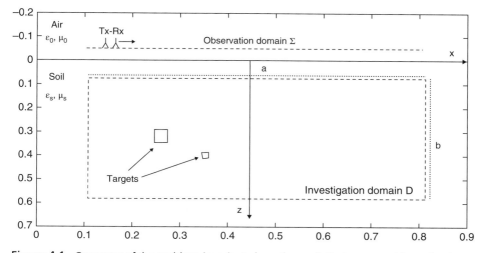

Figure 4.1. Geometry of the problem, invariant along the y-axis that comes out from the sheet. The axes are in meters.

and the inverse Fourier transform is given by

$$f(t) = \frac{1}{2\pi} \int_{-\infty}^{+\infty} \hat{f}(\omega) \exp(j\omega t) \, d\omega \tag{4.4}$$

Let us consider a spatially isotropic time-invariant medium, possibly dispersive, given by two homogeneous half-spaces separated from a plane interface. The upper half-space $(z < 0)$ is built up of free space. The Maxwell's equations, in their differential form in frequency domain, are given by

$$\nabla \times \vec{E} = -j\omega\mu \vec{H} - \vec{J}_m,$$

$$\nabla \times \vec{H} = j\omega\varepsilon \vec{E} + \vec{J},$$

$$\nabla \cdot \varepsilon\vec{E} = \rho, \tag{4.5}$$

$$\nabla \cdot \mu\vec{H} = \rho_m$$

where $\vec{E}$, $\vec{H}$, $\vec{J}$, and $\vec{J}_m$ are the electric field, the magnetic field, the electric impressed density of current, and the magnetic impressed density of current. Beyond the impressed currents, there are induced electric currents too, which are proportional to the electric field by means of the conductivity, and so they have been implicitly considered within the term $j\omega\varepsilon\vec{E}$, because ε is the complex equivalent dielectric permittivity. The quantities ρ and ρ_m are the density of electric and magnetic charges, respectively.

Actually, magnetic currents and magnetic charges do not exist in nature, but some sources can be conveniently described by means of *equivalent* magnetic current and charges (Franceschetti, 1997). In the following we also will make use of this possibility, and so we consider already from now the most complete form of the Maxwell's equations. The charge densities ρ and ρ_m are scalar quantities and are linked to the homologous current densities by means of the continuity equations

$$\nabla \cdot J + j\omega\rho = 0, \tag{4.6}$$

$$\nabla \cdot J_m + j\omega\rho_m = 0 \tag{4.7}$$

The continuity equations descend from the Maxwell's equations. In fact, in order to achieve them, it is sufficient to consider the divergence of the first two Maxwell's equations together with the vector identity $\nabla \cdot \nabla \times v = 0$ $\forall v$. Physically, the continuity equations mean that the charges are neither created nor destroyed. So, if in a given volume the charges augment (or decrease), it means that the difference entered from outside (exited from inside). All the considered quantities are, in general, functions of the spatial point and of the frequency.

4.2   DERIVATION OF THE SCATTERING EQUATIONS WITHOUT CONSIDERING THE EFFECT OF THE ANTENNAS

Let us now consider as source a filamentary current, directed along the y-axis and concentrated at the point (x_0, z_0) in air. Moreover, let us consider the case when no magnetic anomaly is present ($\chi_m = 0$) but dielectric anomalies can be looked for ($\chi_e \neq 0$). The source is given by

$$\vec{J} = I_0 \delta(x-x_0)\delta(z-z_0)i_y \tag{4.8}$$

where I_0 is the level of the current. In this case, the symmetries of the system allow us to state a priori that the electric field has the form

$$\vec{E} = E(x,z)i_y \tag{4.9}$$

Consequently, the magnetic field, from the first Maxwell's equation, can be expressed as

$$\vec{H} = H_x(x,z)i_x + H_z(x,z)i_z \tag{4.10}$$

The permittivity is equal to ε_0 in air and is a generic function of the point $\varepsilon(x,z)$ in the soil, whereas the magnetic permeability is a piecewise constant function given by

$$\mu_b = \begin{cases} \mu_0, & z<0, \\ \mu_s, & z>0 \end{cases} \tag{4.11}$$

In this case, the quantity $\varepsilon(x,z)\,\vec{E}\,(x,z)$ is immediately recognized to be solenoidal—that is, with a null divergence. Thus, the Maxwell equations can be rewritten as

$$\begin{aligned} \nabla \times \vec{E} &= -j\omega\mu_b\,\vec{H}, \\ \nabla \times \vec{H} &= j\omega\varepsilon\vec{E} + I_0\delta(x-x_0)\delta(z-z_0)i_y, \\ \nabla \cdot \varepsilon\vec{E} &= 0, \\ \nabla \cdot \mu_b\vec{H} &= 0 \end{aligned} \tag{4.12}$$

At this point, let us write

$$\varepsilon(x,z) = \varepsilon_b + \Delta\varepsilon(x,z) \tag{4.13}$$

where ε_b is the background permittivity, given by the piecewise constant function

$$\varepsilon_b = \begin{cases} \varepsilon_0, & z<0, \\ \varepsilon_s, & z>0 \end{cases} \tag{4.14}$$

Clearly, the quantity $\Delta\varepsilon(x,z)$ is the contribution to the permittivity function given by the buried objects; that is, it somehow represents the targets.

At this point, let us write

$$\vec{E} = \vec{E}_{inc} + \vec{E}_s = E_{inc}(x,z)i_y + E_s(x,z)i_y, \tag{4.15}$$

$$\vec{H} = \vec{H}_{inc} + \vec{H}_s = H_{incx}(x,z)i_x + H_{incz}(x,z)i_z + H_{sx}(x,z)i_x + H_{sz}(x,z)i_z \tag{4.16}$$

that is, let us decompose the field as the sum of two contributions: the *incident* field plus the *scattered* field. The incident field $(\vec{E}_{inc}, \vec{H}_{inc})$, or unperturbed field, is defined as the field that we would have in absence of buried targets—that is, the field that we would have in correspondence of $\Delta\varepsilon(x,z)=0$. Consequently, the scattered field, or anomalous field $(\vec{E}_s, \vec{H}_s)$, is the contribution due to the buried targets. With these substitutions, the Maxwell's equations (4.12) can be rewritten as

$$\nabla \times \vec{E}_{inc} + \nabla \times \vec{E}_s = -j\omega\mu_b\vec{H}_{inc} - j\omega\mu_b\vec{H}_s,$$

$$\nabla \times \vec{H}_{inc} + \nabla \times \vec{H}_s = j\omega\varepsilon_b\vec{E}_{inc} + j\omega\varepsilon_b\vec{E}_s + j\omega\Delta\varepsilon\,\vec{E} + I_o\delta(x-x_o)\delta(z-z_o)i_y,$$

$$\nabla \cdot \varepsilon_b\vec{E}_{inc} + \nabla \cdot \varepsilon_b\vec{E}_s + \nabla \cdot \Delta\varepsilon\vec{E} = 0, \tag{4.17}$$

$$\nabla \cdot \mu_b\vec{H}_{inc} + \nabla \cdot \mu_b\vec{H}_s = 0$$

Now, each *piece* of the third equation in Eqs. (4.17) is a null quantity, given the direction of the electrical field and its functional dependence [see Eqs. (4.9)–(4.13)]. From Eqs. (4.17), we can extract the Maxwell's equations for the incident field, which are

$$\nabla \times \vec{E}_{inc} = -j\omega\mu_b\vec{H}_{inc},$$

$$\nabla \times \vec{H}_{inc} = j\omega\varepsilon_b\vec{E}_{inc} + I_o\delta(x-x_o)\delta(z-z_o)i_y,$$

$$\nabla \cdot \varepsilon_b\vec{E}_{inc} = 0, \tag{4.18}$$

$$\nabla \cdot \mu_b\vec{H}_{inc} = 0$$

Consequently, the scattered field satisfies the following equations:

$$\nabla \times \vec{E}_s = -j\omega\mu_b\vec{H}_s,$$
$$\nabla \times \vec{H}_s = +j\omega\varepsilon_b\vec{E}_s + j\omega\Delta\varepsilon\vec{E},$$
$$\nabla \cdot \varepsilon_b\vec{E}_s = 0, \tag{4.19}$$
$$\nabla \cdot \mu_b\vec{H}_s = 0$$

In Eqs. (4.19) the quantity $\nabla \cdot \Delta\varepsilon\vec{E}$ has not been considered because, as already said, it is a null function.

At this point, we can recognize that the scattered field is "really" an electromagnetic field, because it is a solution of the Maxwell's equations, relative to the equivalent source

$$\vec{J}_{eq} = j\omega\Delta\varepsilon\vec{E} \tag{4.20}$$

By means of Eqs. (4.19)–(4.20), the buried targets become equivalent to buried sources instead of discontinuities of the medium. So, formally, the scattered field is generated by the equivalent buried sources (also called secondary sources) and not directly by the actual (also called primary) sources. The secondary sources radiate in a homogeneous half-space. Of course, the secondary sources depend, in turn, on the primary ones.

The fact that the secondary sources are solenoidal makes it possible to have a non-null density of current without any accumulation of electrical charges anywhere. Physically, this is possible because of the infinite length of the filamentary current, which makes it an infinite tank for the charges.

Let us now deal with Maxwell's equations relative to the scattered field. To do this, let us use the sampling property of Dirac's function and let us express the secondary sources as follows:

$$\vec{J}_{eq}(x,z) = \iint_D \vec{J}_{eq}(x',z')\delta(x-x')\delta(z-z') \, dx'dz' \tag{4.21}$$

Equation (4.21) expresses the distributed source as an integral sum of concentrated sources. Due to the linearity of Maxwell's equations and also due to the homogeneous radiation conditions that we will impose, we can calculate the field relative to the elementary source $\vec{J}_{eq}(x',z')\delta(x-x')\delta(z-z') \, dx'dz'$ and then integrate all over the investigation domain. So, let us rewrite Maxwell's equations for the scattered field, this time relative to the elementary source $\vec{J}_{eq}(x',z')\delta(x-x')\delta(z-z') \, dx'dz'$:

$$\nabla \times d\vec{E}_s = -j\omega\mu_b \, d\vec{H}_s,$$
$$\nabla \times d\vec{H}_s = +j\omega\varepsilon_b \, d\vec{E}_s + J_{eq}(x',z')\delta(x-x')\delta(z-z') \, dx'dz' \, i_y,$$
$$\nabla \cdot \varepsilon_b \, d\vec{E}_s = 0, \tag{4.22}$$
$$\nabla \cdot \mu_b \, d\vec{H}_s = 0$$

In Eqs. (4.22), the obvious substitution $\vec{J}_{eq}(x',z') = J_{eq}(x',z')i_y$ has been done.

Since both the upper $(z < 0)$ and lower $(z > 0)$ half-spaces are homogeneous, we can look for two solutions in the two half spaces separately and then match them at the interface. So, considering the curl of the first equation and substituting it in the second one, we have

$$\nabla \times \nabla \times d\vec{E}_s = -j\omega\mu_b \nabla \times d\vec{H}_s = -j\omega\mu_b \left(j\omega\varepsilon_b \ d\vec{E}_s + J_{eq}(x',z')\delta(x-x')\delta(z-z') \ dx'dz'i_y \right)$$
$$= k_b^2 \ d\vec{E}_s - j\omega\mu_b J_{eq}(x',z')\delta(x-x')\delta(z-z') \ dx'dz'i_y \qquad (4.23)$$

Let us now use the definition of Laplacian vector: $\nabla^2 \vec{v} = \nabla(\nabla \cdot \vec{v}) - \nabla \times \nabla \times \vec{v}$. Applied in Eq. (4.23), this becomes

$$\nabla\left(\nabla \cdot d\vec{E}_s\right) - \nabla^2 \ d\vec{E}_s = k_b^2 \ d\vec{E}_s - j\omega\mu_b J_{eq}(x',z')\delta(x-x')\delta(z-z') \ dx'dz'i_y \qquad (4.24)$$

However, $\nabla\left(\nabla \cdot d\vec{E}_s\right) = 0$, because of the third of the Maxwell equations (4.22) (the permittivity ε_b is constant in both the upper and lower half-spaces). Consequently, Eq. (4.24) can be rewritten as

$$\nabla^2 \ d\vec{E}_s + k_b^2 \ d\vec{E}_s = j\omega\mu_s J_{eq}(x',z')\delta(x-x')\delta(z-z') \ dx'dz'i_y \qquad (4.25)$$

In Eq. (4.25), the background magnetic permeability has been substituted by the absolute magnetic permeability of the soil, because the secondary sources are null in air. Equation (4.25) is a Helmholtz vector equation. This equation can be projected along the Cartesian axes to produce, in general, three scalar Helmholtz equations. In our case, the electric field has only one component, so we just have a scalar Helmholtz equation along the y-axis. The component of the Laplacian vector operator along the Cartesian axes are just the scalar Laplacians (defined as the divergence of the gradient $\nabla^2\Phi = \nabla \cdot \nabla\Phi$) of the homologous components. Thus, in a Cartesian system, the scalar Laplacian is given by the sum of the second-order nonmixed partial derivatives of the function. Therefore, considering the only y-directed component of the electric scattered field, we have

$$\frac{\partial^2 \ dE_s}{\partial x^2} + \frac{\partial^2 \ dE_s}{\partial z^2} + k_b^2 \ dE_s = j\omega\mu_s J_{eq}(x',z')\delta(x-x')\delta(z-z') \ dx'dz' \qquad (4.26)$$

At this point, we consider the Fourier transform of Eq. (4.26) with respect the x variable, for which we will follow the convention

$$\hat{f}(u,z) = \int_{-\infty}^{+\infty} f(x,z)\exp(-jux) \ dx \qquad (4.27)$$

and consequently the inverse Fourier transform will be given by

$$f(x,z) = \frac{1}{2\pi} \int_{-\infty}^{+\infty} \hat{f}(u,z)\exp(jux) \ du \qquad (4.28)$$

Let us also remember the formula of the Fourier transform of the first derivative $f'(x,z) = \partial f / \partial x$:

$$\hat{f}'(u,z) = \int_{-\infty}^{+\infty} \frac{\partial f}{\partial x}(x,z) \exp(-jux)\, dx = [f(x,z)\exp(-jux)]_{x=-\infty}^{x=+\infty}$$

$$\tag{4.29}$$

$$- \int_{-\infty}^{+\infty} f(x,z)(-ju)\exp(-jux)\, dx = ju \int_{-\infty}^{+\infty} f(x,z)\exp(-jux)\, dx = ju\hat{f}(u,z)$$

where the quantity in square brackets vanishes at infinity because f is a square integrable function. As is customary, we will denote the Fourier transform of the usual operators translated into the (u,z) plane with the same name they had in the (x,z) plane: For example, the gradient of a function $\hat{f}(u,z)$ will be expressed as $\nabla \hat{f} = ju\hat{f} i_x + (\partial \hat{f}/\partial z) i_z$ and, coherently, all the other common differential operators. Actually, this is a loose, even common, language, because (for example) the gradient in the plane (u,z) would be rigorously equal to $\nabla \hat{f} = (\partial \hat{f}/\partial u) i_u + (\partial \hat{f}/\partial z) i_z$. Due to the implicit different semantic of the operators, in general we don't have, in the (u,z) domain, the validity of all the results relative to Maxwell's equations in the (x,z) domain. Notwithstanding, the continuity of tangent component of both the electric and magnetic fields at the air–soil interface holds both in the (x,z) and in the (u,z) domain, because of the continuity of the Fourier operator.

At this point, considering the Fourier transform of Eq. (4.26) with respect to x, we have

$$\frac{\partial^2 d\hat{E}_s}{\partial z^2} + k_z^2 d\hat{E}_s = j\omega\mu_b J_{eq}(x',z')\exp(-jux')\delta(z-z')\, dx'\, dz' \tag{4.30}$$

with

$$k_z^2 = k_b^2 - u^2 = \begin{cases} k_{z0}^2 = k_0^2 - u^2, & z < 0, \\[2mm] k_{zs}{}^2 = k_s^2 - u^2, & 0 < z \end{cases} \tag{4.31}$$

Let us now consider, in the plane (u,z), the three regions $z < 0$, $0 < z < z'$, and $z' < z$, that is, the air half-space, the interval between the air–soil interface and the secondary elementary source, and all the depth range below, respectively.

In each of these areas we don't have sources and the propagation medium is homogeneous. So, we can look for a solution of the homogeneous equation

$$\frac{\partial^2 d\hat{E}_s}{\partial z^2} + k_z^2\, d\hat{E}_s = 0 \tag{4.32}$$

Equation (4.32) is a harmonic equation, whose solution is well known. Therefore, accounting also for the radiation condition to infinity, we can write a formal solution of Eq. (4.32) as follows:

$$d\hat{E}_s = \begin{cases} dA\exp(jk_{z0}z), & z<0, \\ dB\exp(jk_{zs}z) + dC\exp(-jk_{zs}z), & 0<z<z', \\ dD\exp(-jk_{zs}z), & z'<z \end{cases} \tag{4.33}$$

where $k_{z0,zs} = \sqrt{k_{0,s}^2 - u^2}$, and the square root is meant as the determination with nonpositive imaginary part, so that the solution will not diverge at infinity. As already said, in Eqs. (4.33) the radiation condition have been accounted for too, because we have a priori excluded contributions that propagate from the infinite toward the source. At this point, we have to determine the four (differential) constants dA, dB, dC, and dD. In order to do that, we need to impose four independent conditions on them. This is possible, remembering that the tangential components of the fields at the air–soil interface have to be continuous. Moreover, Eq. (4.30) has to be satisfied also at $z = z'$, and this is possible only imposing an impulsive behavior to the second derivative of the field in that point. Consequently, the electric field has to be continuous at $z = z'$, whereas its first derivative has to show a discontinuity of the first kind, whose value [integrating Eq. (4.30) between $z = z'^{-}$ and $z = z'^{+}$] is just $j\omega\mu_b J_{eq}(x',z')\exp(-jux')\,dx'dz'$. All this reasoning leads to four mathematical conditions as follows:

1. *Continuity of the tangential electric field at the air–soil interface*: The electric field is directed along the y-axis, and so it is entirely tangential to the interface. Therefore, this condition means $d\hat{E}_s|_{z=0^-} = d\hat{E}_s|_{z=0^+}$, which from Eqs. (4.33) immediately provides $dA = dB + dC$.

2. *Continuity of the tangential component of the magnetic field at the interface*: To quantify this condition, we have to express the tangent magnetic field versus the electric one. This can be done from the first of Eqs. (4.22) in the transformed domain. So, we have

$$d\hat{H}_s = \frac{j}{\omega\mu_b}\nabla \times d\hat{E}_s = \frac{j}{\omega\mu_b}\begin{vmatrix} i_x & i_y & i_z \\ ju & 0 & \dfrac{\partial}{\partial z} \\ 0 & d\hat{E}_s & 0 \end{vmatrix} = \frac{j}{\omega\mu_b}\left(-\frac{\partial d\hat{E}_s}{\partial z}i_x + ju\,d\hat{E}_s i_z\right).$$

Consequently, the second condition is

$$\frac{1}{\mu_0}\frac{\partial d\hat{E}_s}{\partial z}\bigg|_{z=0^-} = \frac{1}{\mu_s}\frac{\partial d\hat{E}_s}{\partial z}\bigg|_{z=0^+}$$

which, when expressed in terms of the Eqs. (4.33), becomes

$$\frac{k_{z0}\,dA}{\mu_0} = \frac{k_{zs}}{\mu_s}dB - \frac{k_{zs}}{\mu_s}dC.$$

3. *Continuity of the electric field about the source point*: This condition means $d\hat{E}_s|_{z=z'^-} = d\hat{E}_s|_{z=z'^+}$, which, when substituted in Eqs. (4.33), provides $dB\exp(jk_{zs}z') + dC\exp(-jk_{zs}z') = dD\exp(-jk_{zs}z')$.

4. *Integrability of Eq. (4.30) about the singular point*: This means

$$\left.\frac{\partial d\hat{E}_s}{\partial z}\right|_{z'^+} - \left.\frac{\partial d\hat{E}_s}{\partial z}\right|_{z'^-} = j\omega\mu_s J_{eq}(x',z')\exp(-jux')\,dx'dz'$$

In terms of Eqs. (4.33), this is rewritten as follows:

$$-k_{zs}dD\exp(-jk_{zs}z')-(k_{zs}dB\exp(jk_{zs}z')-k_{zs}dC\exp(-jk_{zs}z'))=\omega\mu_s J_{eq}(x',z')\exp(-jux')\,dx'dz'$$

Thus, in the end we work out the following algebraic system:

$$dA = dB + dC,$$

$$\frac{\mu_s k_{z0}}{\mu_0 k_{zs}}dA = dB - dC,$$

$$dD - dB\exp(j2k_{zs}z') - dC = 0,$$ (4.34)

$$dD + dB\exp(j2k_{zs}z') - dC = \frac{-\omega\mu_s}{k_{zs}}\exp(jk_{zs}z')\exp(-jux')J_{eq}(x',z')\,dx'dz'$$

Let us now solve the system. From the difference between the fourth and the third equation, we have immediately

$$dB = \frac{-\omega\mu_s}{2k_{zs}}\exp(-jk_{zs}z')\exp(-jux')J_{eq}(x',z')\,dx'dz'$$ (4.35)

From the sum of the first two equations we obtain

$$dA = \frac{2dB}{1+\dfrac{\mu_s k_{z0}}{\mu_0 k_{zs}}} = \frac{2dB\mu_0 k_{zs}}{\mu_0 k_{zs} + \mu_s k_{z0}}$$

$$= \frac{2\mu_0 k_{zs}}{\mu_0 k_{zs} + \mu_s k_{z0}}\frac{(-1)\omega\mu_s}{2k_{zs}}\exp(-jk_{zs}z')\exp(-jux')J_{eq}(x',z')\,dx'dz'$$

$$= \frac{-\omega\mu_0\mu_s}{\mu_0 k_{zs} + \mu_s k_{z0}}\exp(-jk_{zs}z')\exp(-jux')J_{eq}(x',z')\,dx'dz'$$ (4.36)

Then, from the first equation we get

$$dC = dA - dB = \frac{2dB\mu_0 k_{zs}}{\mu_0 k_{zs} + \mu_s k_{z0}} - dB$$

$$= \frac{\mu_0 k_{zs} - \mu_s k_{z0}}{\mu_0 k_{zs} + \mu_s k_{z0}} dB = \frac{-\omega\mu_s}{2k_{zs}} \frac{\mu_0 k_{zs} - \mu_s k_{z0}}{\mu_0 k_{zs} + \mu_s k_{z0}} \exp(-jk_{zs}z')\exp(-jux')J_{eq}(x',z')\,dx'\,dz'$$

(4.37)

Finally, from the third equation we obtain

$$dD = dB\exp(j2k_{zs}z') + dC$$

$$= \frac{-\omega\mu_s}{2k_{zs}} \left[\left(\frac{\mu_0 k_{zs} - \mu_s k_{z0}}{\mu_0 k_{zs} + \mu_s k_{z0}} \right) \exp(-jk_{zs}z') + \exp(jk_{zs}z') \right] \exp(-jux')J_{eq}(x',z')\,dx'\,dz'$$

(4.38)

Substituting these values in Eqs. (4.33), we have

$$d\hat{E}_s = \begin{cases} \dfrac{-\omega\mu_0\mu_s}{\mu_0 k_{zs} + \mu_s k_{z0}} \exp(-jk_{zs}z')\exp(jk_{z0}z)\exp(-jux')J_{eq}(x',z')\,dx'\,dz', \quad z<0, \\[4mm] \dfrac{-\omega\mu_s}{2k_{zs}} \exp(-jux') \left[\exp(-jk_{zs}(z'-z)) + \dfrac{\mu_0 k_{zs} - \mu_s k_{z0}}{\mu_0 k_{zs} + \mu_s k_{z0}} \exp(-jk_{zs}(z'+z)) \right] \\[2mm] \times J_{eq}(x',z')\,dx'\,dz', \quad 0<z<z', \\[4mm] \dfrac{-\omega\mu_s}{2k_{zs}} \exp(-jux') \left[\exp(-jk_{zs}(z-z')) + \left(\dfrac{\mu_0 k_{zs} - \mu_s k_{z0}}{\mu_0 k_{zs} + \mu_s k_{z0}} \right) \exp(-jk_{zs}(z'+z)) \right] \\[2mm] \times J_{eq}(x',z')\,dx'\,dz', \quad z'<z \end{cases}$$

(4.39)

The second and the third expressions can be joined by considering the absolute value of $z - z'$, thereby providing the more compact formula

$$d\hat{E}_s(u,z;\omega) = \begin{cases} \dfrac{-\omega\mu_0\mu_s}{\mu_0 k_{zs} + \mu_s k_{z0}} \exp(-jk_{zs}z')\exp(jk_{z0}z)\exp(-jux')J_{eq}(x',z')\,dx'\,dz', \quad z<0, \\[4mm] \dfrac{-\omega\mu_s}{2k_{zs}} \exp(-jux') \left[\exp(-jk_{zs}|z-z'|) + \dfrac{\mu_0 k_{zs} - \mu_s k_{z0}}{\mu_0 k_{zs} + \mu_s k_{z0}} \exp(-jk_{zs}(z'+z)) \right] \\[2mm] \times J_{eq}(x',z')\,dx'\,dz', \quad 0<z \end{cases}$$

(4.40)

At this point we can calculate the element of scattered field in the spatial domain (x,z) by back transforming Eq. (4.40). Thus we obtain

$$
dE_s(x,z;\omega) = \begin{cases} \dfrac{-\omega\mu_0\mu_s}{2\pi} J_{eq}(x',z')\,dx'dz' \displaystyle\int_{-\infty}^{+\infty} \dfrac{\exp(-jk_{zs}z')\exp(jk_{z0}z)\exp(-ju(x'-x))}{\mu_0 k_{zs}+\mu_s k_{z0}}\,du, \\[4pt] z<0, \\[10pt] \dfrac{-\omega\mu_s}{4\pi} J_{eq}(x',z')\,dx'dz' \displaystyle\int_{-\infty}^{+\infty} \dfrac{\exp(-ju(x'-x))}{k_{zs}} \\[10pt] \times \left[\exp(-jk_{zs}|z-z'|)+\dfrac{\mu_0 k_{zs}-\mu_s k_{z0}}{\mu_0 k_{zs}+\mu_s k_{z0}}\exp(-jk_{zs}(z'+z))\right]du, \quad 0<z \end{cases}
$$

(4.41)

The overall scattered field is the integral over the investigation domain of the quantity in Eq. (4.41), with the substitution of the value of the equivalent current density, as given in Eq. (4.20). Thus we have:

$$
E_s(x,z;\omega) = \begin{cases} \dfrac{-j\omega^2\mu_0\mu_s}{2\pi}\displaystyle\iint_D \Delta\varepsilon(x',z')E(x',z')\,dx'dz' \displaystyle\int_{-\infty}^{+\infty} \dfrac{\exp(-jk_{zs}z')\exp(jk_{z0}z)\exp(-ju(x'-x))}{\mu_0 k_{zs}+\mu_s k_{z0}}\,du, \\[4pt] z<0, \\[10pt] \dfrac{-j\omega^2\mu_s}{4\pi}\displaystyle\iint_D \Delta\varepsilon(x',z')E(x',z')\,dx'dz' \displaystyle\int_{-\infty}^{+\infty} \dfrac{\exp(-ju(x'-x))}{k_{zs}} \\[10pt] \times \left[\exp(-jk_{zs}|z-z'|)+\dfrac{\mu_0 k_{zs}-\mu_s k_{z0}}{\mu_0 k_{zs}+\mu_s k_{z0}}\exp(-jk_{zs}(z'+z))\right]du, \quad 0<z \end{cases}
$$

(4.42)

At this point, by making use of the dielectric contrast as defined in Eq. (4.1), Eqs. (4.42) can be still rewritten as

$$
E_s(x,z;\omega) = \begin{cases} \dfrac{-jk_s^2\mu_0}{2\pi}\displaystyle\iint_D \chi_e(x',z')E(x',z')\,dx'dz' \displaystyle\int_{-\infty}^{+\infty} \dfrac{\exp(-jk_{zs}(u)z')\exp(jk_{z0}(u)z)\exp(-ju(x'-x))}{\mu_0 k_{zs}(u)+\mu_s k_{z0}(u)}\,du, \\[4pt] z<0 \\[10pt] \dfrac{-jk_s^2}{4\pi}\displaystyle\iint_D \chi_e(x',z')E(x',z')\,dx'dz' \displaystyle\int_{-\infty}^{+\infty} \dfrac{\exp(-ju(x'-x))}{k_{zs}(u)} \\[10pt] \times \left[\exp(-jk_{zs}(u)|z-z'|)+\dfrac{\mu_0 k_{zs}(u)-\mu_s k_{z0}(u)}{\mu_0 k_{zs}(u)+\mu_s k_{z0}(u)}\exp(-jk_{zs}(u)(z'+z))\right]du, \quad 0<z \end{cases}
$$

(4.43)

Adding the incident field to both sides of the second of equations (4.43), eventually we retrieve the scattering equations, in the case of only dielectric anomalies ($\chi_m = 0$):

$$E(x,x_s,z,z_s) = E_{inc}(x,x_s,z,z_s) + k_s^2 \int\int_D \chi_e(x',z')E(x',x_s,z',z_s)G_i(x,x',z,z')\,dx'dz',$$

$$(x,z) \in D, (x_s,z_s) \in \Sigma \tag{4.44}$$

$$E_s(x_0,x_s,z_0,z_s) = k_s^2 \int\int_D \chi_e(x',z')E(x',x_s,z',z_s)G_e(x',x_0,z',z_0)\,dx'dz',$$

$$(x_s,z_s),(x_0,z_0) \in \Sigma \tag{4.45}$$

where $(x,z) \in D$ is the generic point within the investigation domain, $(x_s,z_s) \in \Sigma$ is the source point within the observation domain, and $(x_0,z_0) \in \Sigma$ is the observation point within the observation domain. G_i and G_e are the dielectric internal and external Green's functions, respectively. Their value, based on Eq. (4.43), is given by

$$G_i(x,x',z,z') = \frac{-j}{4\pi} \int_{-\infty}^{+\infty} \frac{\exp(-ju(x'-x))}{k_{zs}} \left[\exp(-jk_{zs}|z-z'|) \right.$$

$$\left. + \frac{\mu_0 k_{zs} - \mu_s k_{z0}}{\mu_0 k_{zs} + \mu_s k_{z0}} \exp(-jk_{zs}(z'+z)) \right] du, \quad (x,z),(x',z') \in D \tag{4.46}$$

$$G_e(x_0,x',z_0,z') = \frac{-j\mu_0}{2\pi} \int_{-\infty}^{+\infty} \frac{\exp(-jk_{zs}z')\exp(jk_{z0}z)\exp(-ju(x'-x_0))}{\mu_0 k_{zs} + \mu_s k_{z0}}\,du,$$

$$(x_0,z_0) \in \Sigma, \quad (x',z') \in D \tag{4.47}$$

The proposed procedure expresses the field in terms of Sommerfeld's integrals (Sommerfeld, 1912) and corresponds to the usual procedure to retrieve the classical scattering equations, with the addition that possible magnetic properties of the soil (usually neglected) have also been accounted for. So, for a correct comparison, these results correspond to those already published [e.g., in Lesselier and Duchene (1996); Chew (1995)] only under the further hypothesis that $\mu_s = \mu_0$.

The exposed calculations have also shown that Green's functions are proportional to the electric field generated by a spatially impulsive source buried in a homogeneous soil. More specifically, the internal Green's function is proportional to the field in the soil, and the external Green's function is proportional to the field in air. This makes the Green's functions square integrable functions, because they are essentially electromagnetic fields generated by finite energy sources.

4.3 CALCULATION OF THE INCIDENT FIELD RADIATED BY A FILAMENTARY CURRENT

In order to calculate the incident field, we have to consider an impulsive source in air instead of an impulsive source in the soil, but the procedure is fully analogous to that just followed, except for the final integration on the investigation domain, which this time is not needed because the source is actually concentrated. The explicit calculation steps are left as an exercise, and the final result is

$$
E_{inc}(x,z;\omega) =
\begin{cases}
\dfrac{-\omega I_o \mu_0}{4\pi} \displaystyle\int_{-\infty}^{+\infty} \dfrac{\exp(-jv(x_s-x))}{k_{z0}(v)} \\[2ex]
\times \left[\exp(-jk_{z0}(v)|z-z_s|) + \dfrac{\mu_s k_{z0}(v)-\mu_0 k_{zs}(v)}{\mu_0 k_{zs}(v)+\mu_s k_{z0}(v)} \exp(-jk_{zs}(v)(z+z_s)) \right] dv, \\[2ex]
0 < z, \\[2ex]
\dfrac{-\omega I_0 \mu_0 \mu_s}{2\pi} \displaystyle\int_{-\infty}^{+\infty} \dfrac{\exp(-jk_{zs}(v)z)\exp(jk_{z0}(v)z_s)\exp(-jv(x_s-x))}{\mu_0 k_{zs}(v)+\mu_s k_{z0}(v)}\, dv, \quad z < 0
\end{cases}
$$

$$(4.48)$$

4.4 THE PLANE WAVE SPECTRUM OF AN ELECTROMAGNETIC SOURCE IN A HOMOGENEOUS SPACE

It is possible, even under a 2D model, to enhance the scattering equation by inserting in them the characteristic of a source more similar to the actual one with respect to the filamentary current adopted in the previous sections. In order to do this, however, we preliminarily need to introduce the concepts of plane wave spectrum.

This is a well-known topic in electromagnetism (Collin, 1985; Franceschetti, 1997). Notwithstanding, a resume is useful both for self-consistency purposes and (above all) in order to have the possibility to outline some subtleties specifically relevant to GPR prospecting.

With regard to the plane wave spectrum, let us initially deal with the most general three-dimensional case, and let us consider a source at height h above the soil and at the coordinatives $x_s = 0$, $y_s = 0$. In a reference system with the z-axis directed downward (as that of Figure 4.1) this means that the source is at a negative height $z_s = -h$. Let us first consider a homogeneous (possibly lossy) medium instead of a half-space. The propagation medium is also assumed to be isotropic and time-invariant. Thus, we have that, for any $z > -h$, the field is the solution of the homogeneous Maxwell's equations.

$$\nabla \times \vec{E} = -j\omega\mu \, \vec{H},$$

$$\nabla \times \vec{H} = j\omega\varepsilon \, \vec{E},$$

$$\nabla \cdot \varepsilon \vec{E} = 0, \tag{4.49}$$

$$\nabla \cdot \mu \vec{H} = 0$$

Following calculation steps very similar to those presented in Section 4.2, it is straightforward to retrieve from Eq. (4.49) a vector Helmholtz equation of the kind

$$\nabla^2 \vec{E} + k^2 \vec{E} = \vec{0} \tag{4.50}$$

where k is the wavenumber of the medium. In scalar terms, Eq. (4.50) is expressed as

$$\frac{\partial^2 E_i}{\partial x^2} + \frac{\partial^2 E_i}{\partial y^2} + \frac{\partial^2 E_i}{\partial z^2} + k^2 E_i = 0 \tag{4.51}$$

where the subscript i indicates any among the three x- y- or z-directed components. At this point, we consider the two-dimensional Fourier transform with respect to the x and y variables, for which we will follow the convention:

$$\hat{f}(u,v,z) = \int_{-\infty}^{+\infty} \int_{-\infty}^{+\infty} f(x,y,z) \exp(-jux) \exp(-jvy) \, dxdy \tag{4.52}$$

Consequently, the inverse Fourier transform is given by

$$f(x,y,z) = \frac{1}{4\pi^2} \int_{-\infty}^{+\infty} \int_{-\infty}^{+\infty} \hat{f}(u,v,z) \exp(jux) \exp(jvx) \, dudv \tag{4.53}$$

The Fourier transform of the partial derivative of the function follows the homologous rule of Eq. (4.29). Consequently, considering the two-dimensional Fourier transform of Eq. (4.51), we obtain

$$-u^2 \hat{E}_i - v^2 \hat{E}_i + \frac{\partial^2 d\hat{E}_i}{\partial z^2} + k^2 \hat{E}_i = 0 \tag{4.54}$$

whose formal solution, accounting for the radiation condition at infinity (which prevents the presence of waves propagating from the infinite toward the sources), can be expressed as

$$\hat{E}_i(u,v,z) = \hat{A}_i(u,v) \exp(-jk_{z1}(u,v)z) \tag{4.55}$$

or, in vector terms,

$$\hat{\vec{E}}_i(u,v,z) = \hat{\vec{A}}(u,v)\exp(-jk_{z1}(u,v)z) \tag{4.56}$$

where

$$k_{z1}(u,v) = \sqrt{k^2 - u^2 - v^2} \tag{4.57}$$

In Eq. (4.56) the determination of the square root with nonpositive imaginary part has to be retained, in order to have a solution where the field vanishes far from the sources (in particular for $z \to +\infty$).[1] We have labeled the quantity in Eq. (4.57) as k_{z1} in order to distinguish it from the quantity k_z previously defined [Eq. (4.31)] in the two-dimensional half-space scalar case. At any rate, physically, both k_z and k_{z1} represent the z-component of the wavevector.

Equation (4.56) tells us that it is sufficient to know the field on any plane $z = \text{const} > -h$ in order to know it (theoretically) everywhere in the half-space $z > -h$. In particular, let us suppose to know the field at the height $z = z_1 > -h$. In this case we can calculate $\vec{E}(u,v,z_1)$ as the two-dimensional Fourier transform of the field on the plane $z = z_1$. Then, according to Eq. (4.56), we have

$$\hat{\vec{E}}(u,v,z_1) = \hat{\vec{A}}(u,v)\exp(-jk_{z1}(u,v)z_1) \Rightarrow \hat{\vec{A}}(u,v) = \hat{\vec{E}}(u,v,z_1)\exp(jk_{z1}(u,v)z_1) \tag{4.58}$$

Consequently, for any other $z = z_2 > -h$, we have

$$\hat{\vec{E}}(u,v,z_2) = \hat{\vec{A}}(u,v)\exp(-jk_{z1}(u,v)z_2) = \hat{\vec{E}}(u,v,z_1)\exp(-jk_{z1}(u,v)(z_2-z_1)) \tag{4.59}$$

At this point, the field in the plane $z = z_2$ is the inverse two-dimensional Fourier transform of $\vec{E}(u,v,z_2)$ according to Eq. (4.53).

Thus, we can refer the plane wave spectrum to any fixed height. Here, we choose the height of the source $z = z_s > -h$,[2] so we calculate the plane wave spectrum as

$$\hat{\vec{E}}(u,v) = \hat{\vec{E}}(u,v,z_s) = \int\limits_{-\infty}^{+\infty}\int\limits_{-\infty}^{+\infty} \vec{E}(x,y,z_s)\exp[-j(xu+yv)]\,dxdy \tag{4.60}$$

The plane wave spectrum expresses the behavior of the sources, acquired either experimentally or theoretically or numerically.

---

[1] Let us remind that the solution looked for is valid only for $z > -h$.

[2] In most texts, there is a default choice at $z = 0$ for the definition of the plane wave spectrum. This is the most natural choice if the propagation medium is a homogeneous medium. However, we will have to pass to the case of a medium composed by two homogeneous half-spaces.

The field for any point with $z > z_s$ is given by

$$\vec{E}(x,y,z) = \frac{1}{4\pi^2} \int\limits_{-\infty}^{+\infty} \int\limits_{-\infty}^{+\infty} \hat{\vec{E}}(u,v) \exp[j(ux + vy - k_{z1}(u,v)(z - z_s))]\, du\, dv \qquad (4.61)$$

Equation (4.61) has been worked out under the condition that $x_s = 0$ and $y_s = 0$. Due to the homogeneity of the considered medium, a shift of the source amounts to the same shift for the field. Thus, if we consider a source placed in a generic (reference) point (x_s, y_s), the field assumes the more general expression:

$$\vec{E}(x,y,z) = \frac{1}{4\pi^2} \int\limits_{-\infty}^{+\infty} \int\limits_{-\infty}^{+\infty} \hat{\vec{E}}(u,v) \exp[j(u(x - x_s) + v(y - y_s) - k_{z1}(u,v)(z - z_s))]\, du\, dv\ z > z_s$$

$$(4.62)$$

The quantity integrated in Eq. (4.62) is a plane wave that propagates along the direction $-ui_x - vi_y + k_{z1}(u,v)i_z$. So, Eq. (4.62) expresses the field as a sum of plane waves (for any fixed frequency).[3] Equivalently, the field can be also expressed as

$$\vec{E}(x,y,z) = \frac{1}{4\pi^2} \int\limits_{-\infty}^{+\infty} \int\limits_{-\infty}^{+\infty} \hat{\vec{E}}(u,v) \exp[-j(u(x - x_s) + v(y - y_s) + k_{z1}(u,v)(z - z_s))]\, du\, dv\ z > z_s$$

$$(4.63)$$

The reader can easily check that the integrals (4.62) and (4.63) provide the same result, but in Eq. (4.63) the generic plane wave component propagates along the direction $ui_x + vi_y + k_{z1}(u,v)i_z$. In the following, we will adopt the expression (4.63).

In the 2D case at hand, it is an easy exercise to check that Eq. (4.63) reduces to

$$\vec{E}(x,z) = \frac{1}{2\pi} \int\limits_{-\infty}^{+\infty} \hat{\vec{E}}(u) \exp[-j(u(x - x_s) + k_z(u)(z - z_s))]\, du\ z > z_s \qquad (4.64)$$

where $k_z(u) = \sqrt{k^2 - u^2}$ and

$$\hat{\vec{E}}(u) = \hat{\vec{E}}(u, z_s) = \int\limits_{-\infty}^{+\infty} \vec{E}(x, z_s) \exp(-jux)\, dx \qquad (4.65)$$

---

[3] Let us outline that the reader can find different expressions of the plane wave spectrum, corresponding to different conventional choices about the definition of the Fourier transform. In particular, Eq. (4.63) might be equivalently expressed with the minus sign before the terms ux and vy in the exponential. Moreover, sometimes the factor $1/4\pi^2$ is included by definition in the plane wave spectrum and thus does not appear explicitly outside the integral.

Moreover, in Eqs. (4.64) and (4.65) both the field and the plane wave spectrum have only the y-component.[4]

Let us now consider a lossless medium. In this case, if the couple (u,v) is internal to the circle of equation $C : u^2 + v^2 = k^2$, then k_{z1} is a real quantity [see Eq. (4.57)] and the plane wave (i.e., a homogeneous plane wave) propagates along a direction with a real component along the z-axis. Instead, if the couple (u,v) is external to the circle C, k_{z1} becomes an imaginary quantity. In this case the plane wave propagates along a direction belonging to the xy plane and vanishes exponentially along the z-axis. The circle C allows us to distinguish the so-called visible plane wave spectrum from the invisible plane wave spectrum. Clearly, at some distance from the sources, only the visible plane wave spectrum gives a meaningful contribution to the field. In the case of a low lossy medium (as in many case the soil is) we cannot distinguish rigorously a visible plane wave spectrum from an invisible one. However, if we define a pseudo-visible plane wave spectrum on the basis of the real part of the wavenumber $k_r = \text{Real}(k)$ (i.e., we consider the circle $C : u^2 + v^2 = k_r^2$), we still recognize that the attenuation along the z-axis increases meaningfully for couples (u,v) outside the pseudo-visible circle. Thus, in any case we have that the field for high values of z contains fewer plane waves than does the field for small values of z.

4.5 THE INSERTION OF THE SOURCE CHARACTERISTICS IN THE SCATTERING EQUATIONS

Let us now come back to the 2D scalar case, but let us consider the case of two homogeneous half-spaces (the upper one made up of free space), and let us start from Eqs. (4.64)–(4.65) written in free space:

$$\vec{E}(x,z) = \frac{1}{2\pi} \int_{-\infty}^{+\infty} \hat{\vec{E}}(u) \exp[-j(u(x-x_s) + k_{z0}(u)(z-z_s))] \, du, \qquad (4.66)$$

$$\hat{\vec{E}}(u) = \hat{\vec{E}}(u,z_s) = \int_{-\infty}^{+\infty} \vec{E}(x,z_s) \exp(-jux) \, dx \qquad (4.67)$$

where $k_{z0}(u) = \sqrt{k_0^2 - u^2}$ is the imaginary part of the nonpositive square root.

Both the electric field and the plane wave spectrum have only one component, y-directed—that is, coming out from the sheet according with the geometry of the problem represented in Figure 4.1.

When the incident field of Eq. (4.66) meets the interface between two homogeneous half-spaces, the comprehensive field can be calculated reflecting and refracting the generic plane wave composing the incident field and then calculating the integral sum of the reflected and refracted waves. The superposition is licit because both the involved

---

[4] The reason why we have considered the more general 3D case instead of directly the 2D case at hand will be clear after reading Section 4.8.

propagation media are linear with respect to the incident field and time invariant. So we have:

$$d\vec{E}_I(x,z) = dE_I(x,z)i_y = \frac{1}{2\pi}\hat{\vec{E}}(u)\exp[j(u(x-x_s)-k_{z0}(u)(z-z_s))]\,du$$

$$= \frac{1}{2\pi}\hat{E}(u)\exp[j(u(x-x_s)-k_{z0}(u)(z-z_s))]\,du i_y \quad (4.68)$$

This incident wave generates both a reflected and a refracted (or transmitted) wave at the air–soil interface, at $z = 0$.

The reflected elementary wave will have the general form

$$dE_R(x,z) = dE_R(x,z)i_y = \frac{1}{2\pi}\hat{\vec{E}}_R(u_1)\exp[j(u_1 x + k_{z0}(u_1)z)]\,du_1$$

$$= \frac{1}{2\pi}\hat{E}_R(u_1)\exp[j(u_1 x + k_{z0}(u_1)z)]\,du_1 i_y \quad (4.69)$$

whereas the elementary transmitted wave will have the general form

$$d\vec{E}_T(x,z) = dE_T(x,z)i_y = \frac{1}{2\pi}\hat{\vec{E}}_T(u_2)\exp[j(u_2 x - k_{zs}(u_2)z)]\,du_2$$

$$= \frac{1}{2\pi}\hat{E}_T(u_2)\exp[j(u_2 x - k_{zs}(u_2)z)]\,du_2 i_y \quad (4.70)$$

where k_{zs} is defined according to Eq. (4.31). Equations (4.69) and (4.70) also account for the fact that Maxwell's equations impose to any plane wave (in any homogeneous isotropic propagation medium) the constraint $u^2 + k_z^2 = k^2$, that is, they impose that the modulus of the wavevector is equal to the wavenumber of the local propagation medium. Moreover, in the expression of the reflected wave, it has been shown that the wavevector has to have its arrow directed in the air, which accounts for the plus sign before $k_{z0}(u_1)z$.

At this point, let us impose the continuity of the tangent components of both the electric and the magnetic field at the air–soil interface, which, as already stated, is also a consequence of Maxwell's equations. With respect to the electric field, we have

$$\frac{1}{2\pi}\hat{E}(u)\exp[j(u(x-x_s)-k_{z0}(u)h)]\,du + \frac{1}{2\pi}\hat{E}_R(u_1)\exp(ju_1 x)\,du_1$$

$$= \frac{1}{2\pi}\hat{E}_T(u_2)\exp(ju_2 x)\,du_2 \quad (4.71)$$

Since the equality has to hold for any value of x, we have as a first result the equality at $u = u_1 = u_2$; that is, the x component of the three wavenumbers is the same. Consequently, Eqs. (4.69)–(4.70) evolve into

$$dE_R(x,z) = dE_R(x,z)i_y = \frac{1}{2\pi}\hat{E}_R(u)\exp[j(ux+k_{z0}(u)z)]\,du$$

$$= \frac{1}{2\pi}\hat{E}_R(u)\exp[j(ux+k_{z0}(u)z)]\,du\,i_y \qquad (4.72)$$

$$dE_T(x,z) = dE_T(x,z)i_y = \frac{1}{2\pi}\hat{E}_T(u)\exp[j(ux-k_{zs}(u)z)]\,du$$

$$= \frac{1}{2\pi}\hat{E}_T(u)\exp[j(ux-k_{zs}(u)z)]\,du\,i_y \qquad (4.73)$$

Equations (4.71)–(4.73) express an equivalent formulation of the Snell's law (Franceschetti, 1997).

However, let us outline that, in the case at hand, part of the incident plane waves are homogeneous and part are inhomogeneous. In other words, the spectral formulation allows us to account for all the involved real and complex incidence, reflection and refraction angles in a compact and smart way. Substituting Eqs. (4.72)–(4.73) into Eq. (4.71), the condition of continuity of the tangent electric field at the interface becomes

$$\hat{E}(u)\exp(-jk_{z0}(u)h)\exp(-jux_s) + \hat{E}_R(u) = \hat{E}_T(u) \qquad (4.74)$$

The magnetic field is expressed versus the electric one from Maxwell's, equations, in the same formal way followed with regard to the scattered field in Section 4.2; thus we have

$$d\vec{H}_I = \frac{j}{\omega\mu_0}\nabla \times d\vec{E}_I = \frac{j}{\omega\mu_0}\begin{vmatrix} i_x & i_y & i_z \\ ju & 0 & \frac{\partial}{\partial z} \\ 0 & d\vec{E}_i & 0 \end{vmatrix} = \frac{j}{\omega\mu_0}\left(-\frac{\partial dE_I}{\partial z}i_x + jud\vec{E}_I i_z\right) \qquad (4.75)$$

As a result, the tangent component of the incident magnetic field is given by

$$dH_{Ix} = -\frac{j}{\omega\mu_0}\frac{\partial dE_I}{\partial z} \qquad (4.76)$$

and analogously

$$dH_{Rx} = -\frac{j}{\omega\mu_0}\frac{\partial dE_R}{\partial z}, \qquad (4.77)$$

$$dH_{Tx} = -\frac{j}{\omega\mu_s}\frac{\partial dE_T}{\partial z} \qquad (4.78)$$

Consequently, the continuity of the tangent component of the magnetic field at the air–soil interface is expressed as

$$\frac{1}{\mu_0}\frac{\partial dE_I}{\partial z}\bigg|_{z=0^-} + \frac{1}{\mu_0}\frac{\partial dE_R}{\partial z}\bigg|_{z=0^-} = \frac{1}{\mu_s}\frac{\partial dE_T}{\partial z}\bigg|_{z=0^+} \qquad (4.79)$$

Substituting Eqs. (4.68), (4.72), and (4.73) into Eq. (4.79), we obtain

$$\frac{k_{z0}(u)}{\mu_0}\hat{E}(u)\exp(-jux_s)\exp(-jk_{z0}(u)h) - \frac{k_{z0}(u)}{\mu_0}\hat{E}_R(u) = \frac{k_{zs}(u)}{\mu_s}\hat{E}_T(u) \qquad (4.80)$$

Equations (4.74) and (4.80) allow us to solve the reflected and the transmitted plane wave spectra versus the incident one. Thus, we have

$$\hat{E}_R(u) = \frac{\mu_s k_{z0}(u) - \mu_0 k_{zs}(u)}{\mu_s k_{z0}(u) + \mu_0 k_{zs}(u)}\hat{E}(u)\exp(-jux_s)\exp(-jk_{z0}(u)h), \qquad (4.81)$$

$$\hat{E}_T(u) = \frac{2\mu_s k_{z0}(u)}{\mu_s k_{z0}(u) + \mu_0 k_{zs}(u)}\hat{E}(u)\exp(-jux_s)\exp(-jk_{z0}(u)h) \qquad (4.82)$$

Equations (4.81)–(4.82) make it clear that the quantity $\dfrac{\mu_s k_{z0}(u) - \mu_0 k_{zs}(u)}{\mu_s k_{z0}(u) + \mu_0 k_{zs}(u)}$ has the physical meaning of an air–soil reflection coefficient, whereas the quantity $\dfrac{2\mu_s k_{z0}(u)}{\mu_s k_{z0}(u) + \mu_0 k_{zs}(u)}$ has the physical meaning of an air–soil transmission coefficient.

Therefore, the (unique) y-component of the field, which is the incident field for the inverse scattering problem related to the reconstruction of buried targets, is given by

$$E_{inc}(x,z) = \begin{cases} \dfrac{1}{2\pi}\displaystyle\int_{-\infty}^{+\infty} \hat{E}(u)\exp(jux)\exp(-jux_s)\exp(-jk_{z0}(u)h) \\[2mm] \times\left[\exp(-jk_{z0}(u)z) + \dfrac{\mu_s k_{z0}(u) - \mu_0 k_{zs}(u)}{\mu_s k_{z0}(u) + \mu_0 k_{zs}(u)}\exp(jk_{z0}(u)z)\right]du, \quad z<0, \\[4mm] \dfrac{1}{2\pi}\displaystyle\int_{-\infty}^{+\infty} \hat{E}(u)\exp(jux)\exp(-jux_s)\exp(jk_{z0}(u)z_s) \\[2mm] \times\dfrac{2\mu_s k_{z0}(u)}{\mu_s k_{z0}(u) + \mu_0 k_{zs}(u)}\exp(-jk_{z0}(u)z)\,du, \quad z>0 \end{cases}$$

$$(4.83)$$

Equation (4.83) provides the incident field everywhere in a medium composed of two homogeneous half-spaces, with interface at $z=0$. Equation (4.83) constitutes the generalization of Eq. (4.48), where the incident field is calculated in the case that the source is a filamentary current. In order to distinguish the characteristics of the source from those of the receiving antenna, from now on we will relabel $\hat{E}$ as $\hat{E}_{Tx}$, where the suffix Tx stands for transmitting antenna. Thus, Eq. (4.83) is rewritten as

$$
E_{inc}(x,z) =
\begin{cases}
\dfrac{1}{2\pi} \displaystyle\int_{-\infty}^{+\infty} \hat{E}_{Tx}(u)\exp(jux)\exp(-jux_s)\exp(-jk_{z0}(u)h) \\[2ex]
\quad \times \left[\exp(-jk_{z0}(u)z) + \dfrac{\mu_s k_{z0}(u) - \mu_0 k_{zs}(u)}{\mu_s k_{z0}(u) + \mu_0 k_{zs}(u)} \exp(jk_{z0}(u)z) \right] du, \quad z < 0, \\[3ex]
\dfrac{1}{2\pi} \displaystyle\int_{-\infty}^{+\infty} \hat{E}_{Tx}(u)\exp(jux)\exp(-jux_s)\exp(jk_{z0}(u)z_s) \\[2ex]
\quad \times \dfrac{2\mu_s k_{z0}(u)}{\mu_s k_{z0}(u) + \mu_0 k_{zs}(u)} \exp(-jk_{z0}(u)z)\, du, \quad z > 0
\end{cases}
$$

$$(4.84)$$

To account for the characteristics of the source in the scattering equation just means to insert the incident field given by Eq. (4.84) in them instead of the incident field related to a filamentary current, given by Eq. (4.48). In particular, the plane wave spectrum of a filamentary current I_0 is given by (Clemmow, 1996; Harrington, 1961)

$$
\hat{E}_{Tx}(v) = \frac{-\omega\mu_0 I_0}{2} \frac{1}{k_{z0}(v)}
\tag{4.85}
$$

which, when substituted in Eq. (4.84), gives back Eq. (4.48).

Before closing this section, a further observation is proposed: Since the calculations are based on Maxwell's equations, the primary source of the field is some given current density, and this is considered unchanged passing from the case of a homogeneous space to the case of a half-space. Actually, the density current that generates the field is provided by the currents that arise on the exploited transmitting antenna. These currents in general are not the same when the antenna radiates in free space and when it radiates near the soil. So, if we somehow measure or simulate the plane wave spectrum of the transmitting antenna in free space, and then we make use of it in Eq. (4.84), then we are implicitly neglecting the different behavior of the antenna in free space and close to the soil.

4.6   THE FAR FIELD IN A HOMOGENEOUS LOSSLESS SPACE IN TERMS OF PLANE WAVE SPECTRUM

Let us now devote attention to the insertion of the characteristics of the receiver into the scattering equations. In order to do that, we need another digression starting from the 3D plane wave spectrum of a transmitting antenna. So, let us consider the polar reference system of Figure 4.2, and let us express the involved vectors by means of their projections along the Cartesian axes, in turn expressed in terms of polar coordinates. So we have

$$\vec{r} = x i_x + y i_y + z i_z = r\sin(\theta)\cos(\varphi)i_x + r\sin(\varphi)\sin(\varphi)i_y + r\cos(\theta)i_z, \tag{4.86}$$

$$\vec{r}_s = x_s i_x + y_s i_y + z_s i_z = r_s \sin(\theta_s)\cos(\varphi_s)i_x + r_s\sin(\theta_s)\sin(\varphi_s)i_y + r_s\cos(\theta_s)i_z, \tag{4.87}$$

$$\vec{k}(u,v) = u i_x + v i_y + k_{z1}(u,v)i_z = w\cos(\varphi_1)i_x + w\sin(\varphi_1)i_y + k_{z1}(w\cos(\varphi_1), w\sin(\varphi_1))i_z$$
$$= k\left(w_n \cos(\varphi_1)i_x + w_n \sin(\varphi_1)i_y + k_{z1n}(w_n)i_z\right) \tag{4.88}$$

where we have introduced the following normalized (with respect to the wavenumber) quantities:

$$w_n = \frac{w}{k},$$

$$k_{z1n}(w_n) = \frac{k_{z1}(w\cos(\varphi_1), w\sin(\varphi_1))}{k} = \frac{\sqrt{k^2 - w^2\cos^2(\varphi_1) - w^2\sin^2(\varphi_1)}}{k} \tag{4.89}$$

$$= \frac{\sqrt{k^2 - w^2}}{k} = \sqrt{1 - w_n^2}$$

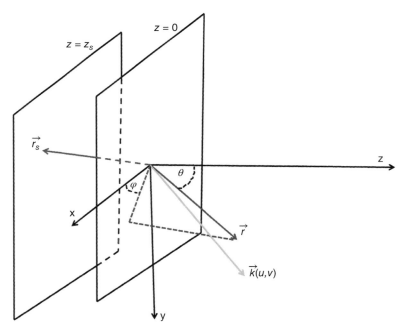

Figure 4.2. Polar and Cartesian reference system at hand. The angular coordinates φ_1, θ_s, and φ_s are easily inferred.

After doing that, expressing the integral of Eq. (4.63) in planar polar variables, we have

$$
\vec{E}(r,\theta,\varphi) = \frac{k^2}{4\pi^2} \int_0^{2\pi} d\varphi_1 \int_0^{+\infty} w_n dw_n \hat{\vec{E}} \left(kw_n \cos(\varphi_1), kw_n \sin(\varphi_1) \right)
$$

$$
\times \exp[-jkr(w_n \cos(\varphi_1) \sin(\theta) \cos(\varphi) + w_n \sin(\phi_1) \sin(\theta) \sin(\varphi) + k_{z1n}(w_n) \cos(\theta))]
$$

$$
\times \exp[jkr_s(w_n \cos(\varphi_1) \sin(\theta_s) \cos(\varphi_s) + w_n \sin(\varphi_1) \sin(\theta_s) \sin(\varphi_s) + k_{z1n}(w_n) \cos(\theta_s))]
$$

$$(4.90)$$

Since the expression is considered only for great values of the distance r, the integral (4.90) can be calculated with the method of the stationary phase method. This method is based on the fact that the exponential term related to the observation point constitutes a strongly oscillating function versus the integration variables, unless the derivatives of the argument of the exponential is equal to zero. Thus, the most important contributions to the results are expected to be related to the neighbors of the roots of derivatives of that argument, which are called "stationary points" (Felsen and Marcuvitz, 1994).

In order to provide the general stationary phase formula for integrals in two variables, let us consider two "slowly varying" functions $f(x,y)$ and $g(x,y)$, where $g(x,y)$ is a real function (instead, $f(x,y)$ can be either a real or a complex function), and let us consider a real parameter $\Omega \rightarrow +\infty$. In this case, for any quadruple of real values X_1, $X_2 > X_1$, Y_1, $Y_2 > Y_1$ we can approximate

$$
\int_{X_1}^{X_2} dx \int_{Y_1}^{Y_2} f(x,y) \exp[j\Omega g(x,y)] \, dy \approx \frac{2\pi j f(x_0,y_0) \exp[j\Omega g(x_0,y_0)]}{\Omega \sqrt{\dfrac{\partial^2 g}{\partial x^2}(x_0,y_0) \dfrac{\partial^2 g}{\partial y^2}(x_0,y_0)}}
\tag{4.91}
$$

where the point (x_0,y_0) is the (supposed unique) stationary point enclosed in the integrated area.

Moreover, in Eq. (4.91), it is implicitly supposed that the stationary point is of the first order; that is, the second-order derivatives do not vanish in it. If more than one first-order stationary points are present, each of them provides an additive contribution of the kind of Eq. (4.91) to the field. If higher-order stationary points are present (i.e., point where the first-, the second- and possibly also higher-order derivatives vanish), they provide further additive contributions, even if these will be not expressed by means of Eq. (4.91).

That said, let us label the phase terms at hand as follows:

$$
\psi(w_n,\varphi_1;\theta,\varphi) = w_n \sin(\theta) \cos(\varphi) \cos(\varphi_1) + w_n \sin(\theta) \sin(\varphi) \sin(\varphi_1) + k_{1zn}(w_n) \cos(\theta)
$$

$$(4.92)$$

and similarly

$$
\psi_s(w,\varphi_1;\theta_s,\varphi_s) = w_n \sin(\theta_s) \cos(\varphi_s) \cos(\varphi_1) + w_n \sin(\theta_s) \sin(\varphi_s) \sin(\varphi_1) - k_{1zn}(w_n) \cos(\theta_s)
$$

$$(4.93)$$

Let us now calculate the first-order derivatives, in order to look for the stationary points:

$$\frac{\partial \psi}{\partial \varphi_1} = -w_n \sin(\theta) \cos(\varphi) \sin(\varphi_1) + w_n \sin(\theta) \sin(\varphi) \cos(\varphi_1)$$

$$= w_n \sin(\theta) \sin(\varphi - \varphi_1),$$

$$\frac{\partial \psi}{\partial w_n} = \sin(\theta) \cos(\varphi) \cos(\varphi_1) + \sin(\theta) \sin(\varphi) \sin(\varphi_1) - \cos(\theta) \frac{w_n}{\sqrt{1-w_n^2}} \qquad (4.94)$$

$$= \sin(\theta) \cos(\varphi - \varphi_1) - \cos(\theta) \frac{w_n}{\sqrt{1-w_n^2}}$$

Looking for the roots of the gradient, the first of Eq. (4.94) admits two distinct solutions (apart from the periodicity, which is not influent here). They are $\varphi_1 = \varphi$ and $\varphi_1 = \varphi + \pi$. However, due to the geometrical condition that the observation point is beyond the source (which means that $0 \le \theta \le \pi/2$ (see Figure 4.2), so that both $\sin(\theta)$ and $\cos(\theta)$ are non-negative quantities), and due to the already quoted radiation conditions, the reader can test that the stationary point with $\varphi_1 = \varphi + \pi$ would lead to a wavevector with negative z-component, which corresponds to an unacceptable wave coming from the infinite toward the sources. Thus, substituting $\varphi_1 = \varphi$ in the second of Eqs. (4.94), we determine that there is a unique acceptable stationary point $(w_n, \varphi_1) = (\sin \theta, \varphi)$. This means that the main contribution is given by the plane wave that propagates in the same direction of the observation point. Let us now calculate the second derivatives of the phase term ψ:

$$\frac{\partial^2 \psi}{\partial \varphi_1^2} = -w_n \sin(\theta) \cos(\varphi - \varphi_1)$$

$$\frac{\partial^2 \psi}{\partial w_n^2} = -\cos(\theta) \frac{\sqrt{1-w_n^2} + \dfrac{w_n^2}{\sqrt{1-w_n^2}}}{1-w_n^2} = -\cos(\theta) \frac{1}{\left(1-w_n^2\right)^{3/2}} \qquad (4.95)$$

which provides the following in the stationary point:

$$\sqrt{\left.\frac{\partial^2 \psi}{\partial w_n^2} \frac{\partial^2 \psi}{\partial \varphi_1^2}\right|_{(w_n,\varphi_1) = (\sin(\theta), \varphi)}} = tg(\theta) \qquad (4.96)$$

Thus, the stationary point is of the first order and we can apply (4.94) thereby obtaining the following in far field:

$$\vec{E}(r,\theta,\varphi) \approx \frac{j}{2\pi} \frac{k\cos(\theta)}{r} \hat{\vec{E}}(k\sin(\theta)\cos(\varphi), k\sin(\theta)\sin(\varphi)) \exp(-jkr) \exp(jkr_s \cdot i_r)$$

$$= \frac{j\exp(-jkr)\exp\left(j\vec{k}\cdot\vec{r}_s\right)}{\lambda r} \cos(\theta) \hat{\vec{E}}(k\sin(\theta)\cos(\varphi), k\sin(\theta)\sin(\varphi)) \qquad (4.97)$$

This equation shows that in far field the angular dependence becomes uncoupled with respect to the radial dependence. In particular, the radial dependence of the far field is independent from the specific source at hand. Let us stress that the hypothesis of the lossless medium was central in the application of the stationary phase. Actually, to speak about far-field pattern in a lossy medium does not have a rigorous sense, even if it is loosely done in some case. However, the plane wave spectrum expresses the characteristic of the source even in a lossy medium.

4.7   THE EFFECTIVE LENGTH OF AN ELECTROMAGNETIC SOURCE IN A HOMOGENEOUS SPACE

Let us now return to a three-dimensional geometry and to a homogeneous medium. The effective length in transmission mode of an antenna is defined as a vector such as the far field radiated by the antenna can be written, in polar coordinates (see Figure 4.2), as

$$\vec{E}(r,\theta,\phi) \approx \frac{jk\varsigma}{4\pi} I_R \vec{h}_T(\theta,\phi) \frac{e^{-jkr}}{r} \qquad (4.98)$$

where ς is the intrinsic impedance of the medium. In the case of free space, it is equal to $\varsigma_0 = \sqrt{\mu_0/\varepsilon_0} \approx 377$ ohms. In the case of wire antennas fed at the gap, I_R is the current at the gap; otherwise it is, more in general, a reference current. In the case that I_R is a reference current, there is some degree of arbitrariness in its choice. However, this will reflect in an inverse proportionality of the value of the effective length so that the electromagnetic field (which is the physical quantity) will not depend on this. The effective length is a complex vector, independent on the distance between the source and the observation point in far field, and has the physical dimension of a length. Let us outline that Eq. (4.98) is valid at a fixed frequency; that is, the effective length (as well as the plane wave spectrum and the far-field pattern) is a function of the frequency too. By comparing Eq. (4.98) with Eq. (4.97), we discover the relationship between the effective length and the plane wave spectrum:

$$\vec{h}_T(\theta,\varphi) = \frac{2\exp\left(j\vec{k}\cdot\vec{r}_s\right)}{I_R\varsigma} \cos(\theta)\hat{\vec{E}}\left(k\sin(\theta)\cos(\varphi), k\sin(\theta)\sin(\varphi)\right) \qquad (4.99)$$

In free space, Eq. (4.99) can be still rewritten as

$$\vec{h}_T(\theta,\phi) = \frac{\lambda_0 \exp\left(j\vec{k}_0\cdot\vec{r}_s\right)}{\pi I_R\varsigma_0} k_0 \cos(\theta)\hat{\vec{E}}\left(k_0\sin(\theta)\cos(\phi), k_0\sin(\theta)\sin(\phi)\right) \qquad (4.100)$$

If we consider only homogeneous waves, then Eq. (4.99) is meant with θ varying in the range $(0,\pi/2)$ and φ varying in the range $(0, 2\pi)$. If we also consider inhomogeneous waves, attenuating along the z-axis, then complex angles should be accounted for; or, alternatively, Eq. (4.100) can be smartly expressed versus the planar component of the wavevector as follows:

$$\vec{h}_T(u,v) = \frac{\lambda_0 \exp\left(j\vec{k}\cdot\vec{r}_s\right)}{\pi I_R \varsigma_0} k_{z1}(u,v)\hat{E}(u,v) = \frac{\lambda_0 \exp\left(j\vec{k}\cdot\vec{r}_s\right)}{\pi I_R \varsigma_0}\sqrt{k_0^2 - u^2 - v^2}\hat{E}(u,v) \quad (4.101)$$

Let us now consider the antenna in reception mode, on which impinges an incident plane wave, which we will express versus its propagation direction as $\vec{E}_{inc}(u,v)$, and let us label as V_0 the open-circuit voltage at the clamps[5] of the gap. The effective length in reception mode of the antenna is defined by means of the equality

$$V_0(x,y,z) = -\vec{h}_R(u,v)\cdot\vec{E}_{inc}(u,v) = -\vec{h}_R(u,v)\cdot\vec{E}_0 \exp[j(ux+vy\pm k_{z1}(u,v)z)] \quad (4.102)$$

The sign before the third term in the exponential depends on whether the plane wave at hand propagates toward the positive or the negative z-axis. Let us outline that $\vec{E}_{inc}(u,v)$ is not the electric field at the gap but instead the incident field—that is, the field that we would have at that point in absence of the receiving antenna. The presence of the receiving antenna, of course, influences and modifies the field—in particular, reirradiating part of the impinging energy. Now, if the incident field is not given by a single plane wave, then it can be expressed as a superposition of plane waves. In this case, due to the linearity of the problem at hand and due to the time-invariant behavior of the receiving antenna, the open-circuit voltage will be expressed as the integral sum of the contribution of all the plane waves composing the incident field, so that we have

$$V_0(x,y,z) = -\frac{1}{4\pi^2}\int_{-\infty}^{+\infty}\int_{-\infty}^{+\infty}\vec{h}_R(u,v)\cdot\hat{E}(u,v)\exp[j(ux+vy\pm k_{z1}(u,v)z)]\,dudv \quad (4.103)$$

At this point, a well-known result in electromagnetism, which is a consequence of the reciprocity theorem (Franceschetti, 1997), is that the effective length in reception mode is the same as the effective length in transmission mode. Therefore, this quantity will now be referred to as just the effective length of the antenna:

$$\vec{h}(u,v) = \vec{h}_T(u,v) = \vec{h}_R(u,v) \quad (4.104)$$

and so we can rewrite Eq. (4.103) as

$$V_0(x,y,z) = -\frac{1}{4\pi^2}\int_{-\infty}^{+\infty}\int_{-\infty}^{+\infty}\vec{h}(u,v)\cdot\hat{E}(u,v)\exp[j(ux+vy\pm k_{z1}(u,v)z)]\,dudv \quad (4.105)$$

---

[5] There are antennas where we cannot identify physical clamps, but even in this case it is possible to consider an equivalent voltage at equivalent clamps (Stutzman and Thiele, 1998).

4.8 THE INSERTION OF THE RECEIVER CHARACTERISTICS IN THE SCATTERING EQUATIONS

We can now consider the problem of inserting the receiver characteristics into the scattering operator. To consider the receiver means to consider as output the "scattered voltage" instead of the scattered field—that is, to consider the voltage at the gap (or at some equivalent gap) of the receiving antenna, instead of the scattered field in the same point. Unlike the case of the source characteristics, where the plane wave spectrum could be easily restricted into a 2D framework, the insertion of the receiver characteristics makes it theoretically hard to remain rigorously into a two-dimensional dealing. In particular, let us note that the filamentary current is a physical abstraction and is a source without clamps, either actual or equivalent. This does not prevent the fact that in many cases a two-dimensional model of the incident and scattered field can be satisfying and convenient. In order to solve this theoretical "embarrassment," we can suppose that a two-dimensional source radiates the field and two-dimensional targets scatter it. However, we will consider momentarily a "real" three-dimensional antenna as receiver. That said, let us reconsider the first in Eqs. (4.41), reported here to simplify the reading:

$$
d\vec{E}_s(x,y,z) = i_y \frac{-\omega\mu_0\mu_s}{2\pi} J_{eq}(x',z')\, dx'dz'
$$

$$
\times \int_{-\infty}^{+\infty} \frac{\exp(-jk_{zs}z')\exp(jk_{z0}z)\exp(-ju(x'-x))}{\mu_0 k_{zs} + \mu_s k_{z0}}\, du, \qquad z < 0
\tag{4.106}
$$

With respect to Eq. (4.41), in Eq. (4.106) we have taken implicit the dependence on the frequency and have instead explicitly shown the fact that the elementary contribution to the field is directed along the y-axis. We have also introduced an artful formal dependence along the y-axis, even if the function at hand is by definition constant along the y-axis.

Let us now consider the three-dimensional plane wave spectrum of this electrical field with reference to the air-soil interface. Considering that

$$
\int_{-\infty}^{+\infty} \exp(-jvy)\, dy = 2\pi\delta(v)
\tag{4.107}
$$

and considering that Eq. (4.106) has the structure of an inverse Fourier transform in the variable x, the plane wave spectrum of Eq. (4.106) is given by

$$
d\hat{\vec{E}}_s(u,v) = -i_y 2\pi\omega\mu_0\mu_s J_{eq}(x',z')\, dx'dz' \frac{\exp(-jk_{zs}(u)z')\exp(-jux')}{\mu_0 k_{zs}(u) + \mu_s k_{z0}(u)}\delta(v)
\tag{4.108}
$$

Therefore, the contribution of the elementary buried source to the scattered voltage can be expressed according to Eq. (4.105) as

$$dV_0 = -\frac{\omega\mu_0\mu_s J_{eq}(x',z')\,dx'\,dz'}{2\pi} \int_{-\infty}^{+\infty}\int_{-\infty}^{+\infty} h_y(u,v)\frac{\exp(-jk_{zs}(u)z')\exp(jk_{z0}(u)z)\exp(-ju(x'-x))}{\mu_0 k_{zs}(u)+\mu_s k_{z0}(u)}\delta(v)\,du\,dv$$

$$= -\frac{\omega\mu_0\mu_s J_{eq}(x',z')\,dx'\,dz'}{2\pi} \int_{-\infty}^{+\infty} h_y(u,0)\frac{\exp(-jk_{zs}(u)z')\exp(jk_{z0}(u)z)\exp(-jux')}{\mu_0 k_{zs}(u)+\mu_s k_{z0}(u)}\,du \qquad (4.109)$$

Accounting for Eq. (4.101) [wherein the source point is $\vec{r}_s = (x',z')$], Eq. (4.109) can be expressed versus the plane wave spectrum of the receiving antenna as follows:

$$dV_0 = -\frac{\omega\mu_0\mu_s J_{eq}(x',z')\lambda_0\,dx'\,dz'}{2\pi^2 I_R \varsigma_0}$$
$$\int_{-\infty}^{+\infty} \hat{E}_y(u,0)k_{z0}(u)\frac{\exp(-jk_{zs}(u)z')\exp(jk_{z0}(u)z)\exp(-ju(x'-x))}{\mu_0 k_{zs}(u)+\mu_s k_{z0}(u)}\,du \qquad (4.110)$$

The comprehensive scattered voltage is determined by integrating on x' and z' after substituting the equivalent buried current density according to Eq. (4.20). So we have

$$V_0(x,z) = -\frac{j\omega^2\mu_0\mu_s\varepsilon_s\lambda_0}{2\pi^2 I_R\varsigma_0} \int_{-\infty}^{+\infty}\int_{-\infty}^{+\infty} E(x',z';x_s,z_s)\chi_e(x',z')$$
$$\times \left[\int_{-\infty}^{+\infty} \hat{E}_y(u,0)k_{z0}(u)\frac{\exp(-jk_{zs}(u)z')\exp(jk_{z0}(u)z)\exp(-ju(x'-x))}{\mu_0 k_{zs}(u)+\mu_s k_{z0}(u)}\,du\right]dx'\,dz'$$

$$(4.111)$$

In Eq. (4.111), E is the internal (i.e., underground) field due to a 2D primary source placed in (x_s,z_s), and it is dealt with as a scalar because it is implied that it has the only y-component. Instead, $\hat{E}_y$ is the y-component of the plane wave spectrum of the receiving antenna in air, which in general has also the x- and the z-components. At this point, we define the pseudo-2D plane wave spectrum of the receiver as

$$\hat{E}_{Rx}(u) = \frac{\hat{E}_y(u,0)}{m} \qquad (4.112)$$

where m is a constant dimensional quantity equal to 1 m. After doing this, Eq. (4.111) can be rewritten as

$$V_0(x,z) = -\frac{jmk_s^2\mu_0\lambda_0}{2\pi^2 I_R\varsigma_0} \int_{-\infty}^{+\infty}\int_{-\infty}^{+\infty} E(x',z';x_s,z_s)\chi_e(x',z')$$
$$\times \left[\int_{-\infty}^{+\infty} \hat{E}_{Rx}(u)k_{z0}(u)\frac{\exp(-jk_{zs}(u)z')\exp(jk_{z0}(u)z)\exp(-ju(x'-x))}{\mu_0 k_{zs}(u)+\mu_s k_{z0}(u)}\,du\right]dx'\,dz'$$

$$(4.113)$$

Equation (4.113) can be expressed in the form

$$V_0(x,z) = k_s^2 \int\limits_{-\infty}^{+\infty} \int\limits_{-\infty}^{+\infty} G_{ev}(x',z';x,z)E(x',z';x_s,z_s)\chi_e(x',z') \, dx'dz' \tag{4.114}$$

where, for comparison, we have

$$G_{ev}(x',z';x,z) = -\frac{jm\mu_0\lambda_0}{2\pi^2 I_R\varsigma_0} \int\limits_{-\infty}^{+\infty} \hat{E}_{Rx}(u)k_{zo}(u)$$

$$\times \frac{\exp(-jk_{zs}(u)z')\exp(jk_{z0}(u)z)\exp(-ju(x'-x))}{\mu_0 k_{zs}(u) + \mu_s k_{z0}(u)} \, du \tag{4.115}$$

Equations (4.115) expresses the 2D external Green's function accounting for the characteristics of the receiving antenna.

Eventually, to insert the characteristics of the receiver into the scattering equation means to substitute the expression (4.115) for the external Green's function into the scattering equation (4.45) instead of the expression (4.47). Note that the internal Green's function, instead, remains unchanged and is given by Eq. (4.46). In fact, the internal Greens function represents the field underground resulting from a buried spatially impulsive source and is therefore not influenced by the characteristics of the antennas.

QUESTIONS

1. Can we say that the scattering equations essentially are the mathematical relationships between the buried targets and the electric field in the observation point?

2. Suppose that the shielding of the antennas is perfect, so that no energy is radiated in or received from the upper half-space. Is this sufficient to assimilate the propagation medium to a homogeneous medium instead of a medium composed by two adjacent half-spaces?

3. In the expression of Green's function and of the incident field [Eqs. (4.46–4.48)] the magnetic permittivity of both the air and the soil appears explicitly, whereas the dielectric permittivities of the air and of the soil appear only implicitly by means of the wave-numbers. What is the reason for this dissymmetry? What should happen in order to have a formulation where the quantities that appear explicitly are the permittivities?

4. In order to achieve a filamentary current at microwave frequency, might we make use of a long wire with lossy nonreflective terminations and then detach the wire in the middle and apply there a microwave voltage?

5. In order to implement a two-dimensional receiver, might we follow the same method described in question 4?

6. Should we expect that the central cut of the field radiated by a three-dimensional antenna is coincident with the 2D field radiated by a filamentary current displaced along the same line of the 3D wire antenna?

THE 2D SCATTERING EQUATIONS FOR MAGNETIC TARGETS

5.1 THE SCATTERING EQUATIONS WITH ONLY MAGNETIC ANOMALIES

Let us now consider the case dual to that dealt with in the previous chapter, when no dielectric anomaly is present ($\chi_e = 0$), but instead magnetic anomalies are looked for ($\chi_m \neq 0$). Actually, this is an extremely rare case, but the relative calculations are preliminary to considering the case with both dielectric and magnetic anomalies, which is less "exceptional." At the moment, we consider a filamentary electric current as primary source and consider the scattered field as datum. So, let us start again from Maxwell's equations, repeated here for simplification.

$$\nabla \times \vec{E} = -j\omega\mu\,\vec{H},$$
$$\nabla \times \vec{H} = j\omega\varepsilon_b\vec{E} + I_0\delta(x-x_0)\delta(z-z_0)i_y,$$
$$\nabla\cdot\varepsilon_b\vec{E} = 0,$$
$$\nabla\cdot\mu\vec{H} = 0$$

$$(5.1)$$

Introduction to Ground Penetrating Radar: Inverse Scattering and Data Processing,
First Edition. Raffaele Persico.

The magnetic permeability is schematized as the sum of the background permeability plus some discrepancy, as follows:

$$\mu(x,z) = \mu_b + \Delta\mu(x,z) \tag{5.2}$$

where the background magnetic permeability μ_b is the piecewise constant function described by Eq. (4.11). Instead, the dielectric permittivity is equal to the background permittivity ε_b, which is the piecewise constant function given in Eq. (4.14).

In the same way that we considered the case with only dielectric anomalies, we can decompose the field into an incident field plus a scattered field, and then we can extract Maxwell's equations for the incident field [which are the same as for the previous chapter and thus are given in Eqs. (4.18)] and those for the scattered field, which now are given by

$$
\begin{aligned}
\nabla \times \vec{E}_s &= -j\omega\mu_b\vec{H}_s - j\omega\Delta\mu\vec{H}, \\
\nabla \times \vec{H}_s &= +j\omega\varepsilon_b\vec{E}_s, \\
\nabla \cdot \varepsilon_b\vec{E}_s &= 0, \\
\nabla \cdot \mu_b\vec{H}_s &= -\nabla \cdot \Delta\mu\,\vec{H}
\end{aligned}
\tag{5.3}
$$

Thus, the scattered field can be viewed as generated by magnetic secondary (equivalent) sources, given by

$$\vec{J}_{meq} = j\omega\Delta\mu\,\vec{H} \tag{5.4}$$

The equivalent sources are not directed along the y-axis, but they have an x-directed component and a z-directed component, as does the magnetic field [see Eq. (4.10)]. Consequently, the secondary sources are not solenoidal in general and, coherently, the fourth equation in Eqs. (5.3) is not homogeneous but just states the continuity equation (4.7) with respect to the secondary (magnetic) sources.

This is a crucial point that prevents us from performing the calculations in the same way followed in Section 4.2 and makes them more complicated. Physically, the point is that, in this case, the secondary source is an equivalent continuous array of magnetic Hertzian dipoles instead of a filamentary electric current. The array is infinitely extended along the y-axis but is directed in the transverse plane and is infinitesimally long in this plane. This implies that the equivalent magnetic charges can move only along an infinitesimal path, and consequently some accumulation of them occurs.

After this premix, Eqs. (5.3) can be solved by means of suitable potential functions. In particular, we can start again from the consideration that the electric field is solenoidal. This assures (Franceschetti, 1997) the existence of a potential vector (also called Fitzgerald vector) $\vec{F}$ such as

$$\vec{E}_s = \frac{1}{\varepsilon_b}\nabla \times \vec{F} \tag{5.5}$$

Substituting Eq. (5.5) in the second of Eqs. (5.3), we have

$$\nabla \times \vec{H}_s = +j\omega \nabla \times \vec{F} \Leftrightarrow \nabla \times \left(\vec{H}_s - j\omega \vec{F} \right) = 0 \tag{5.6}$$

This assures that the irrotational quantity $\vec{H}_s - j\omega \vec{F}$ is the gradient of some scalar potential function Φ (Franceschetti, 1997). So we have

$$\vec{H}_s = j\omega \vec{F} + \nabla \Phi \tag{5.7}$$

Equations (5.5) and (5.7) express the fields in terms of two potential functions, so that we can think of recasting the problem in terms of these two potential functions, and then the fields will be derived from them. However, it is to be emphasized that, while the fields are physical quantities, which makes them univocally determined, the potential functions are mathematical quantities not univocally determined. In particular, suppose that we have found a couple of potential functions $\left(\vec{F}_0, \Phi_0 \right)$ that solve the problem, and let us consider any well-behaved scalar function ϕ. It is easy to recognize that the couple $\left(\vec{F}_0 + \nabla\phi, \Phi_0 - j\omega\phi \right)$ gives rise to the same fields as $\left(\vec{F}_0, \Phi_0 \right)$, and so it is another solution. In fact, we have

$$\frac{1}{\varepsilon_b} \nabla \times \left(\vec{F}_0 + \nabla\phi \right) = \frac{1}{\varepsilon_b} \nabla \times \vec{F}_0 + \frac{1}{\varepsilon_b} \nabla \times \nabla\phi = \frac{1}{\varepsilon_b} \nabla \times \vec{F},$$

$$j\omega \vec{F}_0 + j\omega \nabla\phi + \nabla\Phi_0 - j\omega\nabla\phi = j\omega \vec{F}_0 + \nabla\Phi_0 \tag{5.8}$$

In the first Eq. (5.8), we have exploited the well-known vector identity $\nabla \times \nabla f \equiv 0$.

At this point, let us substitute Eqs. (5.5) and (5.7) in the first equation of Eqs. (5.3).

$$\nabla \times \frac{1}{\varepsilon_b} \nabla \times \vec{F}_s = -j\omega\mu_b \left(j\omega \vec{F} + \nabla\Phi \right) - \vec{J}_{meq} \tag{5.9}$$

From Eq. (5.9), extracting the background permittivity (in both the homogeneous half-spaces separately) from the curl symbol and considering again the Laplacian vector, we have

$$\nabla \left(\nabla \cdot \vec{F} \right) - \nabla^2 \vec{F} = k_b^2 \vec{F} - j\omega\varepsilon_b\mu_b \nabla\Phi - \varepsilon_b \vec{J}_{meq}$$

$$\Leftrightarrow \nabla^2 \vec{F} + k_b^2 \vec{F} = \nabla \left(\nabla \cdot \vec{F} + j\omega\varepsilon_b\mu_b \Phi \right) + \varepsilon_b \vec{J}_{meq} \tag{5.10}$$

At this point, we can decide to look for a couple $\left(\vec{F}, \Phi \right)$ such as we obtain the following result:

$$\left(\nabla \cdot \vec{F}\right) + j\omega\varepsilon_b\mu_b\Phi = 0 \tag{5.11}$$

This possibility is related to the general nonuniqueness of the potentials. In particular, suppose again to have found a solution $\left(\vec{F}_0, \Phi_0\right)$. To find another solution that verifies the condition (5.11) amounts to looking for a scalar function ϕ such as

$$\nabla \cdot \left(\vec{F}_0 + \nabla\phi\right) + j\omega\varepsilon_b\mu_b(\Phi_0 - j\omega\phi) = 0 \Leftrightarrow \nabla^2\phi + k_b^2\phi = -\nabla \cdot \vec{F}_0 - j\omega\varepsilon_b\mu_b\Phi_0 \tag{5.12}$$

That is, the couple that verifies Eq. (5.12) can be found solving a scalar Helmholtz equation in ϕ.

Condition (5.11) is known as the gauge of Lorentz (Franceschetti, 1997). Having shown that it is licit, let us impose it on the solution, so that Eq. (5.10) reduces to

$$\nabla^2 \vec{F} + k_b^2 \vec{F} = \varepsilon_b \vec{J}_{meq} \tag{5.13}$$

that is, eventually we have achieved again a vector Helmholtz equation, which allows manipulations that are analogous to those in Section 4.2. However, the equation is not in terms of the unknown "field" but in terms of the unknown potential vector. Moreover, from Eq. (5.11) we have

$$\Phi = -\frac{\nabla \cdot \vec{F}}{j\omega\varepsilon_b\mu_b} \tag{5.14}$$

Substituting Eq. (5.14) into Eq. (5.7), we have

$$\vec{H}_s = j\omega \vec{F} - \frac{\nabla\left(\nabla \cdot \vec{F}\right)}{j\omega\varepsilon_b\mu_b} \tag{5.15}$$

Under the gauge of Lorentz, Eqs. (5.5) and (5.15) express the fields in terms of the only the Fitzgerald vector, which makes it formally redundant to look also for the scalar potential.

Now, from Eq. (5.5), we know a priori that the Fitzgerald vector has the following form:

$$\vec{F}(x,z) = F_x(x,z)i_x + F_z(x,z)i_z \tag{5.16}$$

Due to the linearity of the relationships between the fields and the Fitzgerald vector, we can consider separately the two components of the Fitzgerald vector, as two separated potentials. Each of the them will be associated with an electromagnetic field, and the final solution will be given by the sum of these two electromagnetic fields. So, we will proceed considering separately the two components.

5.2 THE CONTRIBUTION OF THE X-COMPONENT OF THE FITZGERALD VECTOR

Let us now consider the only x-component of the potential vector:

$$\vec{F}_1(x,z) = F_x(x,z)i_x \tag{5.17}$$

It can be seen as generated by the x-component of the magnetic density current. Thus the equivalent magnetic density current to be considered is

$$\vec{J}_{meq1} = j\omega\Delta\mu(x,z)H_x(x,z)i_x = J_{meqx}(x,z)i_x \tag{5.18}$$

This can be decomposed into the sum of concentrated sources as

$$\vec{J}_{meq1} = J_{meqx}(x,z)i_x = i_x \int\int_D J_{meqx}(x',z')\delta(x-x')\delta(z-z') \; dx'dz' \tag{5.19}$$

So, after Fourier transforming Eq. (5.13) along the x-axis and considering the element of Fitzgerald vector generated by an element of magnetic current $d\vec{J}_{meq1} = J_{meqx}(x',z')\delta(x-x')\delta(z-z') \; dx'dz'i_x$, we achieve the scalar equation

$$\frac{\partial^2 d\hat{F}_x}{\partial z^2} + k_z^2 d\hat{F}_x = \varepsilon_b J_{meqx}(x',z')\exp(-jux')\delta(z-z') \; dx'dz' \tag{5.20}$$

Since the secondary sources are buried, Eq. (5.20) can be still rewritten as

$$\frac{\partial^2 d\hat{F}_x}{\partial z^2} + k_z^2 d\hat{F}_x = \varepsilon_s J_{meqx}(x',z')\exp(-jux')\delta(z-z') \; dx'dz' \tag{5.21}$$

And so its formal solution, accounting also for the radiation condition at infinity, is

$$d\hat{F}_x = \begin{cases} dA_1\exp(jk_{z0}z), & z<0, \\ dB_1\exp(jk_{zs}z) + dC_1\exp(-jk_{zs}z), & 0<z<z', \\ dD_1\exp(-jk_{zs}z), & z'<z \end{cases} \tag{5.22}$$

At this point we have to express the same four conditions considered in Section 4.2 in terms of the Fitzgerald vector.

In particular, the first two conditions are given by the continuity of the tangential component of both electric and magnetic field at the interface. So, in order to impose them, we have to express the fields in terms of Eqs. (5.22) using Eqs. (5.5) and (5.15). In particular, the electric field is equal to

$$
d\hat{E}_{s1} = \frac{1}{\varepsilon_b} \nabla \times d\vec{F}_1 = \frac{1}{\varepsilon_b} \begin{vmatrix} i_x & i_y & i_z \\ ju & 0 & \dfrac{\partial}{\partial z} \\ d\hat{F}_x & 0 & 0 \end{vmatrix} = \frac{1}{\varepsilon_b} \frac{\partial d\hat{F}_x}{\partial z} i_y = \begin{cases} \dfrac{1}{\varepsilon_0} \dfrac{\partial d\hat{F}_x}{\partial z} i_y, & z<0 \\[2mm] \dfrac{1}{\varepsilon_s} \dfrac{\partial d\hat{F}_x}{\partial z} i_y, & z>0 \end{cases} \tag{5.23}
$$

So, the continuity of the tangential component of the electric field is expressed as

$$
\frac{1}{\varepsilon_0} \frac{\partial d\hat{F}_x}{\partial z}\bigg|_{z=0^-} = \frac{1}{\varepsilon_s} \frac{\partial d\hat{F}_x}{\partial z}\bigg|_{z=0^+} \quad\Leftrightarrow\quad \frac{\varepsilon_s k_{z0}}{\varepsilon_0 k_{zs}} dA_1 = dB_1 - dC_1 \tag{5.24}
$$

With regard to the condition about the scattered magnetic field, we have

$$
\nabla \cdot d\hat{\vec{F}}_1 = ju\, d\hat{F}_x, \tag{5.25}
$$

$$
\nabla\left(\nabla \cdot d\hat{\vec{F}}_1 \right) = ju\nabla d\hat{F}_x = ju\left(jud\hat{F}_x i_x + \frac{\partial d\hat{F}_x}{\partial z} i_z \right) \tag{5.26}
$$

So, the tangential component of the element of the scattered magnetic field (i.e., its x-component) is given by

$$
d\hat{H}_{sx1} = j\omega d\hat{F}_x - \frac{\nabla\left(\nabla \cdot d\hat{\vec{F}}_1 \right)}{j\omega\varepsilon_b\mu_b} \cdot i_x = j\omega\, d\hat{F}_x + \frac{u^2 d\hat{F}_x}{j\omega\varepsilon_b\mu_b} = \frac{(u^2 - k_b^2)}{j\omega\varepsilon_b\mu_b} d\hat{F}_x = -\frac{k_z^2}{j\omega\varepsilon_b\mu_b} d\hat{F}_x \tag{5.27}
$$

So, the continuity of the tangential scattered magnetic field translates into

$$
\frac{k_z^2}{\varepsilon_0\mu_0} d\hat{F}_x\bigg|_{z=0^-} = \frac{k_z^2}{\varepsilon_s\mu_s} d\hat{F}_x\bigg|_{z=0^+} \quad\Leftrightarrow\quad \frac{\varepsilon_s\mu_s k_{z0}^2}{\varepsilon_0\mu_0 k_{zs}^2} dA_1 = dB_1 + dC_1 \tag{5.28}
$$

The third condition is the continuity of the Fitzgerald vector at the source point $z = z'$ (because the Helmholtz equation (5.20) is expressed versus the Fitzgerald vector). So we have

$$
dD_1 - dB_1 \exp(j2k_{zs}z') - dC_1 = 0 \tag{5.29}
$$

The fourth condition is given by the integration of Eq. (5.20) about the source point $z = z'$, so we have

$$dD_1 + dB_1 \exp(j2k_{zs}z') - dC_1 = -\varepsilon_s \frac{\exp(jk_{zs}z')\exp(-jux')}{jk_{zs}} J_{meqx}(x',z')\,dx'dz' \quad (5.30)$$

So, we retrieve the following algebraic system:

$$\frac{\varepsilon_s k_{z0}}{\varepsilon_0 k_{zs}} dA_1 = dB_1 - dC,$$

$$\frac{\varepsilon_s \mu_s k_{z0}^2}{\varepsilon_0 \mu_0 k_{zs}^2} dA_1 = dB_1 + dC_1, \quad\quad\quad (5.31)$$

$$dD_1 - dB_1 \exp(j2k_{zs}z') - dC_1 = 0,$$

$$dD_1 + dB_1 \exp(j2k_{zs}z') - dC_1 = j\varepsilon_s \frac{\exp(jk_{zs}z')\exp(-jux')}{k_{zs}} J_{meqx}(x',z')\,dx'dz'$$

The system is formally similar to the system (4.34). So, following the same calculation steps, the final result is given by

$$dA_1 = \frac{j\varepsilon_0\mu_0 k_{zs}}{k_{z0}} \frac{\exp(-iux')\exp(-jk_{zs}z')}{\mu_0 k_{zs} + \mu_s k_{z0}} J_{meq}(x',z')\,dx'dz',$$

$$dB_1 = j\varepsilon_s \frac{\exp(-iux')\exp(-jk_{zs}z')}{2k_{zs}} J_{meq}(x',z')\,dx'dz',$$

$$dC_1 = j\varepsilon_s \frac{1}{2k_{zs}} \frac{\mu_s k_{z0} - \mu_0 k_{zs}}{\mu_s k_{z0} + \mu_0 k_{zs}} \exp(-iux')\exp(-jk_{zs}z') J_{meq}(x',z')\,dx'dz', \quad (5.32)$$

$$dD_1 = j\varepsilon_s \frac{\exp(-iux')}{2k_{zs}} \left[\exp(jk_{zs}z') + \frac{\mu_s k_{z0} - \mu_0 k_{zs}}{\mu_s k_{z0} + \mu_0 k_{zs}} \exp(-jk_{zs}z') \right] J_{meq}(x',z')\,dx'dz'$$

Substituting into Eqs. (5.22), we have

$$d\hat{F}_x(u,z) = \begin{cases} -\dfrac{j\varepsilon_0\mu_0 k_{zs}}{k_{z0}} \dfrac{\exp(-iux')\exp(-jk_{zs}z')\exp(jk_{z0}z)}{\mu_0 k_{zs} + \mu_s k_{z0}} J_{meq}(x',z')\,dx'dz', \quad z<0 \\[2em] -j\varepsilon_s \dfrac{\exp(-iux')}{2k_{zs}} \left[\exp(-jk_{zs}|z-z'|) + \dfrac{\mu_s k_{z0} - \mu_0 k_{zs}}{\mu_s k_{z0} + \mu_0 k_{zs}} \exp(-jk_{zs}(z+z')) \right] \\[1.5em] \times J_{meq}(x',z')\,dx'dz', \quad z>0 \end{cases}$$

$$(5.33)$$

Substituting Eq. (5.33) into Eq. (5.23), we achieve the elementary component of the scattered electric field in the transformed domain:

$$d\hat{E}_{s1}(u,z) = \begin{cases} -\mu_0 k_{zs} \dfrac{\exp(-iux')\exp(-jk_{zs}z')\exp(jk_{z0}z)}{\mu_0 k_{zs}+\mu_s k_{z0}} J_{meq}(x',z)\ dx'dz'i_x, & z<0 \\[2ex] \dfrac{\exp(-iux')}{2} J_{meq}(x',z)\ dx'dz'i_x \\[2ex] \times \left[\mathrm{sgn}(z-z')\exp(-jk_{zs}|z-z'|)+\dfrac{\mu_s k_{z0}-\mu_0 k_{zs}}{\mu_s k_{z0}+\mu_0 k_{zs}}\exp(-jk_{zs}(z+z')) \right], & z>0 \end{cases}$$

(5.34)

where sgn stands for the *signum* functions, equal to -1 if its argument is negative and equal to $+1$ if its argument is positive. Note that the signum is the derivative of the absolute value, which will be exploited in the following. After inverse Fourier transformation and integration all over the investigation domain, eventually the contribution to the scattered field is the following [directed along the y-axis accordingly to Eq. (5.23)]:

$$E_{s1}(x,z)$$
$$= \begin{cases} -\dfrac{\mu_0}{2\pi} \displaystyle\iint_D J_{meqx}(x',z')\ dx'dz' \int_{-\infty}^{+\infty} \dfrac{k_{zs}\exp(-iu(x'-x))\exp(-jk_{zs}z')\exp(jk_{z0}z)}{\mu_0 k_{zs}+\mu_s k_{z0}}\ du, & z<0 \\[3ex] \dfrac{1}{4\pi} \displaystyle\iint_D J_{meqx}(x',z')\ dx'dz' \int_{-\infty}^{+\infty} \exp(-iu(x'-x)) \\[2ex] \times \left[\mathrm{sgn}(z-z')\exp(-jk_{zs}|z-z'|)+\dfrac{\mu_s k_{z0}-\mu_0 k_{zs}}{\mu_s k_{z0}+\mu_0 k_{zs}}\exp(-jk_{zs}(z+z')) \right] du, & z>0 \end{cases}$$

(5.35)

Substituting the expression of the magnetic current according to Eq. (5.4), and accounting for the first Maxwell's equation for the total field (5.1),[1] we have

$$J_{meqx}(x',z')=j\omega\Delta\mu(x',z')H_x(x',z')=j\omega\Delta\mu(x',z')\dfrac{1}{-j\omega\mu(x',z')}\nabla\times\vec{E}(x',z')\cdot i_x$$

$$= -\dfrac{\Delta\mu(x',z')}{\mu_b+\Delta\mu(x',z')}\begin{vmatrix} i_x & i_y & i_z \\ \dfrac{\partial}{\partial x'} & 0 & \dfrac{\partial}{\partial z'} \\ 0 & E & 0 \end{vmatrix} \cdot i_x = -\dfrac{\dfrac{\Delta\mu(x',z')}{\mu_b}}{1+\dfrac{\Delta\mu(x',z')}{\mu_b}}\left(-\dfrac{\partial E}{\partial z'}i_x\right)\cdot i_x=\dfrac{\chi_m}{1+\chi_m}\dfrac{\partial E}{\partial z'}$$

(5.36)

---

[1] Let us remind that the considered magnetic current density is a secondary source; there is no primary magnetic source.

Substituting Eq. (5.36) into Eq. (5.35), we have

$E_{s1}(x,z)$

$$
= \begin{cases}
-\dfrac{\mu_0}{2\pi} \displaystyle\iint_D \dfrac{\chi_m}{1+\chi_m} \dfrac{\partial E}{\partial z'} \, dx' dz' \displaystyle\int_{-\infty}^{+\infty} \dfrac{k_{zs}\exp(-iu(x'-x))\exp(-jk_{zs}z')\exp(jk_{z0}z)}{\mu_0 k_{zs}+\mu_s k_{z0}} \, du, & z<0 \\[4mm]
\dfrac{1}{4\pi} \displaystyle\iint_D \dfrac{\chi_m}{1+\chi_m} \dfrac{\partial E}{\partial z'} \, dx' dz' \displaystyle\int_{-\infty}^{+\infty} \exp(-iu(x'-x)) \\[4mm]
\quad \times \left[\mathrm{sgn}(z-z')\exp(-jk_{zs}|z-z'|) + \dfrac{\mu_s k_{z0}-\mu_0 k_{zs}}{\mu_s k_{z0}+\mu_0 k_{zs}}\exp(-jk_{zs}(z+z')) \right] du, & z>0
\end{cases}
$$

$$(5.37)$$

Equation (5.37) expresses the element of electric scattered field in terms of the internal scattered field and in terms of magnetic contrast.

At this point, let us calculate the first partial derivative of the internal and external electric Green's functions, expressed by Eqs. (4.46) and (4.47), versus the variable z':

$$
\begin{aligned}
\frac{\partial}{\partial z'} G_i(x,x',z,z') &= \frac{\partial}{\partial z'} \left\{ \frac{-j}{4\pi} \int_{-\infty}^{+\infty} \frac{\exp(-ju(x'-x))}{k_{zs}} \right. \\
&\qquad \left. \times \left[\exp(-jk_{zs}|z-z'|) + \frac{\mu_0 k_{zs}-\mu_s k_{z0}}{\mu_0 k_{zs}+\mu_s k_{z0}}\exp(-jk_{zs}(z'+z)) \right] du \right\} \\[3mm]
&= \frac{-1}{4\pi} \int_{-\infty}^{+\infty} \exp(-ju(x'-x)) \\
&\qquad \times \left[(-1)\,\mathrm{sgn}(z-z')\exp(-jk_{zs}|z-z'|) + \frac{\mu_0 k_{zs}-\mu_s k_{z0}}{\mu_0 k_{zs}+\mu_s k_{z0}}\exp(-jk_{zs}(z'+z)) \right] du \\[3mm]
&= \frac{1}{4\pi} \int_{-\infty}^{+\infty} \exp(-ju(x'-x)) \\
&\qquad \times \left[\mathrm{sgn}(z-z')\exp(-jk_{zs}|z-z'|) + \frac{\mu_s k_{z0}-\mu_0 k_{zs}}{\mu_0 k_{zs}+\mu_s k_{z0}}\exp(-jk_{zs}(z'+z)) \right] du, \\
&\qquad\qquad \forall (x,z),(x',z') \in D
\end{aligned}
$$

$$(5.38)$$

$$\frac{\partial}{\partial z'} G_e(x,x',z,z')$$

$$= \frac{\partial}{\partial z'} \left\{ \frac{-j\mu_0}{2\pi} \int_{-\infty}^{+\infty} \frac{\exp(-jk_{zs}z')\exp(jk_{z0}z)\exp(-ju(x'-x))}{\mu_0 k_{zs} + \mu_s k_{z0}} du \right\}$$

$$= \frac{-\mu_0}{2\pi} \int_{-\infty}^{+\infty} k_{zs} \frac{\exp(-jk_{zs}z')\exp(jk_{z0}z)\exp(-ju(x'-x))}{\mu_0 k_{zs} + \mu_s k_{z0}} du, \quad (x,z) \in \Omega, \quad (x',z') \in D$$

$$(5.39)$$

Comparing Eqs. (5.38) and (5.39) with Eq. (5.37), we can express the contribution $E_{s1}(x,z)$ as

$$E_{s1}(x,z) = \begin{cases} \iint_D \frac{\chi_m}{1+\chi_m} \frac{\partial E}{\partial z'} \frac{\partial G_e}{\partial z'} dx' dz' = k_s^2 \iint_D \frac{\chi_m}{1+\chi_m} \left(\frac{1}{k_s^2}\right) \frac{\partial E}{\partial z'} \frac{\partial G_e}{\partial z'} dx' dz', & z<0 \\ \iint_D \frac{\chi_m}{1+\chi_m} \frac{\partial E}{\partial z'} \frac{\partial G_i}{\partial z'} dx' dz' = k_s^2 \iint_D \frac{\chi_m}{1+\chi_m} \left(\frac{1}{k_s^2}\right) \frac{\partial E}{\partial z'} \frac{\partial G_i}{\partial z'} dx' dz', & z>0 \end{cases}$$

$$(5.40)$$

5.3   THE CONTRIBUTION OF THE Z-COMPONENT OF THE FITZGERALD VECTOR

Let us consider the second component of the Fitgerald vector. The calculations proceed in the same way as before, so that we can reduce to a minimum the exposed passages.

This time, the potential vector is given by

$$\vec{F}_2(x,z) = F_z(x,z)i_z \qquad (5.41)$$

The equivalent magnetic density of current is given by

$$\vec{J}_{meq2} = j\omega\Delta\mu(x,z)H_z(x,z)i_z = J_{meqz}(x,z)i_z = i_z \iint_D J_{meqz}(x',z')\delta(x-x')\delta(z-z')\,dx'\,dz'$$

$$(5.42)$$

The element of magnetic density of current is given by

$$d\vec{J}_{meq2} = J_{meqz}(x',z')\delta(x-x')\delta(z-z')\,dx'\,dz'\,i_z \qquad (5.43)$$

The scalar Helmholtz equation in the element of potential vector in the transformed domain (u,z) is given by

$$\frac{\partial^2 d\hat{F}_z}{\partial z^2} + k_z^2 d\hat{F}_z = \varepsilon_s J_{meqz}(x',z') \exp(-jux')\delta(z-z') dx' dz' \tag{5.44}$$

whose formal solution is

$$d\hat{F}_z = \begin{cases} dA_2 \exp(jk_{z0}z), & z<0, \\ dB_2 \exp(jk_{zs}z) + dC_2 \exp(-jk_{zs}z), & 0<z<z', \\ dD_2 \exp(-jk_{zs}z), & z'<z \end{cases} \tag{5.45}$$

The element of the electric field versus the element of the Fitzgerald vector is expressed as follows:

$$d\vec{E}_{s2} = \frac{1}{\varepsilon_b} \nabla \times d\vec{F}_2 = \frac{1}{\varepsilon_b} \begin{vmatrix} i_x & i_y & i_z \\ ju & 0 & \frac{\partial}{\partial z} \\ 0 & 0 & d\hat{F}_z \end{vmatrix} = -\frac{ju}{\varepsilon_b} d\hat{F}_z i_y = \begin{cases} -\dfrac{ju}{\varepsilon_0} d\hat{F}_z i_y, & z<0 \\ -\dfrac{ju}{\varepsilon_s} d\hat{F}_z i_y, & z>0 \end{cases} \tag{5.46}$$

The continuity of the unique (and tangent) component of the electric field at the interface is expressed as follows:

$$\frac{1}{\varepsilon_0} d\hat{F}_z \bigg|_{z=0^-} = \frac{1}{\varepsilon_s} d\hat{F}_z \bigg|_{z=0^+} \Leftrightarrow \frac{\varepsilon_s}{\varepsilon_0} dA_2 = dB_2 + dC_2 \tag{5.47}$$

With regard to the continuity of the tangential component of the scattered magnetic field, we have

$$\nabla \cdot d\vec{F}_2 = \frac{\partial d\hat{F}_z}{\partial z} \tag{5.48}$$

$$\nabla \left(\nabla \cdot d\vec{F}_2 \right) = \nabla \frac{\partial d\hat{F}_z}{\partial z} = \left(ju \frac{\partial d\hat{F}_z}{\partial z} i_x + \frac{\partial^2 d\hat{F}_z}{\partial z^2} i_z \right) \tag{5.49}$$

$$d\hat{H}_{sx2} = -\frac{\nabla \left(\nabla \cdot d\vec{F}_2 \right)}{j\omega\varepsilon_b\mu_b} \cdot i_x = -\frac{ju}{j\omega\varepsilon_b\mu_b} \frac{\partial d\hat{F}_z}{\partial z} = -\frac{u}{\omega\varepsilon_b\mu_b} \frac{\partial d\hat{F}_z}{\partial z} \tag{5.50}$$

Thus, the continuity of the tangential scattered magnetic field is expressed as

$$\frac{1}{\varepsilon_0\mu_0} \frac{\partial d\hat{F}_z}{\partial z} \bigg|_{z=0^-} = \frac{1}{\varepsilon_s\mu_s} \frac{\partial d\hat{F}_z}{\partial z} \bigg|_{z=0^+} \Leftrightarrow \frac{\varepsilon_s\mu_s k_{z0}}{\varepsilon_0\mu_0 k_{zs}} dA_2 = dB_2 - dC_2 \tag{5.51}$$

The continuity of the Fitzgerald vector at the source point $z = z'$ is expressed as

$$dD_2 - dB_2 \exp(j2k_{zs}z') - dC_2 = 0 \qquad (5.52)$$

The integration of Eq. (5.44) about the source point $z = z'$ is expressed as

$$dD_2 + dB_2 \exp(j2k_{zs}z') - dC_2 = -\varepsilon_s \frac{\exp(jk_{zs}z')\exp(-jux')}{jk_{zs}} J_{meqz}(x',z')\,dx'\,dz' \qquad (5.53)$$

Thus, we have the algebraic system:

$$\frac{\varepsilon_s}{\varepsilon_0} dA_2 = dB_2 + dC_2,$$

$$\frac{\varepsilon_s \mu_s k_{z0}}{\varepsilon_0 \mu_0 k_{zs}} dA_2 = dB_2 - dC_2,$$

$$dD_2 - dB_2 \exp(j2k_{zs}z') - dC_2 = 0, \qquad (5.54)$$

$$dD_2 + dB_2 \exp(j2k_{zs}z') - dC_2 = j\varepsilon_s \frac{\exp(jk_{zs}z')\exp(-jux')}{k_{zs}} J_{meqz}(x',z')\,dx'\,dz'$$

The solution of the system is

$$dA_2 = j\varepsilon_0 \mu_0 \frac{\exp(-iux')\exp(-jk_{zs}z')}{\mu_0 k_{zs} + \mu_s k_{z0}} J_{meqz}(x',z')\,dx'\,dz',$$

$$dB_2 = j\varepsilon_s \frac{\exp(-iux')\exp(-jk_{zs}z')}{2k_{zs}} J_{meqz}(x',z')\,dx'\,dz', \qquad (5.55)$$

$$dC_2 = j\varepsilon_s \frac{1}{2k_{zs}} \frac{\mu_0 k_{zs} - \mu_s k_{z0}}{\mu_0 k_{zs} + \mu_s k_{z0}} \exp(-iux')\exp(-jk_{zs}z') J_{meqz}(x',z')\,dx'\,dz',$$

$$dD_2 = j\varepsilon_s \frac{\exp(-iux')}{2k_{zs}} \left[\exp(jk_{zs}z') + \frac{\mu_0 k_{zs} - \mu_s k_{z0}}{\mu_s k_{z0} + \mu_0 k_{zs}} \exp(-jk_{zs}z') \right] J_{meqz}(x',z')\,dx'\,dz'$$

Substituting Eqs. (5.55) into Eqs. (5.45), we have

$$d\hat{F}_z(u,z) = \begin{cases} j\varepsilon_0 \mu_0 \dfrac{\exp(-iux')\exp(-jk_{zs}z')\exp(jk_{z0}z)}{\mu_0 k_{zs} + \mu_s k_{z0}} J_{meqz}(x',z')\,dx'\,dz'\,i_z, & z < 0 \\[4mm] j\varepsilon_s \dfrac{\exp(-iux')}{2k_{zs}} \left[\exp(-jk_{zs}|z-z'|) + \dfrac{\mu_0 k_{zs} - \mu_s k_{z0}}{\mu_s k_{z0} + \mu_0 k_{zs}} \exp(-jk_{zs}(z+z')) \right] \\[4mm] \times J_{meqz}(x',z')\,dx'\,dz'\,i_z, & z > 0 \end{cases}$$

$$(5.56)$$

from which, substituting Eq. (5.56) into Eq. (5.46), we have

$$
d\hat{E}_{s2}(u,z) = \begin{cases} u\mu_0 \dfrac{\exp(-jux')\exp(-jk_{zs}z')\exp(jk_{z0}z)}{\mu_0 k_{zs} + \mu_s k_{z0}} J_{meqz}(x',z')\,dx'dz', & z<0, \\[3ex] \dfrac{u\exp(-iux')}{2k_{zs}} \left[\exp(-jk_{zs}|z-z'|) + \dfrac{\mu_0 k_{zs} - \mu_s k_{z0}}{\mu_s k_{z0} + \mu_0 k_{zs}}\exp(-jk_{zs}(z+z')) \right] \\[3ex] \times J_{meqz}(x',z')\,dx'dz', & z>0 \end{cases}
$$

(5.57)

After inverse Fourier transform and integration all over the investigation domain, eventually the contribution to the scattered field can be expressed as

$$
\begin{aligned}
&E_{s2}(x,z) = \\
&\begin{cases} \dfrac{j\mu_0}{2\pi}\displaystyle\iint_D J_{meqz}(x',z')\,dx'dz' \int_{-\infty}^{+\infty} \dfrac{-ju\exp(-iu(x'-x))\exp(-jk_{zs}z')\exp(jk_{z0}z)}{\mu_0 k_{zs} + \mu_s k_{z0}}\,du, & z<0, \\[3ex] = \dfrac{j}{4\pi}\displaystyle\iint_D J_{meqz}(x',z')\,dx'dz' \int_{-\infty}^{+\infty} \dfrac{-ju}{k_{zs}}\exp(-iu(x'-x)) \\[3ex] \times \left[\exp(-jk_{zs}|z-z'|) + \dfrac{\mu_s k_{z0} - \mu_0 k_{zs}}{\mu_s k_{z0} + \mu_0 k_{zs}}\exp(-jk_{zs}(z+z')) \right] du, & z>0 \end{cases}
\end{aligned}
$$

(5.58)

Expressing the z-component of the equivalent magnetic current in terms of internal electric field and magnetic contrast, we have

$$
\begin{aligned}
J_{meqz}(x',z') &= j\omega\Delta\mu(x',z')H_z(x',z') = j\omega\Delta\mu(x',z')\dfrac{1}{-j\omega\mu(x',z')}\nabla\times\vec{E}(x',z')\cdot i_z \\[2ex]
&= -\dfrac{\Delta\mu(x',z')}{\mu_b + \Delta\mu(x',z')}\begin{vmatrix} i_x & i_y & i_z \\ \dfrac{\partial}{\partial x'} & 0 & \dfrac{\partial}{\partial z'} \\ 0 & E_1 & 0 \end{vmatrix}\cdot i_z = -\dfrac{\dfrac{\Delta\mu(x',z')}{\mu_b}}{1+\dfrac{\Delta\mu(x',z')}{\mu_b}}\dfrac{\partial E}{\partial x'}i_z\cdot i_z = -\dfrac{\chi_m}{1+\chi_m}\dfrac{\partial E}{\partial x'}
\end{aligned}
$$

(5.59)

Substituting Eq. (5.59) into Eq. (5.58), we have

$E_{s2}(x,z)$

$$
= \begin{cases}
-\dfrac{j\mu_0}{2\pi} \displaystyle\iint_D \dfrac{\chi_m}{1+\chi_m} \dfrac{\partial E}{\partial x'} \, dx' dz' \int\limits_{-\infty}^{+\infty} \dfrac{-ju\exp(-iu(x'-x))\exp(-jk_{zs}z')\exp(jk_{z0}z)}{\mu_0 k_{zs}+\mu_s k_{z0}} \, du, & z<0, \\[4ex]
-\dfrac{j}{4\pi} \displaystyle\iint_D \dfrac{\chi_m}{1+\chi_m} \dfrac{\partial E}{\partial x'} \, dx' dz' \int\limits_{-\infty}^{+\infty} \dfrac{-ju}{k_{zs}} \exp(-iu(x'-x)) \\[3ex]
\qquad \times \left[\exp(-jk_{zs}|z-z'|) + \dfrac{\mu_s k_{z0}-\mu_0 k_{zs}}{\mu_s k_{z0}+\mu_0 k_{zs}} \exp(-jk_{zs}(z+z')) \right] du, & z>0
\end{cases}
$$

$$(5.60)$$

At this point, let us consider the derivatives of the internal and external Green's functions, as given by Eqs. (4.46) and (4.47), with respect to the variable x':

$$
\frac{\partial}{\partial x'} G_i(x,x',z,z') = \frac{\partial}{\partial x'} \left\{ \frac{-j}{4\pi} \int\limits_{-\infty}^{+\infty} \frac{\exp(-ju(x'-x))}{k_{zs}} \right.
$$
$$
\left. \times \left[\exp(-jk_{zs}|z-z'|) + \frac{\mu_0 k_{zs}-\mu_s k_{z0}}{\mu_0 k_{zs}+\mu_s k_{z0}} \exp(-jk_{zs}(z'+z)) \right] du \right\}
$$
$$
= \frac{-j}{4\pi} \int\limits_{-\infty}^{+\infty} \frac{-ju\exp(-ju(x'-x))}{k_{zs}}
$$
$$
\times \left[\exp(-jk_{zs}|z-z'|) + \frac{\mu_0 k_{zs}-\mu_s k_{z0}}{\mu_0 k_{zs}+\mu_s k_{z0}} \exp(-jk_{zs}(z'+z)) \right] du, \quad (x,z),(x',z') \in D
$$

$$(5.61)$$

$$
\frac{\partial}{\partial x'} G_e(x,x',z,z')
$$
$$
= \frac{\partial}{\partial x'} \left\{ \frac{-j\mu_0}{2\pi} \int\limits_{-\infty}^{+\infty} \frac{\exp(-jk_{zs}z')\exp(jk_{z0}z)\exp(-ju(x'-x))}{\mu_0 k_{zs}+\mu_s k_{z0}} \, du \right\}
$$
$$
= \frac{-j\mu_0}{2\pi} \int\limits_{-\infty}^{+\infty} \frac{-ju\exp(-jk_{zs}z')\exp(jk_{z0}z)\exp(-ju(x'-x))}{\mu_0 k_{zs}+\mu_s k_{z0}} \, du, \quad (x,z)\in\Omega,\ (x',z')\in D
$$

$$(5.62)$$

Comparing Eq. (5.60) with Eqs. (5.61) and (5.62), the contribution to the electric scattered field of the z-component of the Fitzgerald vector can be rewritten as

$$
E_{s2}(x,z) =
\begin{cases}
\displaystyle \iint_D \frac{\chi_m}{1+\chi_m} \frac{\partial E}{\partial x'} \frac{\partial G_e}{\partial x'} = k_s^2 \iint_D \frac{\chi_m}{1+\chi_m} \left(\frac{1}{k_s^2}\right) \frac{\partial E}{\partial x'} \frac{\partial G_e}{\partial x'}, & z<0, \\[3ex]
\displaystyle \iint_D \frac{\chi_m}{1+\chi_m} \frac{\partial E}{\partial x'} \frac{\partial G_i}{\partial x'} = k_s^2 \iint_D \frac{\chi_m}{1+\chi_m} \left(\frac{1}{k_s^2}\right) \frac{\partial E}{\partial x'} \frac{\partial G_i}{\partial x'}, & z>0
\end{cases}
\tag{5.63}
$$

5.4 THE JOINED CONTRIBUTION OF BOTH THE X- AND Z-COMPONENTS OF THE FITZGERALD VECTOR

As already stated, this case can be dealt with as the superposition of the cases considered in Sections 5.2 and 5.3. Therefore, the comprehensive scattered field due to the buried magnetic anomalies is given by

$$
E_s(x,z) =
\begin{cases}
\displaystyle k_s^2 \iint_D \frac{\chi_m}{1+\chi_m} \left(\frac{1}{k_s^2}\right) \left(\frac{\partial E}{\partial x'} \frac{\partial G_e}{\partial x'} + \frac{\partial E}{\partial z'} \frac{\partial G_e}{\partial z'}\right) dx'dz', & z<0, \\[3ex]
\displaystyle k_s^2 \iint_D \frac{\chi_m}{1+\chi_m} \left(\frac{1}{k_s^2}\right) \left(\frac{\partial E}{\partial x'} \frac{\partial G_i}{\partial x'} + \frac{\partial E}{\partial z'} \frac{\partial G_i}{\partial z'}\right) dx'dz', & z>0
\end{cases}
\tag{5.64}
$$

which can be more compactly rewritten as

$$
E_s(x,z) =
\begin{cases}
\displaystyle k_s^2 \iint_D \frac{\chi_m}{1+\chi_m} \left(\frac{1}{k_s^2}\right) \nabla'E\cdot\nabla'G_e\, dx'dz', & z<0, \\[3ex]
\displaystyle k_s^2 \iint_D \frac{\chi_m}{1+\chi_m} \left(\frac{1}{k_s^2}\right) \nabla'E\cdot\nabla'G_i\, dx'dz', & z>0
\end{cases}
\tag{5.65}
$$

where the gradient symbols have been primed to indicate that they are calculated with respect to the variables of integration x' and z'.

Since the internal field is given by the sum of the incident and scattered internal fields, in the end the complete scattering equations for only magnetic targets are given by

$$
E(x,z) = E_{inc} + k_s^2 \iint_D \frac{\chi_m}{1+\chi_m} \left(\frac{1}{k_s^2}\right) \nabla'E\cdot\nabla'G_i dx'\, dz', \qquad z>0,
\tag{5.66}
$$

$$
E_s(x,z) = k_s^2 \iint_D \frac{\chi_m}{1+\chi_m} \left(\frac{1}{k_s^2}\right) \nabla'E\cdot\nabla'G_e dx'\, dz', \qquad z<0
\tag{5.67}
$$

5.5   THE CASE WITH BOTH DIELECTRIC AND MAGNETIC ANOMALIES

In this chapter we have shown that the product of the magnetic field underground times the magnetic contrast of the buried target can be interpreted as an equivalent buried magnetic density of current (apart from the factor $j\omega$) radiating into a homogeneous half-space. Similarly, in Chapter 4 we showed that the product of the underground electric field times the dielectric contrast can be can as an equivalent electric density of current. As already stated, these two equivalent sources are usually labeled as secondary sources, whereas the real source of the field (i.e., the GPR antennas) are usually labeled as primary sources. Now, the scattered field (both underground and in air) depends on the primary sources only through the secondary ones. This implies that if we "forget" the presence of the primary sources for a while, the scattered field is only a function of the secondary sources. Now, any electric field is a linear function of its sources, and thus the comprehensive scattered field in the presence of both a magnetic and a dielectric contrast is given by the sum of the scattered fields obtained by each of the corresponding secondary sources, respectively. In formulas, this means in particular that the internal (underground) scattered field, which is equal to the internal electric field minus the internal incident field, is given by the sum of the contributions to it by the electric secondary sources [contribution given by Eq. (4.44)] and the magnetic secondary sources [contribution given by Eq. (5.66)]. The internal scattered field is therefore given by Eq. (5.68).

$$E_s(x,z) = k_s^2 \iint_D \left[\chi_e E G_i + \frac{1}{k_s^2} \frac{\chi_m}{1+\chi_m} \nabla' E \cdot \nabla' G_i \right] dx' dz', \quad z > 0 \qquad (5.68)$$

The total internal field is achieved by adding the internal incident field to the scattered internal field, thereby obtaining

$$E(x,z) = E_{inc}(x,z) + k_s^2 \iint_D \left[\chi_e E G_i + \frac{1}{k_s^2} \frac{\chi_m}{1+\chi_m} \nabla' E \cdot \nabla' G_i \right] dx' dz', \quad z > 0 \qquad (5.69)$$

Equation (5.69) is the internal scattering equation in the case of both dielectric and magnetic anomalies, or (which is the same) in the presence of targets with both dielectric and magnetic properties different from those of the embedding soil. From Eq. (5.69), we appreciate that, even if the Maxwell's equations are linear with respect to the sources, the relationships between the contrasts and the internal field is not linear, and in particular the dielectric and magnetic contrasts interfere with each other, in the sense that the resulting internal field is not given by the sum of the internal field that we would separately have if only one kind of anomaly (dielectric or magnetic) were present.

   Once the internal field is given (of course implicitly, because Eq. (5.69) is an equation to be solved in order to retrieve the internal field versus the contrasts), the external scattered field can be retrieved. In fact, because of the linearity of the scattered field with respect to the secondary sources, the external scattered field is provided by the sum of the contributions of the electric secondary sources, given by Eq. (4.45), and the contribution of the magnetic secondary sources, given by Eq. (5.67). The result is

$$E_s(x,z) = k_s^2 \iint_D \left[\chi_e E G_e + \frac{1}{k_s^2} \frac{\chi_m}{1+\chi_m} \nabla' E \cdot \nabla' G_e \right] dx' dz', \quad z<0 \tag{5.70}$$

Equation (5.70) constitutes the external scattering equation in the general 2D case of both dielectric and magnetic anomalies.

It is possible to include the characteristics of the antennas also in the case of magnetic targets, and this is done in the same way as revealed in Chapter 4. In fact, the incident field does not depend on the nature of buried anomalies, and the passage from the scattered field to the scattered voltage also does not depend on what kind of anomaly generated the gathered scattered field. Thus, in the more general case the expression of the incident field is given by Eq. (4.84) instead of Eq. (4.48), and the expression of the external Green's function is given by Eq. (4.115) instead of Eq. (4.47), but Eqs. (5.69) and (5.70) remain formally the same.

QUESTIONS

1. Is the electric field solenoidal in any physical case?
2. Is the magnetic field solenoidal in any physical case?
3. Does the magnetic permeability of the targets interact with the dielectric permittivity for determining the incident field?
4. Does the magnetic permeability of the targets interact with the dielectric permittivity for determining the scattered field?
5. Does the magnetic permeability of the targets interact with the dielectric permittivity for determining the total field?
6. Is it theoretically possible to have a 2D geometry where the electric field is not parallel to the axis of invariance?

6

ILL-POSEDNESS AND NONLINEARITY

6.1 ELECTROMAGNETIC INVERSE SCATTERING

Electromagnetic inverse scattering is the branch of the science that studies how to reconstruct the electromagnetic characteristics of a given volume (e.g., its dielectric permittivity and/or electrical conductivity and/or magnetic permeability) starting from measurements of electric (and/or the magnetic) scattered field data, gathered outside the probed volume and generated by known sources.

The reason why this problem is called "inverse" is historical, and it is due to the fact that it reverses the point of view with respect to another problem conventionally labeled as the "direct" or "forward" scattering problem. The direct (or forward) problem consists in the calculation of the field scattered from a known dielectric permittivity (and/or electrical conductivity and/or magnetic permeability) profile, under the radiation of electromagnetic waves radiated by known sources.

The definitions of direct and inverse scattering problem, within the framework of the GPR prospecting, are explained with the aid of Figure 4.1. In particular, with reference to Figure 4.1, the direct scattering problem consists of the calculation of the scattered field if we know the underground scenario, whereas the inverse scattering problem consists in

Introduction to Ground Penetrating Radar: Inverse Scattering and Data Processing,
First Edition. Raffaele Persico.
© 2014 The Institute of Electrical and Electronics Engineers, Inc. Published 2014 by John Wiley & Sons, Inc.

the reconstruction of the underground scenario if we know the scattered field. In both the direct and the inverse problem the characteristics of the antennas and of the soil are assumed to be known.

Indeed, there are several inverse problems of interest, not only in electromagnetism but also in acoustics, in heat conduction and so on, and customarily the inverse problems present some additional difficulties with respect to the corresponding direct ones. In particular, unlike the direct ones, the inverse problems are usually ill-posed (Colton and Kress, 1992). Moreover, if the direct problem is nonlinear, also the corresponding inverse problem is nonlinear. In particular, the electromagnetic inverse scattering problem is both ill-posed and nonlinear.

6.2   ILL-POSEDNESS

Direct problems, and in particular the forward scattering problem, are customarily well-posed. Mathematically, this means that one has the guarantee that the solution exists, is unique, and has a continuous dependence on the data (Hadamard, 1923). The solution's continuous dependence on the data involves the fact that a small (meant in a limit sense, i.e., infinitesimal) error on the data translates into a small error on the retrieved solution. Regarding the issue of the GPR prospecting, this means (for example) that a small error in the evaluation of the permittivity of the underground scenario will produce a small error in the evaluation of the scattered field. This condition is necessary (but not sufficient) in order to have a solution with a physical sense, because it is impossible to have completely error-free data. A problem is said to be "ill-posed" when at least one of the conditions for the well-posedness fails.

With regard to inverse problem of the GPR data processing, the most relevant issue (and the only one that we will focus on) is the solution's noncontinuous dependence on the data. This means that even a small error on the scattered field can reverse in a meaningful error in the reconstruction of the dielectric (conductive or magnetic) characteristics of the background scenario. Such being the case, one is tempted to say that an ill-posed problem does not make any physical sense, because it is impossible to retrieve any physically reliable solution. However, this difficulty is overcome by means of the *regularization*. To regularize a problem substantially means to renounce to look for its ideal "perfect" solution and to look instead for a suitable, not fully detailed but more robust, solution.

The regularization is not a specific protocol written once and for all. There are endless possibilities and "degrees" of regularization, and one has to choose the regularization for the case at hand as a compromise between the exigencies to achieve a refined (not overregularized) and, at the same time, robust (not underregularized) solution. This is a key point, on which we will come back again.

6.3   NONLINEARITY

Any mathematical relationship R between two quantities is said to be linear when, for any two arguments x_1 and x_2 and for any two scalar quantities a and b, the following property holds:

$$R(ax_1 + bx_2) = aR(x_1) + bR(x_2) \tag{6.1}$$

Otherwise it is said to be nonlinear. The inversion of a nonlinear relationship is in general more difficult than the inversion of a linear relationship, because of the possible presence of false solutions or local minima (Persico et al., 2002).

In order to introduce the problem of the local minima, let us consider a generic (nonlinear) equation:

$$F(x) = y \tag{6.2}$$

Suppose also that we know a priori that Eq. (6.2) has a unique ideal (i.e., for error-free data) solution. Even under these hypotheses, in many problems of physical relevance (including the inversion of GPR data) we are not able to find this solution in a closed form, and we can't even be sure to find an exact solution at all, because the actual data are not error-free. So, customarily one looks for a numerical solution, that is, the value of x that makes minimum the quantity $|F(x) - y|^2$.

In order to provide a graphical example regarding local minima, let us consider a case where

$$\begin{aligned} F(x) &= 0.002x^5 + 0.005x^4 - 0.0585x^3 - 0.08875x^2 + 0.47725x + 0.25325 \\ y &= 0.05 \end{aligned} \tag{6.3}$$

In this case, Eq. (6.2) reduces to

$$0.002x^5 + 0.005x^4 - 0.0585x^3 - 0.08875x^2 + 0.47725x + 0.25325 = 0.05 \tag{6.4}$$

Equation (6.4) cannot be solved in a closed form.[1] Thus, we look for a least square solution, given by the value of the variable x that provides the minimum value of the quantity

$$|F(x) - 0.05|^2 = \left|0.002x^5 + 0.005x^4 - 0.0585x^3 - 0.08875x^2 + 0.47725x + 0.20325\right|^2 \tag{6.5}$$

In Figure 6.1 the graph of the cost function (6.5) is depicted. Being a function of a single variable, the absolute minimum point is approximately but immediately identified by eye; in the case at hand, it is about equal to −0.4. However, in the more general case of a function of several (possibly thousands) variables it is not possible to identify the global minimum among the local minima by eye.

Thus, in the more general case, one looks for the minimum by making use of the gradient of the function[2] in any point, and he/she "follows" the direction opposite to that of the gradient[3] up to finding a point where the gradient is equal to zero. In this way, at a

---

[1] The Abel–Ruffini theorem states that it is impossible to solve by radicals algebraic equations of the fifth or higher degree.

[2] For a scalar function, the gradient reduces to the first derivative times the unitary vector directed along the x-axis.

[3] More in general a direction depending on the gradient is followed (Press et al., 1987), but we will not include these specific aspects.

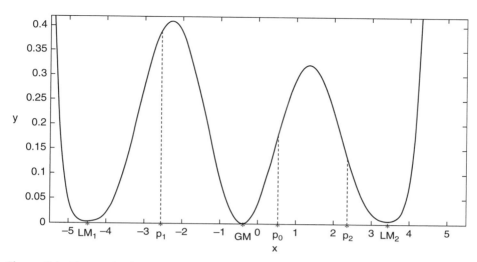

Figure 6.1. The graph of the cost function 6.5. Beyond the global minimum GM, there are two local minima LM$_1$ and LM$_2$.

certain point one reaches a minimum of the function. If the starting point of the procedure is in a suitable position (e.g., p_0 in Figure 6.1), the procedure will produce the actual (global) minimum; but if the starting point is not "suitable" (e.g., p_1 or p_2 in Figure 6.1), the procedure will converge toward a local minimum, and this will be erroneously interpreted as the solution of the problem. The suitability of the starting point is not merely given by the distance between it and the global minimum; the question is whether the starting point lies or not in the "valley" of the global minimum, which is virtually never quantifiable a priori.

The problem of the local minima can be mitigated by repeating the minimization procedure starting from several starting points, chosen and updated by means of stochastic criteria (Pastorino, 2010). In this way, it becomes more probable (at the price of an increased computational burden) that one of the achieved minima (of course the one that provides the smaller value of the cost function) is really the global minimum.

Under either a deterministic or a statistical minimization procedure, a nonlinear approach can be computationally burdening, and this hinders the application of nonlinear approaches toward large-scale problems. In the framework of GPR data processing, often the problem at hand is a large-scale one (with respect to the involved wavelengths), and therefore approximated linear algorithms are exploited in most cases. In particular, the common migration algorithms are based on a linear approximation of the scattering equations, as will be shown in Chapter 10. Notwithstanding, it is important to be aware of the intrinsic nonlinearity of the inverse scattering problem.

Let us explicitly note that the problem of the local minima is intrinsically related to the nonlinearity, because linear problems do not suffer the presence of local minima. To show this in a simple way, let us consider the linear function $F(x) = ax$, where a is some constant. In this case, the cost function to be minimized is quadratic, and its graph is a parabola with the concavity toward the positive side of the y-axis. Such a function does not show any local minimum, and any starting point is suitable in order to look for its minimum numerically.

Let us also note that not necessarily a nonlinear function produces a cost functional showing local minima. An immediate counterexample is given by the case $F(x) = ax + b$, where a and b are two constants, which is not a linear relationship. More in general, the research of conditions for the disappearing of the local minima within a nonlinear approach is itself an issue of scientific interest [see Persico et al. (2002) and references therein] because this makes more reliable (possibly also faster) the result of a numerical nonlinear inversion.

6.4   THE ILL-POSEDNESS OF THE INVERSE SCATTERING PROBLEM

To show the ill-posedness of the electromagnetic inverse scattering problem at hand, let us provide some preliminary observations:

1. At each position, the incident field is a finite energy (square integrable) function, because the sources can radiate only a finite energy.
2. The internal field is a finite energy function too, because it is produced by the interaction of the finite energy incident field with a passive medium.
3. The internal and external Green's functions are finite energy functions too, because (as has been shown in Chapter 4) they correspond to an equivalent field generated by impulsive (with finite energy) buried sources.

Therefore, labeling as f any among the listed quantities, we have that $\int\limits_{-\infty}^{+\infty} \int\limits_{+\infty}^{+\infty} |f(x',z')|^2 dx' dz'$ is a finite quantity. Moreover, let us consider the following specific class of dielectric contrasts:

$$\bar{\chi}_e(x',z') = \begin{cases} \sin(kx)\sin(ky) & |x'| < 1m, \quad 0 < z' < 1m \\ 0 & \text{elsewhere} \end{cases} \tag{6.6}$$

where k is a spatial circular frequency. Let us now consider the limit value of the scattering equations when the oscillations of the contrast become faster and faster. To do this, let us first particularize the internal scattering Equation (4.44) to the case at hand:

$$E(x,x_s,z,z_s) = E_{inc}(x,x_s,z,z_s) + k_s^2 \int\limits_{-1}^{1}\int\limits_{0}^{1} G_i(x,x',z,z')E(x',x_s,z',z_s)\bar{\chi}_e(x',z')\,dx'dz',$$

$$(x,z) \in D, (x_s,z_s) \in \Sigma \tag{6.7}$$

We will now show that, when the oscillations of the contrast become faster and faster (i.e., for $k \to \infty$), the integral term in Eq. (6.7) vanishes. To accomplish this task, let us define a sequence of functions as follows (the functional dependence of the integrands are left implicit):

$$E_1 = E_{inc},$$

$$E_2 = E_{inc} + k_s^2 \int_{-1}^{1}\int_{0}^{1} G_i E_1 \bar{\chi}_e \, dx' dz',$$

$$E_3 = E_{inc} + k_s^2 \int_{-1}^{1}\int_{0}^{1} G_i E_2 \bar{\chi}_e \, dx' dz', \qquad (6.8)$$

$$\cdots$$

$$E_n = E_{inc} + k_s^2 \int_{-1}^{1}\int_{0}^{1} G_i E_{n-1} \bar{\chi}_e \, dx' dz'$$

Clearly, if this sequence converges, its limit is the internal field. In this case, the limit value of the sequence (6.8) can be also rewritten as the sum of the following series:

$$E = E_{inc} + k_s^2 \int_{-1}^{1}\int_{0}^{1} G_i E_{inc} \bar{\chi}_e \, dx' dz'$$

$$+ k_s^4 \int_{-1}^{1}\int_{0}^{1} G_i \bar{\chi}_e \, dx' dz' \int_{-1}^{1}\int_{0}^{1} G_i E_{inc} \bar{\chi}_e \, dx'' dz''$$

$$+ k_s^6 \int_{-1}^{1}\int_{0}^{1} G_i \bar{\chi}_e \, dx' dz' \int_{-1}^{1}\int_{0}^{1} G_i \bar{\chi}_e \, dx'' dz'' \int_{-1}^{1}\int_{0}^{1} G_i E_{inc} \bar{\chi}_e \, dx''' dz''' +$$

$$\cdots + k_s^{2n} \int_{-1}^{1}\int_{0}^{1} G_i \bar{\chi}_e \, dx' dz' \int_{-1}^{1}\int_{0}^{1} G_i \bar{\chi}_e \, dx'' dz'' \cdots \int_{-1}^{1}\int_{0}^{1} G_i E_{inc} \bar{\chi}_e \, dx^n dz^n + \cdots \qquad (6.9)$$

For simplicity of notation, the functional dependence has been left implicit in Eq. (6.9). In order to specify it, let us note that the observation point of the subsequent nested integral becomes the integration variable of the previous integral. For example, the second integral term is extendedly written as

$$k_s^4 \int_{-1}^{1}\int_{0}^{1} G_i(x,z,x',z') \bar{\chi}_e(x',z') \, dx' dz' \int_{-1}^{1}\int_{0}^{1} G_i(x',z',x'',z'') E_{inc}(x'',z'') \bar{\chi}_e(x'',z'') \, dx'' dz'' \qquad (6.10)$$

and so on

Equation (6.9) expresses the so-called Born series of the field (Born and Wolf, 1999) in the case at hand.

Let us now consider the first integral term $k_s^2 \int\limits_{-1}^{1}\int\limits_{0}^{1} G_i E_{inc}\bar{\chi}_e \, dx' dz'$. Its integrand is composed of the contrast times a square integrable function independent from it. Therefore, this integral vanishes for $k \to \infty$ (it's the same principle underlying the stationary phase method illustrated in Chapter 4). Consequently, for any small positive and dimensionless ε we can find a value of k high enough to guarantee that $\left| k_s^2 \int\limits_{-1}^{1}\int\limits_{0}^{1} G_i E_{inc}\bar{\chi}_e \, dx' dz' \right| < \varepsilon \times 1\frac{V}{m}$.

Let us now consider the integral term $k_s^4 \int\limits_{-1}^{1}\int\limits_{0}^{1} G_i \bar{\chi}_e \, dx' dz' \int\limits_{-1}^{1}\int\limits_{0}^{1} G_i E_{inc}\bar{\chi}_e \, dx'' dz''$. It's modulus is smaller than $\varepsilon \left| k_s^2 \int\limits_{-1}^{1}\int\limits_{0}^{1} G_i \bar{\chi}_e \, dx' dz' \right| \times 1\frac{V}{m}$, and we can find a value of k that (further than the previous inequality) also guarantees the inequality $\left| k_s^2 \int\limits_{-1}^{1}\int\limits_{0}^{1} G_i \bar{\chi}_e \, dx' dz' \right| < \varepsilon$. This guarantees that $\left| k_s^4 \int\limits_{-1}^{1}\int\limits_{0}^{1} G_i \bar{\chi}_e \, dx' dz' \int\limits_{-1}^{1}\int\limits_{0}^{1} G_i E_{inc}\bar{\chi}_e \, dx'' dz'' \right| < \varepsilon^2 \times 1\frac{V}{m}$. At this point, it is easily shown iteratively that the same k guarantees that

$$|E_n - E_{inc}| < 1\frac{V}{m} \times \left(\varepsilon + \varepsilon^2 + \varepsilon^3 + \cdots + \varepsilon^n \right) \tag{6.11}$$

Passing to the limit value, we have

$$|E_\infty - E_{inc}| \leq 1\frac{V}{m} \times \left(\varepsilon + \varepsilon^2 + \varepsilon^3 + \cdots \right) = 1\frac{V}{m} \times \varepsilon \sum_{n=0}^{+\infty} \varepsilon^n = 1\frac{V}{m} \times \frac{\varepsilon}{1-\varepsilon} \tag{6.12}$$

where in Eq. (6.12) we have substituted the sum of the geometrical series of common ratio ε. Given the arbitrary level of ε, Eq. (6.12) shows that, in the limit case at hand, the Born series converges uniformly and the final result in the limit for $k \to +\infty$ is $E = E_{inc}$, which is to say that very fast oscillating and limited contrasts are transparent to the radiation or, equivalently, the internal field cannot "follow" any contrast variation. The reasoning could be repeated also with a contrast given by the sum of a "slow varying" contrast and a fast oscillating contrast, and in the limit we would determine that the internal field is influenced only by the "slowly varying" portion of the contrast profile.

Substituting $E = E_{inc}$ in the external scattering equation [see Eq. (4.45)], we have

$$E_s \approx k_s^2 \int\limits_{-1}^{1} \int\limits_{0}^{1} G_e E_{inc} \bar{\chi}_e \, dx' dz' \qquad (6.13)$$

At this point, since both the external Green's function and the incident field square integrable functions are independent from the contrast, we have that in the limit for $k \to \infty$ the contrast $\bar{\chi}_e$ produces a null scattered field; that is, it represents an invisible target. Now, any square integrable contrast target can be expanded, within the investigation domain D, along its Fourier series, and so any target can be generically seen as composed by a slow varying part plus a fast varying part. We have shown that at a certain point the fast oscillating part becomes more and more transparent, and in the considered limit completely transparent, and consequently it becomes more and more difficult (because of the uncertainties on the data) and eventually impossible to retrieve it. This demonstrates that the inverse scattering problem at hand is ill-posed, and it also shows that the ill-posedness involves the impossibility to retrieve all the details of the buried scenario. For a wider and deeper dealing about the ill-posedness, the interested reader is referred to Colton and Kress, (1992), Tikhonov and Arsenine (1977), and Kirsh (1996). In particular, let us clearly say that the invisible targets are not necessarily only those oscillating in a fast way, even if this is probably the aspect of main interest in the framework of GPR prospecting.

6.5 THE NONLINEARITY OF THE INVERSE SCATTERING PROBLEM

The nonlinearity of the problem can immediately be shown. In fact, the electric field inside the investigation domain depends on the contrast, due to Eq. (4.44). Consequently, if in Eq. (4.44) the contrast is multiplied times any constant, the scattered field will be not just be multiplied for the same constant, because the internal field has changed too, and so Eq. (6.1) does not hold.

Physically, the nonlinearity is due to the electromagnetic interferences among the scattering objects embedded in the soil, which makes the internal field change in accordance with them; we will explain this more completely in Chapter 8.

QUESTIONS

1. In this chapter the Born series has been introduced. Does a truncation of the series at the second order necessarily provide a better approximation of the field with respect to one at the first order? And if this happens, does the truncation at the third order necessarily provide a still better approximation of the field?

2. Is the scattered field due to two imposed density currents radiating at the same time given by the sum of the scattered fields achieved under the radiation of any of the sources independently from the other one?

3. Is the scattered field due to two antennas radiating at the same time given by the sum of the scattered fields achieved under the radiation of any of the sources independently from the other one?

4. Is the scattered field due to two targets illuminated at the same time given by the sum of the scattered fields achieved from each of the two targets independently from the other one?

5. Does the distance between two targets make weaker the nonlinear effects?

6. Do the losses in the soil make weaker the nonlinear effects?

7

EXTRACTION OF THE SCATTERED FIELD DATA FROM THE GPR DATA

7.1 ZERO TIMING

Before extracting the scattered field data, we have to first discuss the problem of the zero timing. In particular, the choice of the zero time is a problem arising from a physical constraint, namely the fact that any GPR system has a finite band and, consequently, it cannot radiate or receive "correctly" an impulse with an immediate rising up. Moreover, there is necessarily some propagation of the signal inside the instrument before the impulse is "launched" outside. So, there is a delay between the time instant when the generator begins to produce the impulse and the time instant when the propagation of the signal in the external environment begins.

This problem involves both real and synthetic pulses—this is, both impulsive and stepped frequency systems. The practical consequence is that we have to choose, on the basis of the received data, the moment when we assume that the signal was launched outside from the GPR system, which is the also the reference origin of the time for each trace (Yelf, 2004). This operation is called zero timing and necessarily involves some degree of arbitrariness. The zero timing is important, because it influences the retrieving of the time depth of the targets and formally also the evaluation of the propagation

Introduction to Ground Penetrating Radar: Inverse Scattering and Data Processing,
First Edition. Raffaele Persico.
© 2014 The Institute of Electrical and Electronics Engineers, Inc. Published 2014 by John Wiley & Sons, Inc.

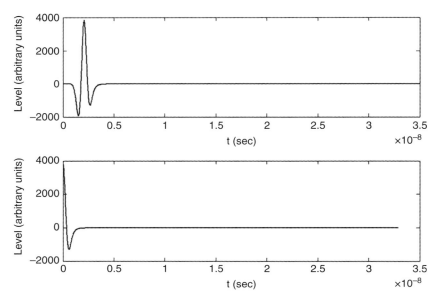

Figure 7.1. Example of zero timing on a GPR trace simulated with the code GPRmax. The upper panel shows the trace as it comes out, and the lower panel shows the zero timed trace. The bottom scale is 35 ns, but it reduces to 32.9 for the zero timed trace. The source is a Ricker pulse with central frequency 500 MHz. The source is in contact with a soil that shows a relative permittivity $\varepsilon_{sr} = 5$ and an electric conductivity $\sigma = 0.001$ S/m and with no magnetic properties.

velocity of the waves. In particular, a different choice of the zero time corresponds to a rigid shift of the diffraction curves along the t-axis, which amounts to an alteration of the retrieved value of propagation velocity, as is easily understood from Chapter 2.

In turn, an incorrect evaluation of the electromagnetic characteristics of the soil affects the retrieved depth of the targets as well as the result of any focusing algorithm, because these characteristics enter as parameters in the scattering equations (see Chapters 4 and 5). A common (and generally reasonable) choice is to fix the zero time at the first maximum modulus of the received signal, so that we gate out all the energy received before this instant. Actually, the value of this maximum changes from trace to trace, even if the height of the antennas is kept constant along the B-scan, because of clutter and noise. However, the relative discrepancies are usually minimal, and a heuristic average value of the time of this maximum is sufficiently accurate in most cases. Figure 7.1 shows the result of such a gating on simulated data.

7.2 MUTING OF INTERFACE CONTRIBUTIONS

After the preliminary zero timing, the datum observed in the observation point is the voltage related to the total field in that point, which is roughly proportional to the total field in that point.

Eventually, we can therefore write the datum in the time domain as

$$E(x,t) = E_{inc}(x,t) + E_s(x,t) \tag{7.1}$$

or equivalently we can write the datum in frequency domain as

$$E(x,\omega) = E_{inc}(x,\omega) + E_s(x,\omega) \tag{7.2}$$

Equation (7.1) or (7.2) accounts for the fact that the height of the data is usually a fixed parameter, not varied along the scan. The data in time and in frequency domain are linked by Fourier direct or inverse transforms; but we have not made use of hat symbols, which have been reserved for the Fourier transforms with respect to the spatial variables.

Now, the data that we want to process are the scattered field data, that is, the E_s quantities. In fact, as shown by means of the scattering equations, the contrasts (i.e., the buried targets) are specifically related to the scattered field data and not to the total field data. In Chapter 4 we have provided expressions for the incident field both in the case of a filamentary current and in the case of a source characterized by a given plane wave spectrum. So, one might think of subtracting the calculated incident field data from the measured total field data. This operation is mathematically correct, but is not robust against the uncertainties present with regard to both the properties of the soil and the characteristics of the transmitting antenna (Persico and Soldovieri, 2006). In order to show this, an exercise is now proposed, with data simulated with GPRMAX2D. In particular, a $30 \times 30 \text{ cm}^2$ cavity is buried at the depth of 1 m in a lossless nonmagnetic soil with relative permittivity $\varepsilon_s = 5\varepsilon_0$. The source is a Ricker pulse with central frequency 500 MHz; it is moved along a path 2.5 m long, centered on the target, at height 30 cm above the air–soil interface, with a spatial step of 5 cm. The offset between the source and the observation point is equal to zero. Figure 7.2a shows the total field result; as can be seen, the diffraction curve is almost invisible. In fact, as often happens, the contribution of the incident field is quite stronger than that of the scattered field and masks it. The result of the subtraction of the incident field from the total one without any uncertainties on the data is shown in Figure 7.2b. To do this, the total and the incident field have been calculated in two distinct simulations, the second of which differs from the first one only for the absence of the buried target. This result shows the diffraction curve in a clear fashion. However, the procedure has worked because we have implicitly assumed that we have a perfect knowledge of the characteristics of the soil, of the source, and of the receiver, so that we could model the incident field perfectly. In real cases we don't have a perfect knowledge of any among the involved parameters, and so a reliable and precise calculation of the incident field is not a trivial matter. In order to simulate the effect of the uncertainties, the incident field has been calculated a second time for a soil with relative permittivity equal to 5.35 instead of 5. Moreover, the amplitude of the Ricker source has been changed times a factor 1.05 and the central frequency has been shifted at 502 MHz. Then, we have subtracted this incident field to the total field data of Figure 7.2a. In this way, we have simulated an uncertainty about the permittivity, the intensity of the source, and the central frequency. The result is shown in Figure 7.2c, and it clearly shows that the involved uncertainties meaningfully

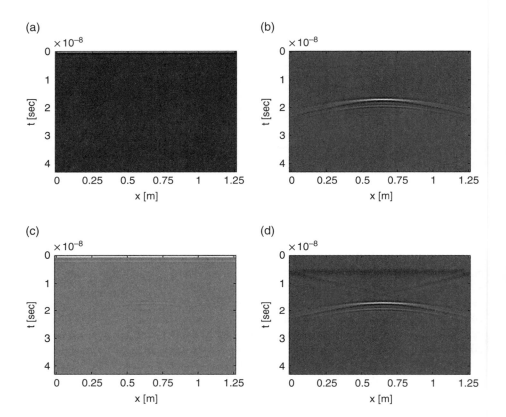

Figure 7.2. **Panel a**: Total field data. **Panel b**: Scattered field data worked out by the difference between the total and the incident field data with no parametric uncertainty. **Panel c**: Scattered field data worked out by the difference between the total and the incident field data with some parametric uncertainty. **Panel d**: Scattered field data worked out by muting. In all the panels the data have been previously zero-timed at the first maximum of the received total field data.

mask the buried targets. Let us also emphasize that the introduced errors are minimal with respect to realistic situations: and, in particular, no clutter due to the surface roughness has been synthesized, and there is no uncertainty about the kind of radiated pulse and about the antenna pattern, which further enforces the statement.

In Figure 7.2d, instead, we have muted (i.e., we have put equal to zero) the received signal up to the second horizontal band visible in Figure 7.2a. In fact, the first (shallower) horizontal band is due to the direct coupling between the source and the receiver, whereas the second (deeper) one is due to the reflection of the signal from the air–soil interface. If the antennas are moved in common offset at the air–soil interface (in the presented example they are instead at 30 cm from the air–soil interface), these two signals are generally superposed and indistinguishable from each other.

In the absence of buried targets, the gathered datum would consist exclusively of these horizontal bands, which is to say that the two horizontal belts are essentially the incident field data. More in general, in the case of any layered medium (and in particular in the case of a masonry) the incident field data amount to several quasi-horizontal bands, caused either by horizontal physical interfaces or by obstacles that move synchronically with the antennas (e.g., a trolley or a car or the same human operator). Finally, a further possible cause of horizontal bands is some undesired reflection of the currents within the arms of the antenna, which is known as ringing (Daniels, 2004).

Whatever the cause, these bands might be erased from the data by trivially muting them. However, in this way, we pay the price to erase, partially or totally, also the signal scattered by targets located within those belts. So, there is some trade-off between the "pureness" of the scattered field worked out and the range of depths filtered out. The optimal choice for the thickness of the erased time-belts is, in general, case-dependent and, in particular, depends on the band of the antennas and on the depth range of the targets of interest. In the end, the muting of the interfaces can be seen as an imperfect filtering of the incident field. This muting damps down a little share of the scattered field and preserves some "tail" of the incident field; this is a price to pay in order to have a more robust procedure with respect to a formal subtraction of the incident field form the total one. As said, in Figure 7.2d the result of the muting is shown. As can be seen, the results of Figures 7.2d and 7.2b differ only by a weak residual horizontal band.[1]

Another theoretical possibility might be to gather a B-scan in an area where no meaningful buried target is registered. In this way, we would have a measured version of the incident field to be subtracted from the total field. However, also this procedure is not robust against the uncertainties on the characteristics of the soil. In particular, it is virtually impossible to have at one's disposal (unless in a test site built up on purpose) a B-scan where one gathers only the interfaces (and *the same interfaces* of a B-scan that has to be processed) under the same characteristics of the soil, the same surface roughness, and so on.

Indeed, in microwave imaging there are cases where the scattered field data can be reliably achieved experimentally by subtraction of an incident field measured apart. This happens, in particular, when the structure to be probed is a mobile opaque body. In fact, in such a situation we can perform two measurements with the antennas in the same position, one in the presence of the body and another one after removing it, which provides experimentally and rigorously the incident field data (Otto and Chew, 1994). The "trick" is that in such a case we are assuming that the air is the (homogeneous) background medium, and so we don't have any meaningful uncertainty about it (in GPR applications it can be assimilated to the free space all the times). In the case of GPR prospecting, instead, the background medium is commonly chosen as the air–soil couple of half-spaces, and so we cannot remove the investigation domain in order to measure the incident field. Going on in this theoretical speculation, at this point let us also say that we might think of assimilating the underground zone to be investigated to a "big box"containing the targets of interest and comprehensively embedded in air.

---

[1] Actually, in Figure 7.2d, also two oblique "X"-shaped lines are visible, but this a mere numerical effect due to the limitedness of the simulated domain.

So, we might retrieve the incident field, measuring it in air—that is, making the antennas radiate and receive in the same relative positions but without any obstacle. There is nothing formally incorrect in assuming that the air (instead of the air–soil couple) is the background medium also in the case of GPR data. However, under such a choice, the quantity to be reconstructed would not be constituted by the buried targets, but rather by the investigated "soil box" with inside the targets of interest. In other words, the soil itself would compose the contrast (with respect to the air) to be reconstructed. It is intuitive (but in the next chapter this will be formalized more completely), that in this way the quantity to be retrieved is somehow "farther" from the background medium with respect to the quantity achieved under the choice of the air–soil couple as background medium. This can cause problems in relationship with the intrinsic nonlinearity of the scattering phenomenon, as will be shown in the next chapter. Therefore, the common praxis in GPR data processing is to refer to a background medium composed by a couple of half-spaces.

That said, in order to give a rough quantification of the targets erased by the muting, we have to quantify the thickness of the "horizontal bands" related to the incident field. In general, they depend on the frequency band of the signal and on the propagation velocity of the waves. In particular, physically, the thickness of the horizontal band is essentially the "spatial duration" of the propagating pulses. Therefore, it is given by the temporal duration of the pulse times the propagation velocity in the soil divided by two (because of the round-trip). The temporal duration of the pulse is roughly given by the inverse of its band or, still more roughly (see Chapter 3), by the inverse of its central frequency. So, by restricting our attention to the case of a homogeneous soil, all the targets shallower than

$$D_{min} = \frac{c}{2B} \tag{7.3}$$

will be meaningfully erased, where B is the band of the antennas. It is often reported that a large band is desirable in order to have a good resolution. From Eq. (7.3), we can appreciate that a large band can also reduce the drawbacks related to the muting of the interface.

7.3 THE DIFFERENTIAL CONFIGURATION

There are alternative possibilities to the muting of the horizontal bands, in order to extract scattered field data from the total field data. One of them is to make use of a differential configuration (Gurel and Oguz, 2003; Persico and Soldovieri, 2006). It consists in assuming as datum the difference between the total field gathered at two different positions with an opposite offset with respect to the source. In other words, if x_s is the position of the source and Δ is a fixed offset, then the datum is given by

$$\Delta E = E(x_s + \Delta, t) - E(x_s - \Delta, t) \tag{7.4}$$

or, in the frequency domain

$$\Delta E = E(x_s + \Delta, \omega) - E(x_s - \Delta, \omega) \qquad (7.5)$$

at variance of x_s. The differential configuration can be also implemented dually with two transmitting antennas and a receiving one in the middle point between them, with the two transmitted signals equal in shape and amplitude but reversed in phase.

In a layered medium, and in particular in a homogeneous half-space, the incident field is constant along x_s, and so it is immediate to see that the difference between the total field in the two symmetrical points reduces to the difference between the scattered fields in those two points.[2] In this way, we have erased the contribution of the incident field without needing any muting. This has the advantage that we don't erase the targets embedded in the horizontal belts, and in particular we don't erase the shallowest targets.

However, the achieved datum is not any longer the scattered field but the differential scattered field in the observation point, given by

$$\Delta E = E_s(x_s + \Delta, t) - E_s(x_s - \Delta, t) \qquad (7.6)$$

or in the frequency domain

$$\Delta E = E_s(x_s + \Delta, \omega) - E_s(x_s - \Delta, \omega) \qquad (7.7)$$

This implies that the mathematical model to achieve the reconstruction should change too. In other words, the scattering operator at hand changes and has different mathematical properties. In Chapter 9, we will show that this somehow worsens the ill-posedness of the inverse scattering problem.

7.4   THE BACKGROUND REMOVAL

Another possibility is that of applying a background removal (BKGR). The background removal is a well-known procedure in geophysics (Daniels, 2004; Conyers, 2004; Jol, 2009). However, it is rarely thought that, in terms of physics-mathematics, it can be viewed as a heuristic way to filter out the contribution of the incident field from the data (Persico and Soldovieri, 2008; Persico and Soldovieri, 2010). The reasoning that shows this is substantially the same shown with regard to the differential configuration. In fact, let us consider a BKGR with moving average: It consists in taking as datum the difference between the current trace and the average value of a set of traces symmetrically centered around the current one. So, let us suppose that the averaging is performed on $2N + 1$ traces, starting from the Nth traces before the current one and ending at the Nth traces after the current one. Let us label as s the spatial step of the measurements. So, the datum is given by

---

[2] Indeed, the roughness of the interface does not make this statement rigorous, but the discrepancy is usually marginal on the spatial scale of a reasonable offset (expected to be of the order of 10 cm)

$$E_b(x,t) = E(x,t) - \frac{1}{2N+1} \sum_{n=-N}^{N} E(x-ns,t) \qquad (7.8)$$

In the time domain, or alternatively

$$E_b(x,\omega) = E(x,\omega) - \frac{1}{2N+1} \sum_{n=-N}^{N} E(x-ns,\omega) \qquad (7.9)$$

in the frequency domain. The subscript b stands for background removal. Again, considering the incident field invariant along the horizontal abscissa, the contribution of the incident field to the average trace and to the current trace are the same and so they erase each other, and the result is rewritten as

$$E_b(x,t) = E_s(x,t) - \frac{1}{2N+1} \sum_{n=-N}^{N} E_s(x-ns,t) \qquad (7.10)$$

In the time domain or

$$E_b(x,\omega) = E_s(x,\omega) - \frac{1}{2N+1} \sum_{n=-N}^{N} E_s(x-ns,\omega) \qquad (7.11)$$

in the frequency domain. BKGR is automatically implementable by several commercial software. Often, one can choose either the number of traces or the "horizontal extension" to be averaged—that is, the distance between the first and the last averaged traces. The two quantities are related to each other, because this distance is equal to $2Ns$.

The definition of BKGR presents a problem with regard to both the initial trace and to final trace. In fact, for example, when considering the first trace of the GPR scan, of course we don't have N traces *before* it, and when considering the last one we don't have N further traces *beyond* it. So, an artificial prolongation (implicit or explicit) of the B-scan is needed both before its beginning and beyond its end.[3] To this pros, several choices are possible, and in general the best choice depends on the situation at hand.

One of the possible choices is to place M times the comprehensive average trace of the whole B-scan before the B-scan and M times the same average trace after the B-scan, where M is the total number of traces of the gathered B-scan. In this way, we create an equivalent B-scan three times longer than the actual one. At this point, if one performs a mobile averaging on $2M+1$ traces on the real B-scan, it can be checked that the result is equal to the well-known background removal procedure performed on all the traces—this is, a background removal where the datum is given by the current traces minus the average of all the traces. This shows that the background removal on all

---

[3] Alternatively, one should avoid reconstructing the scenario under the first and last part of the gathered B-scan, but this is never convenient.

the traces is substantially a particular case of the BKGR with moving average, and so the properties that we will investigate about the moving average background removal procedure hold also in the case of BKGR on all the traces.

Similar to the differential configuration, BKGR can be implemented in either the time domain or the frequency domain, and it is a way to filter out the incident field without erasing the targets embedded in the horizontal bands related to the layers present in the raw data. Moreover, again similar to the differential configuration, BKGR erases the constant returns from any target that moves together with the antennas. Of course, also this time the returned datum is not the scattered field in the observation point but rather a function of the scattered field in the observation point, and this changes the mathematical properties of the scattering operator, as will be shown in Chapter 9. In particular, we will show that the properties of the scattering operator get somehow worsened also in this case (again similarly to the differential configuration). In the end, this means that the possibility to erase horizontal belts without erasing the targets inside them has some price in terms of resolution capabilities of the scattering operator.

An important difference between the differential datum and the BKGR datum is that the first one is conceived as associated to a specific hardware, whereas BKGR is a procedure that can be applied on the usual common offset data. In particular, we can think of applying the background removal only within certain time intervals (which is impossible with a differential configuration), in order to erase the horizontal band while "saving" the targets embedded in them and, at the same time, keeping to a heuristic minimum the modifications of the scattering operator. An example of BKGR is shown in Figure 7.3, where the diffraction hyperbola relative to a small reflector is shown. The target is a square object that is built with a perfect electric conductor, sized 5×5 cm^2 and buried at the depth of 1 m. The medium is a homogeneous soil with relative permittivity equal to 5 and electrical conductivity equal to 0.01 S/m. The data have been simulated with the GPRMAX code. The observation line is 2.5 m long and is at a height of 1 cm. The spatial step is 2.5 cm. The offset between source and observation point is zero. The source is a Ricker pulse with central frequency 500 MHz. In particular, in panel a, the diffraction hyperbola achieved after muting of the first 5 ns is represented; in panel b, the homologous quantity achieved from BKGR on all the traces is shown. In panel c the homologous quantity achieved from BKGR with mobile averaging on 9 traces is depicted; in panel d the homologous quantity achieved from BKGR performed on all the traces but only on the first 5 ns is shown. Figure 7.3 shows that the background removal can be done in several ways, each of which can have some pros and some cons. In particular, the background removal on all the traces (panel 2) erases the spurious interface generated by an abrupt muting but generates a spurious flat interface in correspondence to the top of the diffraction hyperbola. This is because the high level of the signal in the target point is able to change the level of the comprehensive average trace at that time instant. This can also be viewed as an effect of the limitedness of the observation line. In fact, if the BKGR is done on a longer and longer observation line, this effect is expected to be progressively weaker. This effect is usually easily recognized, because the lines generated by it are too straight to be ascribable to a real target. Notwithstanding, this effect might mask some weak target at the same time depth of the strong reflection. A BKGR with a moving average doesn't show this drawback, because the limitedness of the averaging line confines to a local effect this possible spreading. This is clear from panel c, where a moving BKGR on nine traces (corresponding to an averaging length of 20 cm) is shown. However, from panel c, it is also evident that the top of the diffraction hyperbola gets partially erased. This is becaues the top is the most flat part of the curve, so that the averaged traces are quite similar to each other and erase with the central one while performing the moving average BKGR. Finally, panel d shows the diffraction hyperbola achieved by a BKGR performed only on the first 5 ns of all the traces. In this case the result is very similar to that of panel a. However, unlike the simple "muting," if we had had a target in the first 5 ns, in the case of Figure 7.3d the BKGR would not erase it. So, in the case at hand, the solution of panel d seems to be a good compromise between the contrasting exigencies to minimize the distortion of the data and to avoid the erasing of shallow targets.

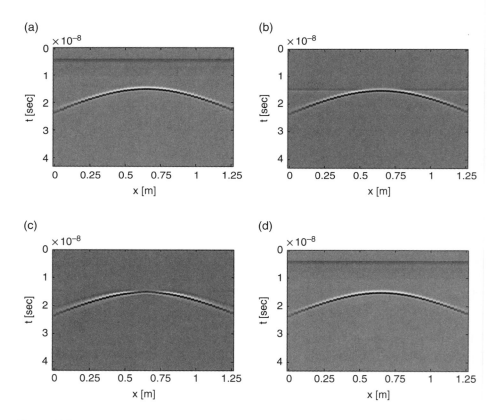

Figure 7.3. **Panel a:** The diffraction hyperbola achieved after muting of the first 5 ns. **Panel b:** The same quantity achieved from BKGR on all the traces. **Panel c:** The same quantity achieved from BKGR with mobile averaging on 9 traces. **Panel d:** The same quantity achieved from BKGR performed on all the traces but only on the first 5 ns.

QUESTIONS

1. Is it necessarily well-advised to process data gathered with a stepped-frequency GPR system entirely in frequency domain?

2. Can the interface muting be performed in frequency domain?

3. Let us suppose that we have to perform a background removal on N traces, for any fixed N. Is a positive effect expected if the spatial step is made progressively narrower?

4. Let us suppose that we have to perform a background removal on N traces, for any fixed N. Is a positive effect expected if the spatial step is made progressively larger?

5. Does the interface muting provide a rigorous calculation of the scattered field?

6. In the case of a homogeneous soil, do the background removal and/or the differential configuration provide a quantity rigorously independent from the incident field?

8

THE BORN APPROXIMATION

8.1 THE CLASSICAL BORN APPROXIMATION

In Chapter 6, we have introduced the Born series of the scattering operator, even if in a specific case and not in its more general form. The basis for the construction of this Born series was given in Eq. (6.8) with regard to the case of nonmagnetic anomalies. The sequence provided in Eq. (6.8) can be written in a more general (i.e., for any dielectric contrast) and complete form as follows:

$$
\begin{aligned}
E_1(x_s,x,z) &= E_{inc}(x_s,x,z), \\
E_2(x_s,x,z) &= E_{inc}(x_s,x,z) + k_s^2 \iint_D G_i(x,z,x',z')E_1(x_s,x',z')\chi_e(x',z')\,dx'dz', \\
E_3(x_s,x,z) &= E_{inc}(x_s,x,z) + k_s^2 \iint_D G_i(x,z,x',z')E_2(x_s,x',z')\chi_e(x',z')\,dx'dz', \\
&\cdots \\
E_n(x_s,x,z) &= E_{inc}(x_s,x,z) + k_s^2 \iint_D G_i(x,z,x',z')E_{n-1}(x_s,x',z')\chi_e(x',z')\,dx'dz', \quad z>0
\end{aligned}
\tag{8.1}
$$

Introduction to Ground Penetrating Radar: Inverse Scattering and Data Processing,
First Edition. Raffaele Persico.
© 2014 The Institute of Electrical and Electronics Engineers, Inc. Published 2014 by John Wiley & Sons, Inc.

where D is the investigation domain and x_s is the source point. If the Born sequence (8.1) converges, its limit point is the internal field by virtue of Eq. (4.44). Analogously to the particular case examined in chapter 6, the Born sequence for the internal field, as any sequence, can be put in the form of a series as follows:

$$
E = E_{inc} + k_s^2 \iint_D G_i E_{inc} \chi_e \, dx' dz'
$$

$$
+ k_s^4 \iint_D G_i E_{inc} \chi_e \, dx' dz' \iint_D G_i E_{inc} \chi_e \, dx'' dz''
$$

$$
+ k_s^6 \iint_D G_i E_{inc} \chi_e \, dx' dz' \iint_D G_i E_{inc} \chi_e \, dx'' dz'' \iint_D G_i E_{inc} \chi_e \, dx''' dz''' + \cdots
$$

$$
\cdots + k_s^{2n} \iint_D G_i E_{inc} \chi_e \, dx' dz' \iint_D G_i E_{inc} \chi_e \, dx'' dz'' \cdots \iint_D G_i E_{inc} \chi_e \, dx^n dz^n + \cdots, \qquad z > 0
$$

$$
(8.2)
$$

Analogously to Eq. (6.9), in Eq. (8.2) all the functional dependences of the nested integrands have been dropped out in order to not over-prolong the expression. Being bi-univocally related to the Born sequence, the Born series for the internal field does not necessarily converge. Incidentally, some conditions for the convergence[1] have been studied (D'Urso et al., 2007), but they are beyond the purposes of this text and will not be dealt with here. Substituting the Born series for the internal field into the external scattering equation [see Eq. (4.45)], we achieve the Born series for the scattered field, given by

$$
E_s = k_s^2 \iint_D \chi_e E_{inc} G_e \, dx' dz' + k_s^4 \iint_D \chi_e G_e \, dx' dz' \iint_D G_i E_{inc} \chi_e \, dx'' dz'' +
$$

$$
+ k_s^{2n+2} \iint_D \chi_e G_e \, dx' dz' \iint_D G_i E_{inc} \chi_e \, dx'' dz'' \cdots \iint_D G_i E_{inc} \chi_e \, dx^{n+1} dz^{n+1} + \cdots, \quad z < 0
$$

$$
(8.3)
$$

Analogously to the Born series for the internal field, there is no general guarantee for the convergence of the Born series for the scattered field. In any case, the first term of this sequence provides the first-order Born Approximation (BA) for the scattered field, which is commonly just reported as the BA (Chew, 1995; Born and Wolf, 1999). Expanding the expression in order to make clear the functional dependences, under the BA the external scattering equation becomes

$$
E_s(x_s, \omega) = k_s^2 \iint_D \chi_e(x', z') E_{inc}(x', x_s, z'; \omega) G_e(x', x_0 = x_s + \Delta, z'; \omega) \, dx' dz', \qquad (x_s, x_o) \in \Sigma
$$

$$
(8.4)
$$

---

[1] The convergence is meant in the least square sense.

Equation (8.4) also accounts for a common offset configuration where the observation point and the source point are separated by a fixed offset Δ and the height of the data is a fixed parameter too. Under the BA, it is immediate that the scattering is approximated as a linear phenomenon, which poses the basis for faster and easier processing algorithms with respect to the actual nonlinear model, and of course a linear model is immune from the problem of the local minima. Moreover, as will be shown, BA allows a noticeable insight about the characteristics of the expected results, essentially based on the spatial filtering properties of the linear scattering operator.

From a physical point of view, the BA amounts to neglect the mutual interactions between any two different buried targets or (which is mathematically the same) between any two different parts of the same target. The situation can be pictorially explained by means of Figure 8.1.

In particular, let us regard A and B as two very small targets, so that we can assume that each of them, in the absence of the other one, would provide a scattered field given by Eq. (8.4), where the contrast is a function that describes each of the two targets in turn. In fact, if the target is very small, the integral term in Eq. (4.44) can be regarded as a negligible perturbation with respect to the incident field, because the contrast is a low norm function. However, if the two targets are present at the same time (so that the contrast is given by the sum of the contrasts relative to each of the two targets separately considered), they interact with each other, and consequently the comprehensive scattered field is not simply given by the sum of the scattered field that each of them would produce in absence of the other. This interaction is the physical genesis of the nonlinearity, because it makes the internal total field different from the internal incident field.

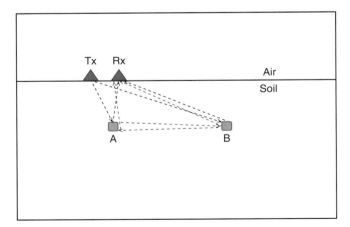

Figure 8.1. Schematic for the physical reason of the nonlinearity of the scattering. The black arrows represent the direct contributions of the two small buried targets separately considered. The red arrows represent the contribution of the mutual interaction between A and B to the scattered field. For color detail, please see color plate section.

8.2 THE BORN APPROXIMATION IN THE PRESENCE OF MAGNETIC TARGETS

From the complete internal scattering equations (5.69), we retrieve immediately that the BA in the more general case of both dielectric and magnetic targets requires a formally more restrictive condition with respect to the classical BA for only dielectric targets. In particular, with regard to the internal field, it is now required (Persico and Soldovieri, 2011) that we have

$$E(x,z) \approx E_{inc}(x,z), \quad (x',z') \in D, \tag{8.5}$$

$$\nabla' E(x',z') \approx \nabla E_{inc}(x',z'), \quad (x',z') \in D \tag{8.6}$$

where the symbols are in agreement with those adopted in the scattering equations provided in Chapters 4 and 5.[2] Indeed, an intriguing theoretical question arises whether Eq. (8.6) is a direct consequence of Eq. (8.5) or is a really further independent condition [or whether (8.6) can be derived from (8.5) under some further wide hypotheses]. In particular, if the incident field is regarded as just a function, it is well known that the approximation (8.6) cannot be derived as a consequence of the approximation (8.5), because two quantities might be very close to each other (e.g., in the least square sense) and notwithstanding their derivatives might be very different from each other. However, the incident field is not a generic function, because it is a solution of the Maxwell's equations, which represents a meaningful constraint. The question is open at the moment, to the best of our knowledge, because we have not found any proof that condition (8.6) can be derived directly from condition (8.5), nor have we found a proof of the opposite statement. So, we adopt "conservatively" both conditions.

Substituting Eqs. (8.5) and (8.6) into the external scattering equation (5.68), we achieve the equation of the scattered field under BA for dielectric and magnetic targets:

$$E_s(x_s;\omega) = k_s^2 \iint\limits_D [G_e(x_s + \Delta, x', z';\omega) E_{inc}(x_s, x', z';\omega) \chi_e(x', z')$$

$$+ \frac{1}{k_s^2} \nabla' G_e(x_s + \Delta, x', z';\omega) \nabla' E_{inc}(x_s, x', z';\omega) \chi_{m1}(x', z')] \, dx' dz', \quad x_s, x_o \in \Sigma$$

$$\tag{8.7}$$

where the height of the source and of the observation point have been supposed invariant along the GPR scan and where the quantity $\dfrac{\chi_m(x',z')}{1 + \chi_m(x',z')}$ has been labeled as $\chi_{m1}(x',z')$.

---

[2] Let us stress that the erasing of the incident field discussed in Chapter 7 was referred to the incident field in the observation point (in air), whereas the incident field in Eqs. (8.1) and (8.2) is meant as the incident field in the investigation domain (underground).

Note that these two quantities have the same support[3]; that is, from a practical point of view, they approximately "express" two anomalies with the same size, position, and shape (but with different values). The scattering equation (8.7) is in frequency domain, and in particular the contrasts might depend on the frequency too. However, this dependence will be neglected.

That said, under the BA the scattering problem is in any case linear and so any inversion algorithm based on the BA is not affected by local minima and is, in general, computationally less demanding with respect to a nonlinear approach.

Of course, based on the reasoning presented in Chapter 6, the ill-posedness of the problem remains.

As already said, physically the BA amounts to neglecting the mutual electromagnetic interactions among different buried targets and among the different parts of any buried target. In the case of magnetic and dielectric targets, this also amounts to neglecting the interaction between the magnetic contrasts and the dielectric contrasts, so that the scattered field under the BA is the sum of the scattered fields generated by the only dielectric part plus that generated by the only magnetic part of the buried targets.

8.3   WEAK AND NONWEAK SCATTERING OBJECTS

Theoretically, to neglect the mutual interactions between the buried targets is licit for weak scattering targets—that is, buried objects that make negligible the integral term on the right-hand side of Eq. (5.69). That term is the difference between the internal actual field and the internal incident field; that is, it is by definition the internal scattered field.

The definition of weak scattering target is often accepted as an object that that modifies negligibly (conventionally $\pm \pi/8$ at most) the phase of an electromagnetic wave propagating through it (Slaney et al., 1984). However, this is a definition essentially thought to be appropriate for free space cases. Actually, an inhomogeneous background can influence the degree of nonlinearity of the problem (Persico and Soldovieri, 2004). Therefore, here "weak scatterer" is defined as just an object for which the internal field, and (in case of magnetic targets) its first-order spatial derivatives can be approximated with those of the incident field. The "weakness," in particular, is a feature not only related to the maximum level of the contrast, but also to the electrical size of the buried target (which implies a dependence on the frequency), its shape, and the nature of the of background medium (homogeneous or layered, lossless or lossy).

It is important to emphasize that, independently from the validity of the BA in the current situation (with GPR field data, customarily BA is not valid), an aspect worth emphasizing is that the secondary sources that generate the scattered field under BA have the same support of the actual secondary sources, and this support is just the extension of the buried targets. In fact (see Chapter 4), the actual secondary source in case of dielectric

---

[3] The support of a function is defined as the closure of the set of the points where a function assumes non-null values.

targets are given by $j\omega\varepsilon_s\chi_e E$ [see Eq. (4.20)], whereas the secondary source under the BA are provided by $j\omega\chi_e\varepsilon_s E_{inc}$. Now, assuming that both the incident and the actual internal fields are supported throughout the entire investigation domain, it is clear that either under the exact model or under the BA the support of the secondary source is equal to the support of the dielectric contrast, which is just the extension of the buried targets. Therefore, it is licit to expect that in many cases the position, the size, and (under certain limits) the shape of the buried targets can be satisfyingly retrieved under BA, even in cases when the targets are not weak (Slaney et al. 1984; Idemen and Ackduman, 1990). This has also been widely shown experimentally (Meincke, 2001; Catapano et al., 2006; Persico and Sala, 2011). Let us also state that "satisfyingly" is a voluntarily generic term: It means that the achievable results are useful for some applications but does not mean that the geometrical reconstruction is "perfect" or very good (e.g., in the least square sense). In particular, the achievable reconstruction is affected by the filtering properties of the linear scattering operator, examined in the next chapter. In any case, in situations beyond the limits of the BA, we will not achieve a quantitative reconstruction of the electromagnetic characteristics of the buried object. Analogous statements can be straightforwardly worked out in the case of magnetic anomalies.

At this point, we can propose a brief flashback on the question of the background medium, prompted in Chapter 7: Indeed, making a rough parallel between the Born series for the scattered field and the Taylor series for an ordinary function, we recognize that the background medium plays the role of the starting point of the series. The sense of the incident field, as well as the sense of the contrast function, actually depends on the chosen background medium. Consequently, to chose a background medium somehow "closer" to the physical situation is helpful in order to mitigate the nonlinearity of the problem. In particular, the common praxis for GPR data is to refer to a half-space geometry (or even to a homogeneous medium with the electromagnetic characteristics of the soil) because a "big box" model embedded in air would make quite more critical the intrinsic nonlinearity of the problem.

QUESTIONS

1. Consider a given weak scattering target embedded in a homogeneous soil. Let us embed this same target in a masonry composed of a material with the same characteristics of the previous soil. Is the scattering target "weak" in the same way? Namely, might the ratio between the norms of the internal incident field and the internal total field meaningfully change? Provide a qualitative answer considering the target under the radiation of the same sources placed in the same relative positions with respect to it in the two cases.

9

DIFFRACTION TOMOGRAPHY

9.1 INTRODUCTION TO DIFFRACTION TOMOGRAPHY

Diffraction Tomography is a topic that has been dealt with for a long time (Lesselier and Duchene, 1996; Meincke, 2001; Tabbra et al., 1988; Witten et al., 1996; Cui and Chew, 2002). In general terms, a DT relationship is an algebraic relationship between the spectrum of the data and the spectrum of the unknown function. There are several kinds of DT relationships in relationship with the measurement configuration (Persico et al., 2005; Persico, 2006), but here we will focus only on the common offset configuration. In general, a DT relationship requires more approximations than does the linearization provided by the BA, as will be shown in the following. In particular, two general hypotheses assumed for any DT relationship in common offset configuration are the following:

1. The soil or more in general the propagation medium is lossless.
2. The targets are not close to the sources in terms of wavelength.

Introduction to Ground Penetrating Radar: Inverse Scattering and Data Processing,
First Edition. Raffaele Persico.
© 2014 The Institute of Electrical and Electronics Engineers, Inc. Published 2014 by John Wiley & Sons, Inc.

9.2   DIFFRACTION TOMOGRAPHY FOR DIELECTRIC TARGETS

In order to introduce the first DT relationship of this chapter, let us specify some further hypotheses, beyond BA and the assumptions 1 and 2 of the previous section. In particular, in this section we will assume the following:

1. There is no magnetic target
2. The observation line is at the air–soil interface.
3. The observation line is infinitely long.
4. The source is a filamentary electrical current.
5. The datum is constituted by the scattered field in the observation point.

In this case, the linear external scattering equation [see Eq. (8.4)] is particularized into

$$E_s(x;\omega) = k_s^2 \iint\limits_D G_e(x,x',0,z';\omega) E_{inc}(x+\Delta,x',0,z';\omega)\chi_e(x',z')\,dx'dz', \quad x\in\Sigma \quad (9.1)$$

where Δ is the fixed offset between the source and the observation point. After substituting the expressions of the external Green's function given in Eq. (4.47) and the expression of the incident field given in Eq. (4.48) in Eq. (9,1), we obtain

$$E_s(x;\omega) = \frac{j\omega I_0 \mu_0^2 \mu_s k_s^2}{2\pi} \int_{-\infty}^{+\infty}\int_{-\infty}^{+\infty}\int_{-\infty}^{+\infty}\int_{-\infty}^{+\infty} \frac{\exp(-j(k_{zs}(u)+k_{zs}(v))z')\exp(j(u+v)x)}{(\mu_0 k_{zs}(u)+\mu_s k_{z0}(u))(\mu_0 k_{zs}(v)+\mu_s k_{z0}(v))}$$

$$\times \exp(jv\Delta)\exp(-j(u+v)x')\chi_e(x',z')\,dx'dz'dudv, \quad x\in\Sigma \quad (9.2)$$

In Eq. (9.2), the integral on the investigation domain D has been replaced with an integral on the entire plane, because the dielectric contrast is a null function outside the investigation domain. Let us now write $p = u+v \Rightarrow u = p-v$. The integrals can be rearranged as (Persico et al., 2005)

$$E_s(x;\omega) = \frac{j\omega I_0 \mu_0^2 \mu_s k_s^2}{2\pi} \int_{-\infty}^{+\infty}\int_{-\infty}^{+\infty}\int_{-\infty}^{+\infty}\int_{-\infty}^{+\infty} \frac{\exp(-j(k_{zs}(p-v)+k_{zs}(v))z')\exp(jpx)}{(\mu_0 k_{zs}(p-v)+\mu_s k_{z0}(p-v))(\mu_0 k_{zs}(v)+\mu_s k_{z0}(v))}$$

$$\times \exp(jv\Delta)\exp(-jpx')\chi_e(x',z')\,dx'dz'dpdv, \quad x\in\Sigma \quad (9.3)$$

In the four nested integrals, we can recognize a Fourier transform of the contrast with respect to x' (the conjugate variable being p) and an inverse Fourier transform with respect to the variable p (the conjugate variable being the observation point x). So, by Fourier transforming the data, from Eq. (9.3) we obtain

$$\hat{E}_s(p;\omega) = j\omega I_0 \mu_0^2 \mu_s k_s^2 \int_{-\infty}^{+\infty} dz' \hat{\chi}_e(p,z') \times$$

$$\times \int_{-\infty}^{+\infty} \frac{\exp(-j(k_{zs}(p-v) + k_{zs}(v))z')\exp(jv\Delta)}{(\mu_0 k_{zs}(p-v) + \mu_s k_{z0}(p-v))(\mu_0 k_{zs}(v) + \mu_s k_{z0}(v))} dv, \quad x \in \Sigma \quad (9.4)$$

Now, the integral in dv can be solved under the stationary phase approximation, because we have assumed high values of z', that is, deep targets (see hypothesis 2 above). The rationale of the stationary phase method has already been introduced in Chapter 4 and thus will be not repeated now. However, in that chapter the two-dimensional stationary phase approximation was dealt with. Here, instead, we need the one-dimensional stationary phase approximation, which is provided by (Felsen and Marcuvitz, 1994)

$$\int_{X_1}^{X_2} f(x) \exp[j\Omega g(x)] \, dx \approx \frac{\sqrt{2\pi} \exp\left(j\frac{\pi}{4}\right) f(x_0) \exp[j\Omega g(x_0)]}{\sqrt{\Omega}\sqrt{\dfrac{d^2 g}{dx^2}(x_0)}} \quad (9.5)$$

where the symbols are homologous to those of Eq. (4.91). If the second derivative in the denominator is negative, the determination of the square root to be retained is $\sqrt{\dfrac{d^2 g}{dx^2}(x_0)} = -j\sqrt{\left|\dfrac{d^2 g}{dx^2}(x_0)\right|}$. In the case at hand, the phase function to consider is

$$\psi(v) = -k_{zs}(p-v) - k_{zs}(v) = -\sqrt{k_s^2 - (p-v)^2} - \sqrt{k_s^2 - v^2} \quad (9.6)$$

Thus

$$\frac{d\psi}{dv} = -\frac{p-v}{k_{sz}(p-v)} + \frac{v}{k_{sz}(v)} \quad (9.7)$$

$$\frac{d^2\psi}{dv^2} = \frac{-1}{(k_{sz}(p-v))^2}\left(\frac{(p-v)^2}{k_{sz}(p-v)} + k_{sz}(p-v)\right) + \frac{1}{(k_{sz}(v))^2}\left(\frac{-v^2}{k_{sz}(v)} - k_{sz}(v)\right)$$

$$= -\frac{k_s^2}{(k_{sz}(p-v))^3} - \frac{k_s^2}{(k_{sz}(v))^3} \quad (9.8)$$

It easy to check that there is a unique first-order (see Section 4.6) stationary point at $v = p/2$, and in the stationary point we have

$$\frac{d^2\psi}{dv^2}\Big|_{v=P/2} = \frac{-2k_s}{\left(k_{sz}\left(\frac{p}{2}\right)\right)^3} \Rightarrow \sqrt{\frac{d^2\psi}{dv^2}\Big|_{v=P/2}} = -j\frac{\sqrt{2k_s}}{\left(k_{sz}\left(\frac{p}{2}\right)\right)^{3/2}} \qquad (9.9)$$

Substituting into Eq. (9.4), we obtain

$$\hat{E}_s(p;\omega) = -\frac{\omega I_0 \mu_0^2 \mu_s k_s \exp\left(j\Delta\frac{p}{2}\right) \exp\left(j\frac{\pi}{4}\right)}{\left(\mu_0 k_{zs}\left(\frac{p}{2}\right) + \mu_s k_{z0}\left(\frac{p}{2}\right)\right)^2} \sqrt{\pi\left(k_{zs}\left(\frac{p}{2}\right)\right)^3}$$

$$\times \int_{-\infty}^{+\infty} \frac{\hat{\chi}_e(p,z')}{\sqrt{z'}} \exp\left(-2jk_{zs}\left(\frac{p}{2}\right)z'\right) dz' \qquad (9.10)$$

The integral term in Eq. (9.10) is once again a Fourier transform, so that eventually we obtain

$$\hat{E}_s(p;\omega) = f(p;\omega)\hat{\chi}_{e1}(\eta(p,\omega),\varsigma(p,\omega)) \qquad (9.11)$$

where

$$f(p;\omega) = -\frac{\omega I_0 \mu_0^2 \mu_s k_s \exp\left(j\Delta\frac{p}{2}\right) \exp\left(j\frac{\pi}{4}\right)}{\left(\mu_0 k_{zs}\left(\frac{p}{2}\right) + \mu_s k_{z0}\left(\frac{p}{2}\right)\right)^2} \sqrt{\pi\left(k_{zs}\left(\frac{p}{2}\right)\right)^3} \qquad (9.12)$$

$$\chi_{e1}(x',z') = \frac{\chi_e(x',z')}{\sqrt{z'}} \qquad (9.13)$$

$$\eta(p,\omega) = p \qquad (9.14)$$

$$\varsigma(p,\omega) = 2k_{zs}\left(\frac{p}{2}\right) = 2\sqrt{k_s^2 - \left(\frac{p}{2}\right)^2} = 2\sqrt{\left(\frac{\omega}{c}\right)^2 - \left(\frac{p}{2}\right)^2} = \sqrt{\frac{4\omega^2}{c^2} - p^2}$$

The reader can check that hypotheses 1–5 listed at the beginning of this section and also hypotheses 1 and 2 of the previous section have all been used in the performed calculations. In particular, the last Fourier relationship and the application of the stationary phase method require the hypothesis of a lossless soil.

Equation (9.11) also requires that $k_{zs}(p/2)$ be a real quantity; otherwise the integral in Eq. (9.10) is not a Fourier transform but instead a Laplace transform (and this involves an exponential attenuation of the scattered waves). This restricts the admitted range of variability of p to the so-called "visible" interval $I_v = (-2k_s, 2k_s)$. Physically, this means that the secondary sources, as any electromagnetic source, generate both a propagating field (related to the visible interval) and a reactive field (related to the values of p beyond the visible interval). The retrievable spectral set is provided by the image of Eq. (9.14) in the plane (η, ς), on condition that the spectral weighting function given in Eq. (9.12) is a relatively smooth function. In particular, the neighbors of the zeroes of the spectral weight should be excluded from the retrievable spectral set, because in those points the reconstruction of the spectrum of the contrast becomes unreliable. However, in order to introduce progressively the difficulties, let us neglect the effect of the spectral weight in a first moment.

In this case, the mapping equation (9.14) is easily seen to be the parametric equations of a curve in the plane (η, ς). The corresponding Cartesian equation is easily retrieved as follows:

$$\varsigma = 2k_{zs}\left(\frac{\eta}{2}\right) = 2\sqrt{k_s^2 - \left(\frac{\eta}{2}\right)^2} = \sqrt{4k_s^2 - \eta^2} \Rightarrow \varsigma^2 + \eta^2 = 4k_s^2 \qquad (9.15)$$

Equation (9.15) is the equation of a circumference centred in the origin and with ray $2k_s$. Moreover, it is immediate to check that we have the following correspondences, according to Eq. (9.14):

$$p = -2k_s \rightarrow (\eta, \varsigma) = (-2k_s, 0),$$

$$p = 0 \rightarrow (\eta, \varsigma) = (0, 2k_s), \qquad (9.16)$$

$$p = 2k_s \rightarrow (\eta, \varsigma) = (2k_s, 0)$$

This means that, while p ranges from $-2k_s$ to $2k_s$, the corresponding point in the plane (η, ς) describes in the clockwise direction a half-circumference in the half-plane $\varsigma \geq 0$, from the point $(-2k_s, 0)$ to the point $(2k_s, 0)$. Consequently, multifrequency data in a limited band ranging between some f_{min} and some f_{max} "cover" a half-annulus in the plane (η, ς). In Figure 9.1, the theoretically retrievable multifrequency spectral set is quantitatively depicted in four cases.

Figure. 9.1 shows the fact that, even in the ideal case of a continuous amount of data gathered on an infinitely long observation line, the amount of retrievable information is limited and only a spatially filtered version of the object function is retrievable. As said, the spectral weight can provide a restriction of the ideal maximum spectral set of Figure 9.1 (and actually does). In order to get an insight on this effect, let us examine the behavior of the modulus of the spectral weight. In particular, in Figure 9.2 it is shown the modulus of the weighting function, normalized to its maximum, in four cases.

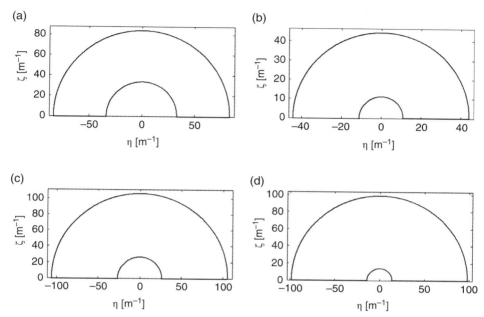

Figure 9.1. The theoretical retrievable spectral set in relationship with an infinitely long B-scan.
Panel a: $\varepsilon_s = 4\varepsilon_0$, $\mu_s = 4\mu_0$, $f_{min} = 200$ MHz, $f_{max} = 500$ MHz. **Panel b:** $\varepsilon_s = 7\varepsilon_0$, $\mu_s = \mu_0$, $f_{min} = 100$ MHz, $f_{max} = 400$ MHz. **Panel c:** $\varepsilon_s = 5\varepsilon_0$, $\mu_s = 2\mu_0$, $f_{min} = 200$ MHz, $f_{max} = 800$ MHz. **Panel d:** $\varepsilon_s = 15\varepsilon_0$, $\mu_s = 3\mu_0$, $f_{min} = 50$ MHz, $f = 350$ MHz.

As can be seen, the behavior of the function is case-dependent, but some general features are easily identified. In particular, the function is even and vanishes at $p = \pm 2k_s$, as immediately understood from its expression (9.12). This means that the visible interval should not be considered equal to the entire interval $(-2k_s, 2k_s)$, but rather an interval of the kind $I_v = (-2k_s + \varepsilon, 2k_s - \varepsilon)$, where ε is a positive quantity frequency dependent and case dependent (in particular, noise-dependent). Moreover, as it can be appreciated from Figure 9.3, the shape of the modulus of the spectral weight normalised to its maximum value is somehow frequency invariant. Therefore, ε is expected roughly proportional to the frequency and the shape of the really retrievable spectral set is expected of the kind in Figure 9.4. Also note that the restriction is more severe if the soil shows magnetic properties, because this makes concave the graph (see Figure 9.2, and this feature has been confirmed by further not reported numerical results).

Let us now consider the effect of the level of the spectral weight versus the frequency. From Eq. (9.12), it is easy to work out that we have an increasing of the level of the spectral weight versus the frequency, asymptotically proportional to $f\sqrt{f}$ for any fixed p. In particular, in Figure 9.5, the same spectral weights of Figure 9.3 are

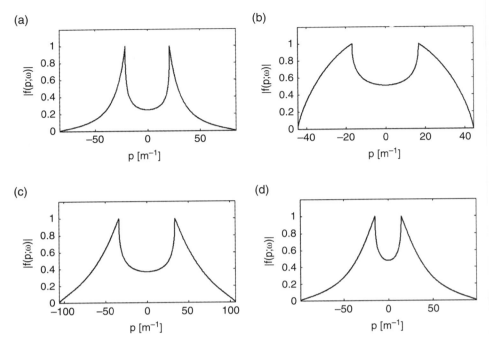

Figure 9.2. The normalized modulus of the spectral weighting function in four cases. **Panel a:** $\mu_s = 4\mu_0$, $\varepsilon_s = 4\varepsilon_0$, $f = 500$ MHz. **Panel b:** $\mu_s = \mu_0$, $\varepsilon_s = 7\varepsilon_0$, $f = 400$ MHz. **Panel c:** $\mu_s = 2\mu_0$, $\varepsilon_s = 5\varepsilon_0$, $f = 800$ MHz. **Panel d:** $\mu_s = 3\mu_0$, $\varepsilon_s = 15\varepsilon_0$, $f = 350$ MHz; in all the cases the variable p ranges from $-2k_s$ to $2k_s$ and the modulus of the spectral weight is normalized to its maximum.

shown. However, this time the four functions are normalized not with respect to their own maxima, but rather with respect to the maximum value over all of them, which is achieved at the maximum considered frequency.

Physically, the result shown in Figure 9.5 means that a filamentary current, with a fixed (maximum or rms) value of the current flowing in it, radiates more power at higher frequencies. This should be accounted when considering very large band data (e.g., one decade or more) processed under a filamentary current model. In particular, in these cases a model based on a filamentary current might penalize too much the lower frequencies and should be somehow compensated.

In conclusion, the retrievable spectral set is a limited set, even in the ideal case of an infinite observation line, and this is due to the ill-posedness of the problem.

Due to its analytic properties, the spectrum of a square integrable function can be theoretically prolonged in a unique fashion all over the plane, starting from the retrievable spectral set. However, the prolonging of the Fourier transform is in turn an ill-posed problem. Indeed, the prolongation of the spectrum has been widely studied in one-dimensional cases, where some results in closed form have been achieved too

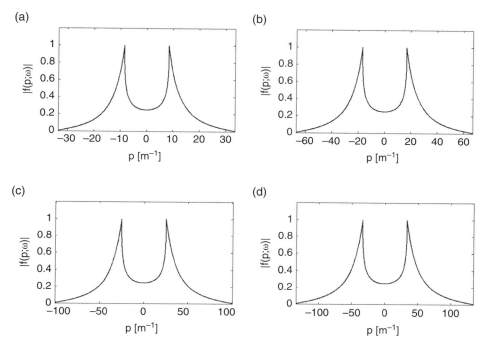

Figure 9.3. The normalized modulus of the spectral weighting function in four cases. **Panel a:** $f = 200$ MHz. **Panel b:** $f = 400$ MHz. **Panel c:** $f = 600$ MHz. **Panel d:** $f = 800$ MHz; in all the cases $\varepsilon_s = 4\varepsilon_0$, $\mu_s = 4\mu_0$, p ranges from $-2k_s$ to $2k_s$ and the modulus of the spectral weight is normalized to its own maximum.

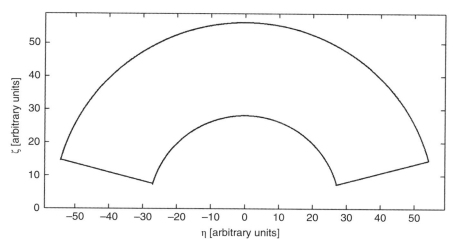

Figure 9.4. Qualitative reduction of the spectral set due to the vanishing of the spectral weight at the two extremes of the visible interval. The unities are arbitrary, but the shape of the set is achieved supposing $f_{max} = 2f_{min}$.

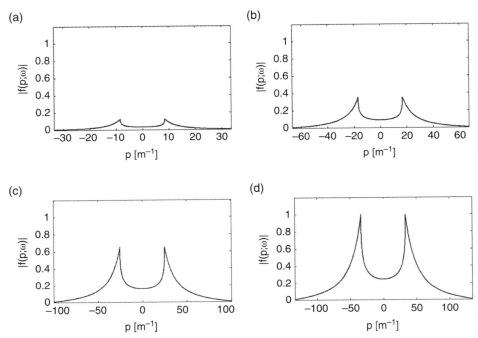

Figure 9.5. The normalized modulus of the spectral weighting function in four cases. **Panel a:** $f = 200$ MHz. **Panel b:** $f = 400$ MHz. **Panel c:** $f = 600$ MHz. **Panel d:** $f = 800$ MHz. In all the cases $\varepsilon_s = 4\varepsilon_0$, $\mu_s = 4\mu_0$, p ranges from $-2k_s$ to $2k_s$ and the modulus of the spectral weight is normalized to the highest value among the maximum levels of the four functions.

(Slepian and Pollack, 1961; Landau and Pollak, 1961; Landau and Pollak, 1962). Under the highlight provided by these studies, we can even say that the prolongation of the spectrum is the ill-posed problem "per antonomasia" and in general is not a suitable strategy to enhance the DT relationships.

9.3   DIFFRACTION TOMOGRAPHY FOR DIELECTRIC TARGETS SEEN UNDER A LIMITED VIEW ANGLE

Among the five hypotheses assumed in Section 9.2, there was the infinite length of the observation line. This hypothesis allowed us to recast the problem as an algebraic relationship between spectra. However, in any practical case the observation line is limited. So, let us now consider a finite-length observation line, leaving unchanged the other hypotheses assumed in Sections 9.1 and 9.2. The finite length of the observation line corresponds to a limited-view angle range, smaller than $(-\pi/2, \pi/2)$, which is the range of view angles available with an infinite observation line, as illustrated in Figure 9.6 with regard to a single point-like target. In Figure 9.6, the angles are referred to the

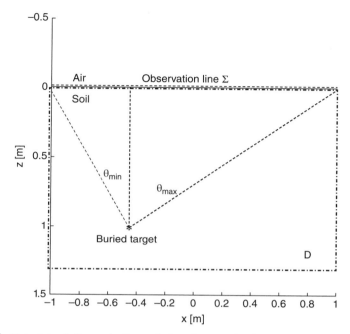

Figure 9.6. Pictorial of the reduction of the view angle due to the limitedness of the observation line. In particular, $\theta_{min} \rightarrow -\pi/2$ and $\theta_{max} \rightarrow \pi/2$ if and only if the observation is infinitely long on both sides. The length units are arbitrary.

vertical line passing on the target and are counted clockwise. Therefore, they are negative if the observation point is on the left-hand side with respect to the target and positive when the observation point is on the right-hand side with respect to the target.

In order to calculate the DT relationship in the case at hand, let us re-start from Eq. (9.1). However, in order to make it symmetrical, let us refer the scattered field not to the observation point, but to the intermediate point between the observation point x and source point $x + \Delta$. This intermediate point is given by

$$x_m = x + \frac{\Delta}{2} \qquad (9.17)$$

Thus the observation and the source points can be expressed versus it as

$$x = x_m - \frac{\Delta}{2} \qquad (9.18)$$

$$x_s = x + \Delta = x_m + \frac{\Delta}{2} \qquad (9.19)$$

Under this change of variable, after substituting the expressions of the external Green's function [Eq. (4.47)] and of the incident field [Eq. (4.48)] in Eq. (9.1), we obtain

$$E_s(x;\omega) = \frac{j\omega I_0 \mu_0^2 \mu_s k_s^2}{2\pi} \int_{-\infty}^{+\infty} \int_{-\infty}^{+\infty} \int_{-\infty}^{+\infty} \int_{-\infty}^{+\infty} \frac{\exp(-j(k_{zs}(u)+k_{zs}(v))z') \exp(j(u+v)x_m)}{(\mu_0 k_{zs}(u)+\mu_s k_{z0}(u))(\mu_0 k_{zs}(v)+\mu_s k_{z0}(v))}$$

$$\times \exp\left(jv\frac{\Delta}{2}\right) \exp\left(-ju\frac{\Delta}{2}\right) \exp(-j(u+v)x')\chi_e(x',z')\, dx'dz'dudv, \quad x \in \Sigma$$

$$(9.20)$$

The integrals can be reordered as

$$E_s(x;\omega) = \frac{j\omega I_0 \mu_0^2 \mu_s k_s^2}{2\pi} \int_{-\infty}^{+\infty} \int_{-\infty}^{+\infty} \chi_e(x',z')dx'dz' \int_{-\infty}^{+\infty} \exp\left(-ju\frac{\Delta}{2}\right)$$

$$\times \frac{\exp(-j(k_{zs}(u))z') \exp(jux_m) \exp(-jux')}{(\mu_0 k_{zs}(u)+\mu_s k_{z0}(u))} du \int_{-\infty}^{+\infty} \exp\left(jv\frac{\Delta}{2}\right)$$

$$\times \frac{\exp(-jk_{zs}(v)z') \exp(jvx_m) \exp(-jvx')}{(\mu_0 k_{zs}(v)+\mu_s k_{z0}(v))} dv, \quad x \in \Sigma \qquad (9.21)$$

After some straightforward manipulations, Eq. (9.21) can be still rewritten as

$$E_s(x;\omega) = \frac{j\omega I_0 \mu_0^2 \mu_s k_s^2}{2\pi} \int_{-\infty}^{+\infty} \int_{-\infty}^{+\infty} \chi_e(x',z')\, dx'dz' \times$$

$$\times \left[\left(\int_{-\infty}^{+\infty} \cos\left(u\frac{\Delta}{2}\right) \frac{\exp(-j(k_{zs}(u))z') \exp(-ju(x_m-x'))}{(\mu_0 k_{zs}(u)+\mu_s k_{z0}(u))} du \right)^2 + \right.$$

$$\left. + \left(\int_{-\infty}^{+\infty} \sin\left(u\frac{\Delta}{2}\right) \frac{\exp(-j(k_{zs}(u))z') \exp(-ju(x_m-x'))}{(\mu_0 k_{zs}(u)+\mu_s k_{z0}(u))} du \right)^2 \right], \quad x \in \Sigma$$

$$(9.22)$$

Let us now write

$$r' = \sqrt{(x_m-x')^2 + z'^2},$$

$$\cos(\theta') = \frac{z'}{r'},$$

$$\sin(\theta') = \frac{x_m-x'}{r'}$$

$$(9.23)$$

These positions express the vector of the relative position between the observation and the integration point in polar coordinates. Under these positions, the integral (9.22) is expressed as

$$E_s(x;\omega) = \frac{j\omega I_0 \mu_0^2 \mu_s k_s^2}{2\pi} \int\limits_{-\infty}^{+\infty} \int\limits_{-\infty}^{+\infty} \chi_e(x',z')\, dx'\, dz'$$

$$\times \left[\left(\int\limits_{-\infty}^{+\infty} \cos\left(u\frac{\Delta}{2}\right) \frac{\exp(-j(k_{zs}(u))r'\cos(\theta'))\exp(-jur'\sin(\theta'))}{(\mu_0 k_{zs}(u)+\mu_s k_{z0}(u))}\, du \right)^2 \right.$$

$$\left. + \left(\int\limits_{-\infty}^{+\infty} \sin\left(u\frac{\Delta}{2}\right) \frac{\exp(-j(k_{zs}(u))r'\cos(\theta'))\exp(-jur'\sin(\theta'))}{(\mu_0 k_{zs}(u)+\mu_s k_{z0}(u))}\, du \right)^2 \right], \quad x \in \Sigma$$

(9.24)

where the dependence of r' and θ' from x', z', and x_m is taken implicitly. Both the integrals in parentheses can be solved, for high values of r', making use of the stationary phase approximation. There is a unique first-order stationary point at $u = k_s \sin(\theta')$, and the result is (Soldovieri et al., 2007)

$$E_s(x;\omega) = \frac{j\omega I_0 \mu_0^2 \mu_s k_s^2}{2\pi} \int\limits_{-\infty}^{+\infty} \int\limits_{-\infty}^{+\infty} \chi_e(x',z')\, dx'\, dz'$$

$$\times \left[\cos^2\left(\frac{k_s \Delta \sin(\theta')}{2}\right) \frac{2\pi j \exp(-j2r'k_s)\cos^3(\theta')k_s}{\left(\mu_0 k_s \cos(\theta')+\mu_s \sqrt{k_0^2 - k_s^2 \sin^2(\theta')}\right)^2 z'} \right.$$

$$\left. + \sin^2\left(\frac{k_s \Delta \sin(\theta')}{2}\right) \frac{2\pi j \exp(-j2r'k_s)\cos^3(\theta')k_s}{\left(\mu_0 k_s \cos(\theta')+\mu_s \sqrt{k_0^2 - k_s^2 \sin^2(\theta')}\right)^2 z'} \right]$$

$$= -\omega I_0 \mu_0^2 \mu_s k_s^3 \int\limits_{-\infty}^{+\infty} \int\limits_{-\infty}^{+\infty} \chi_e(x',z') \frac{\exp(-j2r'k_s)\cos^3(\theta')}{\left(\mu_0 k_s \cos(\theta')+\mu_s \sqrt{k_0^2 - k_s^2 \sin^2(\theta')}\right)^2 z'}\, dx'\, dz', \quad x \in \Sigma$$

(9.25)

In Eq. (9.25), we have applied the phase stationary method with respect to r'. Formally there is a dependence on r' also in the residual phase term. In particular, the phase term was given by

$$g(u,r') = -k_{zs}(u)\cos(\theta') - u\sin(\theta')$$

(9.26)

This quantity formally depends also on r' because r' and θ' are linked by Eqs. (9.23). We find comfortable, for a reason that will be clear in the following, to solve this question by generalizing the stationary phase formula (9.5) into

$$\int_{X_1}^{X_2} f(x)\exp[j\Omega g(x;\Omega)]\,dx \approx \frac{\sqrt{2\pi}\exp\left(j\frac{\pi}{4}\right)f(x_0)\exp[j\Omega g(x_0;\Omega)]}{\sqrt{\Omega}\sqrt{\dfrac{\partial^2 g}{\partial x^2}(x_0;\Omega)}} \tag{9.27}$$

which is not a correct operation in general. However, in the case at hand the application of (9.27) is licit, because the phase term of Eq. (9.26) does not vanish when r' increases, which makes the rationale of the stationary phase method still valid, because it remains true that the exponential term $\exp(jr'(-k_{zs}(u)\cos(\theta') - u\sin(\theta')))$ in Eq. (9.24) is more and more oscillating when r' increases.

Equation (9.25) shows that, in the current observation point, the first contribution to the scattered field is given by a plane wave coming from the target to the observation point. The phase factor $2r'k_s$ expresses the round-trip of the wave from the source to the target and vice versa.[1]

At this point, let us consider (with regard to Figure 9.6) the case that the contrast is provided by a point like target. So, let us focus the attention on the following case:

$$\chi_e(x',z') = \chi_0\delta(x'-x_0)\delta(z'-z_0) \tag{9.28}$$

Where, implicitly, z_0 is supposed electrically large. Substituting Eq. (9.28) in Eq. (9.25), we achieve

$$E_s(x;\omega) = -\omega I_0\mu_0^2\mu_s k_s^3\chi_0 \frac{\exp(-j2r_0k_s)\cos^3(\theta_0)}{\left(\mu_0 k_s\cos(\theta_0) + \mu_s\sqrt{k_0^2-k_s^2\sin^2(\theta_0)}\right)^2 z_0}, \qquad x\in\Sigma \tag{9.29}$$

where

$$r_0 = \sqrt{(x_m-x_0)^2 + z_0^2},$$
$$\cos(\theta_0) = \frac{z_0}{r}, \tag{9.30}$$
$$\sin(\theta_0) = \frac{x_m-x_0}{r}$$

Let us now consider the spectrum of the scattered field with respect to x_m:

---

[1] Actually, the physical optical path of the signal is from the source point to the target and from the target back to the observation point, but for long distances (in particular distances much longer than the offset between the antennas) it can be approximated by the round-trip from the source–receiver midpoint to the target.

$$
\hat{E}_s(p;\omega) = -\omega I_0 \mu_0^2 \mu_s k_s^3 \chi_0 \int_{-\infty}^{+\infty} \frac{\exp(-j(px + 2r_0 k_s))\cos^3(\theta_0)}{\left(\mu_0 k_s \cos(\theta_0) + \mu_s \sqrt{k_0^2 - k_s^2 \sin^2(\theta_0)}\right)^2 z_0} \, dx_m
$$

$$
= -\omega I_0 \mu_0^2 \mu_s k_s^3 \chi_0 \int_{-\infty}^{+\infty} \frac{\exp(-j(p(x_m - x_0) + 2r_0 k_s))\exp(-jpx_0)\cos^3(\theta_0)}{\left(\mu_0 k_s \cos(\theta_0) + \mu_s \sqrt{k_0^2 - k_s^2 \sin^2(\theta_0)}\right)^2 z_0} \, dx_m
$$

$$
= -\omega I_0 \mu_0^2 \mu_s k_s^3 \chi_0 \int_{-\infty}^{+\infty} \frac{\exp\left(-jz_0\left(-p\operatorname{tg}(\theta_0) + \frac{2k_s}{\cos(\theta_0)}\right)\right)\exp(-jpx_0)\cos^3(\theta_0)}{\left(\mu_0 k_s \cos(\theta_0) + \mu_s \sqrt{k_0^2 - k_s^2 \sin^2(\theta_0)}\right)^2 z_0} \, dx_m
$$

$$(9.31)$$

Let us solve the integral (9.31) with the stationary phase method with respect to z_0. In doing this, we are again applying the generalized stationary formula (9.27). In particular, the phase term is now given by

$$
g(x_m, z_0) = -p\tan(\theta_0) + \frac{2k_s}{\cos(\theta_0)} = p\frac{x_m - x_0}{z_0} + \frac{2k_s}{z_0}\sqrt{(x_m - x_0)^2 + z_0^2} \tag{9.32}
$$

whose first derivative with respect to the integration variable is

$$
\frac{\partial g}{\partial x_m} = -\frac{p}{z_0} + \frac{2k_s}{z_0}\frac{(x_m - x_0)}{\sqrt{(x_m - x_0)^2 + z_0^2}} \tag{9.33}
$$

so that there is a unique first-order stationary point given by

$$
x_{mo} = x_0 + \frac{pz_0}{\sqrt{4k_s^2 - p^2}} \Rightarrow \left(\frac{x_{m0} - x_0}{z_0}\right)^2 = \frac{p^2}{4k_s^2 - p^2}
$$

$$
\Rightarrow \left(\frac{x_{m0} - x_0}{z_0}\right)^2 + 1 = \frac{(x_{m0} - x_0)^2 + z_0^2}{z_0^2} = \frac{r_{m0}^2}{z_0^2} = \frac{p^2}{4k_s^2 - p^2} + 1 = \frac{4k_s^2}{4k_s^2 - p^2} \tag{9.34}
$$

Based on Eq. (9.34), in this stationary point x_{m0} we have

$$
\cos(\theta_{m0}) = \frac{z_0}{r_{m0}} = \frac{\sqrt{4k_s^2 - p^2}}{2k_s},
$$

$$
\sin(\theta_{m0}) = \sqrt{1 - \cos^2(\theta_{m0})} = \frac{p}{2k_s}, \tag{9.35}
$$

$$
r_{m0} = \frac{2k_s z_0}{\sqrt{4k^2 - p^2}}
$$

The second derivative of the phase term is given by

$$\frac{\partial^2 g}{\partial x_m^2} = \frac{2k_s z_0}{\left[(x_m - x_0)^2 + z_0^2\right]^{3/2}} \tag{9.36}$$

and its value in the stationary point is equal to

$$\frac{\partial^2 g}{\partial x_m^2}\bigg|_{x_m = x_{m0}} = \frac{2k_s z_0}{\left[\dfrac{p^2 z_0^2}{4k_s^2 - p^2} + z_0^2\right]^{3/2}} = \frac{2k_s z_0 \left(4k_s^2 - p^2\right)^{3/2}}{\left(4k_s^2 z_0^2\right)^{3/2}} = \frac{\left(4k_s^2 - p^2\right)^{3/2}}{4k_s^2 z_0^2} \tag{9.37}$$

Consequently, the final result is

$$
\begin{aligned}
\hat{E}_s(p;\omega) &= \frac{-\omega I_0 \mu_0^2 \mu_s k_s^3 \chi_0 \sqrt{2\pi} \exp\left(j\dfrac{\pi}{4}\right) \exp(-jpx_0)\left(4k_s^2 - p^2\right)^{3/2} \exp\left(-jz_0\sqrt{4k_s^2 - p^2}\right) 2k_s z_0}{8k_s^3 \sqrt{z_0}\left(\mu_0\sqrt{k_s^2 - \dfrac{p^2}{2}} + \mu_s\sqrt{k_0^2 - \dfrac{p^2}{2}}\right)^2 \left(4k_s^2 - p^2\right)^{3/4} z_0} \\[2ex]
&= \frac{-k_s\omega I_0 \mu_0^2 \mu_s \chi_0 \sqrt{2\pi} \exp\left(j\dfrac{\pi}{4}\right) \exp(-jpx_0) 4^{3/4}\left(k_s^2 - \dfrac{p^2}{4}\right)^{3/4} \exp\left(-2jz_0\sqrt{k_s^2 - \dfrac{p^2}{4}}\right)}{4\sqrt{z_0}\left(\mu_0\sqrt{k_s^2 - \dfrac{p^2}{4}} + \mu_s\sqrt{k_0^2 - \dfrac{p^2}{4}}\right)^2} \\[2ex]
&= \frac{-k_s\omega I_0 \mu_0^2 \mu_s \chi_0 \sqrt{\pi} \exp\left(j\dfrac{\pi}{4}\right) \exp(-jpx_0)\left(k_s^2 - \dfrac{p^2}{4}\right)^{3/4} \exp\left(-2jz_0\sqrt{k_s^2 - \dfrac{p^2}{4}}\right)}{\sqrt{z_0}\left(\mu_0\sqrt{k_s^2 - \dfrac{p^2}{4}} + \mu_s\sqrt{k_0^2 - \dfrac{p^2}{4}}\right)^2} \\[2ex]
&= \frac{-k_s\omega I_0 \mu_0^2 \mu_s \chi_0 \exp\left(j\dfrac{\pi}{4}\right) \exp(-jpx_0)\exp\left(-j2k_{sz}\left(\dfrac{p}{2}\right)z_0\right)\sqrt{\pi k_{sz}^3\left(\dfrac{p}{2}\right)}}{\sqrt{z_0}\left(\mu_0 k_{sz}^2\left(\dfrac{p}{2}\right) + \mu_s k_{s0}^2\left(\dfrac{p}{2}\right)\right)^2}
\end{aligned}
\tag{9.38}
$$

At this point, let us note that if we particularize the result of Eq. (9.11) to the point-like target at hand, we obtain

$$\hat{E}_s(p;\omega) = -\frac{\omega I_0 \mu_0^2 \mu_s k_s \chi_0 \exp\left(j\Delta\frac{p}{2}\right)\exp\left(j\frac{\pi}{4}\right)}{\left(\mu_0 k_{zs}\left(\frac{p}{2}\right)+\mu_s k_{z0}\left(\frac{p}{2}\right)\right)^2 \sqrt{z_0}}\sqrt{\pi k_{sz}^3\left(\frac{p}{2}\right)}$$

$$\times \exp(-jpx_0)\exp\left(-j2k_{sz}\left(\frac{p}{2}\right)z_0\right) \tag{9.39}$$

which is the same as Eq. (9.38) except for the factor $\exp\left(j\Delta\frac{p}{2}\right)$ (because the Fourier transform of the scattered field is computed with respect to the variable x_m instead of the observation point x).

However, the intermediate passages that have led us to Eq. (9.38) reveal a crucial point, expressed by Eqs. (9.35). In fact, Eqs. (9.35) show a noticeable one-to-one correspondence between the spectral variable and the spatial variable. In particular, the scattered field in the point x_{m0} provides the main contribution to the spectrum of the scattered field in the point $p = 2k_s \sin(\theta_{m0})$. Therefore, we can say that any x_m is the stationary point with regard to the related spectral point $p = 2k_s \sin(\theta_m)$. Now, when x_m ranges the real axis from $-\infty$ to $+\infty$, p ranges the visible interval from $-2k_s$ to $2k_s$. However, if x_m ranges from a finite x_{min} to a finite x_{max}, then p ranges through a subset of the visible interval from $2k_s \sin(\theta_{min})$ to $2k_s \sin(\theta_{max})$, where $\sin(\theta_{min}) = \dfrac{x_{min}-x_0}{\sqrt{(x_{min}-x_0)^2 + z_0^2}}$ and $\sin(\theta_{max}) = \dfrac{x_{max}-x_0}{\sqrt{(x_{max}-x_0)^2 + z_0^2}}$ are the sines of the view angles at the extreme points of the observation line, as illustrated in Figure 9.6.

This means that, for a single point-like target, the visible interval with a limited view angle is restricted from $I_v = (-2k_s, 2k_s)$ to $I_{vl} = (2k_s \sin(\theta_{min}), 2k_s \sin(\theta_{max}))$, where the extra subscript l stands for "limited" observation line. So, the visible interval is in general asymmetric and becomes theoretically symmetric if and only if the target point is centered with respect to observation line, in which case $\theta_{min} = -\theta_{max}$. The retrievable spectral set in the case of an observation line with finite extension is the spectral set corresponding to this reduced visible interval. In particular, the mapping between the point (p,ω) and the point (η,ς) is still given by Eq. (9.14) and the spectral weight is still given by Eq. (9.12), of course restricted to the available reduced visible interval.

In general, we don't have a single point-like target to focus but rather an entire investigation domain to reconstruct. Therefore, the achieved result means that the quality of the reconstruction is not constant throughout the investigation point within the investigation domain. In fact, as can be seen from Figure 9.6, the minimum and maximum view angles vary when ranging a point within the investigation domain. In particular, the retrievable spectral set reduces for deeper targets and gets particularly asymmetric for targets close to the lateral edges.[2] In order to achieve a global-average characterization of the retrievable spectral set, one possibility is to refer it to the deepest central point of the investigation domain (Soldovieri et al., 2007). This is a worst-case evaluation with

---

[2] This is actually related to the choice of an investigation domain of the same length and centered with respect to the observation line, which is the most customary and reasonable choice, as will be shown.

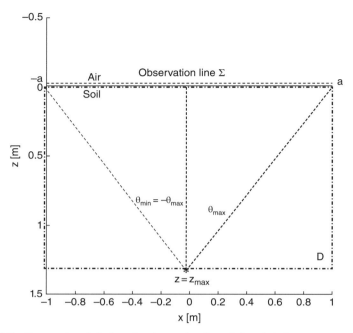

Figure 9.7. Conventional choices for the evaluation of the global-average spectral set.

respect to the depth of the target, but does not consider the asymmetry of the spectral set for lateral targets. With this conventional choice, we have $\sin(\theta_{min}) = -\sin(\theta_{max})$, and therefore the visible interval becomes $I_{vl} = (-2k_s \sin(\theta_{max}), 2k_s \sin(\theta_{max}))$. Consequently, if the observation line at the air–soil interface is extended from $x = -a$ to $x = a$ and if the maximum investigated depth is z_{max}, then we have (see Figure 9.7)

$$\sin(\theta_{max}) = \frac{a}{\sqrt{a^2 + z_{max}^2}} \tag{9.40}$$

At this point, we can represent the spectral sets relative to a limited observation line. Four (multifrequency) cases are quantitatively depicted in Figure 9.8. The four corner points in Figure 9.8 are $P_1 = (2k_{s\,min} \sin(\theta_{max}), 2k_{s\,min} \cos(\theta_{max}))$, $P_2 = (-2k_{s\,min} \sin(\theta_{max}), 2k_{s\,min} \cos(\theta_{max}))$, $P_3 = (-2k_{s\,max} \sin(\theta_{max}), 2k_{s\,max} \cos(\theta_{max}))$, and $P_4 = (2k_{s\,max} \sin(\theta_{max}), 2k_{s\,max} \cos(\theta_{max}))$. The points P_1 and P_2 are connected by an arch of circumference centered in the origin and with ray $2k_{s\,min}$, where the $k_{s\,min}$ is the minimum involved wavenumber in the soil; the points P_3 and P_4 are connected by an arch of circumference centered in the origin and with ray $2k_{s\,max}$, where $k_{s\,max}$ is the maximum involved wavenumber in the soil. The points P_1 and P_4, as well as the points P_2 and P_3, are connected by two straight line segments.

The global-average spectral set is useful for a global estimation of the resolution capabilities of the linear scattering operator, but it cannot account for the variable

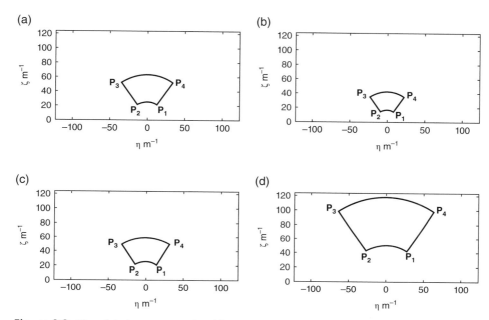

Figure 9.8. The global-average retrievable spectral set under a limited view angle. In all cases $a = 1$ m and $z_{max} = 1.5$ m. **Panel a:** $f_{min} = 200$ MHz, $f_{max} = 500$ MHz, $\varepsilon_{rs} = 9$, $\mu_{rs} = 1$. **Panel b:** $f_{min} = 200$ MHz, $f_{max} = 500$ MHz, $\varepsilon_{rs} = 4$, $\mu_{rs} = 1$. **Panel c:** $f_{min} = 300$ MHz, $f_{max} = 700$ MHz, $\varepsilon_{rs} = 4$, $\mu_{rs} = 1$. **Panel d:** $f_{min} = 300$ MHz, $f_{max} = 700$ MHz, $\varepsilon_{rs} = 4$, $\mu_{rs} = 4$.

quality of the focusing for different points within the investigation domain. In particular, the reduction of maximum view angle versus the depth involves a loss of resolution versus the depth (Deming and Devaney, 1996), as will be shown.

The spectral sets of Figure 9.8 express a fundamental characteristic of the reconstruction that we can achieve under a Born model, which is essentially expected low pass filtered with regard to the horizontal frequencies and bandpass filtered with respect to the vertical spatial frequencies. Practically, this means that the horizontal size of the targets is expected to be satisfyingly reconstructed unless the target is quite shorter than the dominant wavelength (we will be more precise in Section 9.5). Instead, with respect to the vertical direction, the reconstruction will essentially image the discontinuities along the depth, associated with "high" vertical spatial frequencies (Higgins, 1996). In other words, substantially we expect to retrieve (at most) the top and the bottom of the buried targets. Actually, in some cases the bottom is not imaged, due to the losses and/or to the strong reflection from the top. In some other cases, instead, when the target is vertically small, the reconstructions of the top and the bottom can superpose to each other, resulting in a unique "spot." The vertical band-pass characteristics of the inverse scattering operator are the mathematical counterpart of the fact that the GPR signal is substantially composed by radar echoes from the buried targets, basically generated by discontinuities along the vertical direction.

Another consequence of the "morphology" of the retrievable spectral set is that we cannot expect to retrieve the buried targets quantitatively but we can only retrieve their position, size, and (under some limits as said) shape. In particular, let us emphasize that the spectral set does not include the Cartesian origin of the plane (η,ς). Since the value of the spectrum of the contrast profile in the origin is equal the integral average value of the function, it is quite hard, under BA, to retrieve the mean value of the contrast, even if the target is a weak scattering object. In particular, the level of the reconstructed contrast is customarily quite lower than the level of the actual contrast.

It is interesting to note that, formally, the "exploding" shape of the spectral set indicates that the amount of information retrievable from higher frequencies is somehow greater than that obtained from lower frequencies. In particular, after easy geometrical calculations, the area of the spectral is equal to

$$A_s = 8\theta_{max} \Delta k_s k_{sm} \tag{9.41}$$

where Δk_s is the difference $k_{s\,max} - k_{s\,min}$ and k_{sm} is the medium wavenumber $(k_{smax} + k_{smian})/2$. This means that, for example, the band 400–800 MHz formally provides more information than the band 50–450 MHz, even if the extension of the two bands is the same. The physical reason resides in the spatial diversity of the data, as will be shown later on in this chapter. However, this should not make us think that the lower frequencies are less important. In particular, as is well known, in order to investigate deeper layers in the soil, due to the losses (not accounted for in the DT relationships but always present in the field), we need low frequencies, as already discussed in Chapter 1. This warns about the fact that the optimality of the available (or needed) amount of information should not be meant in a merely quantitative way, but above all in a qualitatively way and in relationship with the specific case at hand.

9.4 THE EFFECTIVE MAXIMUM AND MINIMUM VIEW ANGLE

Up to now, we have implicitly assumed that the target is really "seen" from any observation point ranging along the observation line. This is a simplifying hypothesis adopted in order to introduce the calculations. However, in real cases, especially when the data are gathered along a long B-scan (e.g., 50 m long or more, which usually means hundreds or thousands central wavelengths in the soil), the GPR antennas will really perceive any buried target only within a subset of the entire observation line. We can account for this introducing the concept of the effective maximum (minimum) view angle, defined as the maximum (minimum) angle at which the GPR receiving antenna perceives some signal from the crossed buried target. The effective maximum (minimum) view angle $\theta_{e\,max}$ ($\theta_{e\,min}$) is, in general, narrower than the geometrical maximum (minimum) view angle because of the geometrical spreading of the power, the losses, and the directivity of the antennas. The effective maximum view angle is difficult to be predicted in a theoretical way, but in general it can be heuristically evaluated from the data. For example, in Figure 9.9 a GPR scan is represented. The data have been gathered on the

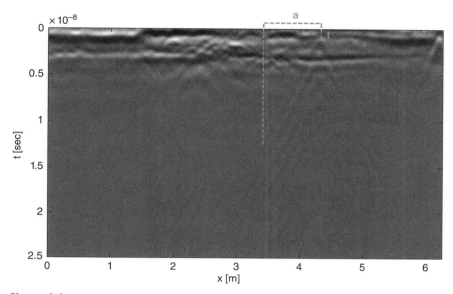

Figure 9.9. Evalutation of the maximum effective view angle from a superficial target.

floor of a baroque building,[3] and the visible anomalies are ascribable to superficial pipes under the floor. In this example, for the evidenced anomaly, we have $a \approx 1$ m and the return time t corresponding to the top (neglecting the offset between the antennas) is about equal to 1 ns. From the shape of the diffraction curve, the propagation velocity has been estimated about equal to $c \approx 1.35 \times 10^8$ m/s, so that $t \approx 1$ ns $\Rightarrow z = ct/2$ ≈ 0.07 m. So, for this anomaly, we can estimate, applying Eq. (9.40) (with $z_{max} = z$), $\sin(\theta_{e\,max}) \approx 0.998$. Clearly, superficial targets are seen under a large angle, but this changes for deeper targets. In particular, in Figure 9.10 a second example is proposed.

The data have been gathered in an archaeological site,[4] and the insulated anomaly is alleged to be ascribable to the basis of a column or to a piece of a column. Also in this case, we have $a \approx 1$ m and the return time t corresponding to the top (neglecting the offset between the antennas) is equal to ~18 ns. From the shape of the diffraction curve, the propagation velocity has been estimated to be $c \approx 0.85 \times 10^8$ m/s, so that $t \approx 18$ ns $\Rightarrow z = ct/2 \approx 0.76$ m. So, for this anomaly, we can estimate, applying Eq. (9.40) (with $z_{max} = z$), $\sin(\theta_{e\,max}) \approx 0.8$.

In general, the evaluation of the maximum (or equivalently the minimum) effective view angle depends on the single target, and so it should be averaged among more targets at the maximum depth of interest. The dealing of Section 9.3 should be referred in general to the effective view angle. Notwithstanding, the quantitative evaluations of Figure 9.8 are meaningful, because in those cases a relatively short observation line was considered,

---

[3] The ex church of Saint Sebastian, nowadays headquarter of the Foundation Palmieri in Lecce, Italy (http://www.fondazionepalmieri.it/).

[4] Hierapolis, a Roman and Byzantine town in Turkey.

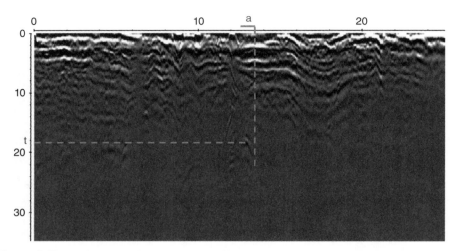

Figure 9.10. Evalutation of the maximum effective view angle from a target less superficial with respect to that considered in Figure 9.9. The abscissas are in meters, and the times are in nanoseconds.

so that the geometrical maximum view angle was not larger than the effective maximum view angle.

Let us also stress the fact that the introduction of the effective view angle shows that the problem of the asymmetry of the spectral set is not so dramatic in real cases: Actually, since the length of the B-scan usually is much longer than the central wavelength, in many cases the peripheral targets represent only a little share of the targets overflown by the GPR.

9.5 HORIZONTAL RESOLUTION

Let now calculate the available horizontal resolution. Actually, we cannot rigorously speak of the horizontal resolution as something separated by the vertical one, because the two quantities are correlated within the DT relationships. However, we are looking for an order of size, and so we will find an approximate relationship where the horizontal resolution is mainly related to the horizontal extent of the retrievable spectral set.

The evaluation of the horizontal resolution proposed here refers (as is customarily done) to the capability to distinguish two "equally strong" scattering targets at the same depth. Of course, if the energy scattered by the two targets is different, one of them might mask the other one (Daniels, 2004).

That said, the horizontal resolution can be defined as the minimum distance at which two electrically small scattering objects, of the same nature and buried at the same depth, can be distinguished from each other. In order to evaluate this distance, we can proceed approximating the spectral sets of Figure 9.8 with the best matching rectangle, as shown in Figure 9.11.

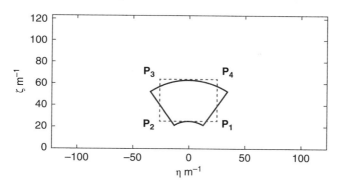

Figure 9.11. The spectral set of Figure 9.8 panel a and the rectangle to which it is assimilated.

The shape of the best matching rectangle depends on the current case, but it is evident from Figure 9.11 that a reasonable matching is achieved with the horizontal frequencies ranging between $-2k_{sc}\sin(\theta_{e\,max})$ and $2k_{sc}\sin(\theta_{e\,max})$ (where $k_{sc} = k_{s\,max} + k_{s\,min}/2$ is the average wavenumber and $\theta_{e\,max}$ is the maximum effective view angle, assumed symmetrical) and with the vertical spatial frequencies ranging between $2k_{s\,min}$ and $2k_{s\,max}$. Consequently, the corner points of the rectangle in Figure 9.11 are $P_1 = (2k_{sc}\sin(\theta_{e\,max}),$ $2k_{s\,min})$, $P_2 = (-2k_{sc}\sin(\theta_{e\,max}), 2k_{s\,min})$, $P_3 = (-2k_{sc}\sin(\theta_{e\,max}), 2k_{s\,max})$, and $P_4 = (2k_{sc}\sin(\theta_{e\,max}), 2k_{s\,max})$. In particular, the spectral set is approximately large $4k_{sc}\sin(\theta_{e\,max})$ along the η-axis and $2(k_{s\,max} - k_{s\,min})$ along the ζ-axis.

Now, let us consider a point-like target placed in the point (x_0,z_0), expressed by the contrast function $\chi_e(x',z') = \chi_0\delta(x' - x_0)\delta(z' - z_0)$, so that its spatial spectrum is given by $\hat{\chi}_e(\eta,\varsigma) = \chi_0\exp(-j\eta x_0)\exp(-j\varsigma z_0)$. Neglecting the spectral weight, the reconstruction of such a target through DT is essentially the filtered version of the contrast achieved considering its spectrum only within the retrievable spectral set, in its turn approximated with the dashed rectangle in Figure 9.11. In this way, we will achieve a result in closed form. The reconstruction available for a point like target is also known as point spread function (Moghaddam and Chew, 1992). In the case at hand, it is given by

$$\chi_{er}(x',z') = \frac{\chi_0}{4\pi^2}\iint_R \exp(j(x'-x_0)\eta)\exp(j(z'-z_0)\varsigma)\,d\eta d\varsigma$$

$$= \frac{2\chi_0}{\pi^2}k_{sc}\sin(\theta_{e\,max})(k_{s\,max} - k_{s\,min})\exp(j(k_{s\,max} + k_{s\,min})(z'-z_0))$$

$$\times \mathrm{sinc}(2k_{sc}\sin(\theta_{e\,max})(x'-x_0))\mathrm{sinc}((k_{s\,max} - k_{s\,min})(z'-z_0))$$

$$= \frac{4\chi_0 k_{sc}\sin(\theta_{e\,max})B\exp(2jk_{sc}(z'-z_0))}{\pi c}$$

$$\times \mathrm{sinc}(2k_{sc}\sin(\theta_{e\,max})(x'-x_0))\mathrm{sinc}\left(\frac{2\pi B}{c}(z'-z_0)\right) \qquad (9.42)$$

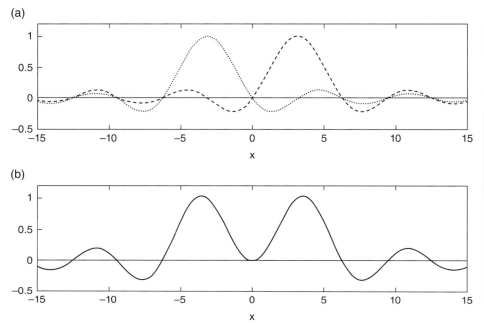

Figure 9.12. **Panel a**: The graph of two functions sinc($x - \pi$) and sinc($x + \pi$), for which the distance between the maxima is equal to the extent of the main lobe. **Panel b**: The graph of the sum sinc($x - \pi$) + sinc($x + \pi$). The axes are in arbitrary units.

where B is the frequency band and R indicates the rectangle in Figure 9.11. The size of the main lobe of this sinc-like function along the horizontal direction is $\pi/k_{sc}\sin(\theta_{emax}) = \lambda_{sc}/2\sin(\theta_{emax})$. It is found reasonable that two targets will be reliably distinguishable if their distance is equal to this value, which guarantees that the main lobes of the two reconstructions do not overlap to each other. So we have

$$HR \approx \frac{\lambda_{sc}}{2\sin(\theta_{e\,\max})} \tag{9.43}$$

In Figure 9.12, two equal sinc functions shifted of the main lobe with respect to each other are represented; this makes clear the rationale of the choice of Eq. (9.43).

Equation (9.43) derives from several assumptions and approximations. So, it is no surprise that different expressions of the horizontal resolution can be found in the literature. In particular, the literature often reports the value $\lambda_s/2$ (or $\lambda_s/\sqrt{2}$) where λ_s is defined as the central wavelength. This evaluation, however, does not account for the loss of resolution versus the depth. Jol (2009) reports the value $HR = \sqrt{d\lambda_s/2}$, where d the depth. This accounts for the loss of resolution versus the depth but drives to an infinitesimal resolution for very shallow targets. Another reported value

(Sheriff, 1980) is $HR = \sqrt{(d + \lambda_s/4)^2 - d^2}$, which reduces to the previous estimation for d $>> \lambda_s$ and becomes $HR \approx \lambda_s/4$ for shallow targets. Equation (9.43) accounts for the loss of resolution versus the depth by means of $\sin(\theta_{e\,max})$. In particular, it accounts for the fact that the available resolution depends on the size of the portion of the observation line under which the buried target is "seen" by the GPR antennas. In fact [see Eq. (9.40)], $\sin(\theta_{emax})$ accounts both for the depth of the target and the wideness of the effective observation line. This means, for example, that if for some reason (e.g., the presence of superficial obstacles) the buried targets can be crossed over only along a short B-scan, then the horizontal resolution is expected to degrade. This implicitly also means that the available resolution is to be evaluated after the processing. Actually, the resolution is evaluated sometime on the basis of the raw data and sometimes it is thought that it is a mere hardware attribute of the antennas. Actually, the focusing capabilities of the processing [sometimes referred to as synthetic aperture radar SAR effect (Daniels, 2004)] influence the achievable resolution meaningfully, even if their effectiveness depends on the current case (in particular on the amount of losses).

Equation (9.43) can be also applied on low-loss soils, by means of a "perturbative" approach (Franceschetti, 1997), where the wavelength is evaluated on the basis of the real part of the wavenumber. However, for a more refined evaluation, it should be also accounted the fact that, in most cases [even if some exceptions are reported (Daniels, 2004; Jol, 2009)] the losses attenuate the higher frequencies versus the depth more strongly than the lower frequencies. Therefore, in Eq. (9.43), λ_{sc} should be meant as the central received frequency from the range of depth of interest, which is in most cases greater or equal to the central radiated wavelength. So, we can generalize Eq. (9.43) by introducing an effective central received wavelength in the soil λ_{sec} and writing

$$HR \approx \frac{\lambda_{sec}}{2 \sin(\theta_{emax})} \tag{9.44}$$

Likewise θ_{emax}, also λ_{sec} is not easily predictable, but can be evaluated from the GPR data by windowing suitably (i.e., without introducing strong discontinuities) the range of depths of interest and considering the spectrum of the windowed data, as done (for example) in Sala and Linford (2010).

So, in general the losses degrade the horizontal resolution, because they make λ_{sec} increase and θ_{emax} decrease.

9.6 VERTICAL RESOLUTION

Likewise the horizontal resolution, also the vertical resolution can be approximately estimated from the retrievable spectral set. The vertical resolution can be defined as the maximum distance between two electrically small and vertically piled targets distinguishable from each other. This time, however, beyond the same assumptions assumed in the previous section, we have to account for two further elements:

1. The upper target can partially mask the lower one, even if the two targets are of the same nature. So, two piled targets with the same electromagnetic character-istics do not, in general, provide echoes of the same level. Therefore, we will make the further assumption that the two targets are relatively weak targets, so that the masking effect of the upper one on the lower one is negligible.

2. The phase term $\exp(j(k_{s\,max} + k_{s\,min})(z' - z_0)) = \exp(2jk_{sc}(z' - z_0))$ in Eq. (9.42), deriving from the fact that the spectral set is band-pass along the vertical spatial frequency axis, can influence the reconstruction. In particular, in the case of two small piled objects, the two relative phase terms can sum in a constructive or destructive way, depending on the electrical distance between the source and the targets at the central frequency.

That said, an order of magnitude for the vertical resolution can be formally calculated in the same way followed for the horizontal resolution—that is, from the wideness of the vertical main lobe of the point spread function. Consequently, based on Eq. (9.42), the vertical resolution in lossless cases is given by

$$VR = \frac{c}{B} = \frac{c}{(f_{max} - f_{min})} \tag{9.45}$$

As said, the band of a GPR antenna is often of the same order of the central frequency. So a rougher evaluation can be derived from Eq. (9.45) as follows:

$$VR = \frac{c}{B} \approx \frac{c}{f_c} = \lambda_{sc} \tag{9.46}$$

It is sometime reported that the vertical resolution is equal to one half of the central wave-length in the soil instead of just the wavelength in the soil. Of course, the two quantities are of the same order, and so this difference is not a big trouble. At any rate, Eq. (9.46) is coherent with the fact that, under the geometrical approximations performed on the retrievable spectral set and under the hypothesis of large band (in particular with the max-imum frequency much larger than the minimum one) the retrievable spectral set is essen-tially twice larger along the direction of the horizontal spatial frequencies than it is along the direction of the vertical spatial frequencies. This makes the theoretical horizontal res-olution for shallow targets equal to one half of the corresponding vertical resolution. In lossy cases, likewise in the case of the horizontal resolution, we can introduce the effective band, the effective central frequency, and the effective central wavelength. Thus, a rough extension of Eqs. (9.45) and (9.46) can be

$$VR = \frac{c}{B_e} \approx \frac{c}{f_{ce}} = \lambda_{sce} \tag{9.47}$$

As said, the higher frequencies attenuates more than the lower ones versus the depth, so that the effective band is customarily narrower than the radiated band. This means that in real cases some degradation of the vertical resolution versus the depth is expected, even if in an ideal lossless soil the vertical resolution is constant versus the depth.

9.7 SPATIAL STEP

DT also provides an approximated but powerful tool to calculate the spatial step needed for taking GPR measurements correctly. In particular, based on the previous dealing, the scattered field should be sampled in a way that allows the correct evaluation of its spectrum within the effective visible interval $I_v = (2k_s \sin(\theta_{emin}), 2k_s \sin(\theta_{emax}))$. Of course, the effective visible interval is a function of the target position. So, in order to choose the spatial step needed for the data in a "conservative" way, the best choice is to estimate the visible interval in relationship to the shallowest part of the investigation domain, because this will provide the narrowest step. Let us outline that this is the opposite criterion with respect to that adopted for the evaluation of the resolution. At this point, we should identify from the data a maximum view angle for the shallowest targets in the investigation domain and, from this, the visible interval for the evaluation of the needed spatial step. Actually, in most cases the investigation domain starts just from the air–soil interface, and in this case the maximum (minimum) view angle is equal to $\theta_{emax} \approx \pi/2 (\theta_{emin} \approx -\pi/2)$. The visible interval is large $4k_s \sin(\theta_{emax})$. Based on the Nyquist criterion, the required spatial step s_{sf} at a single frequency is therefore given by

$$s_{sf} = \frac{\pi}{2k_s \sin(\theta_{emax})} = \frac{\pi}{2(2\pi/\lambda_s)\sin(\theta_{emax})} = \frac{\lambda_s}{4\sin(\theta_{emax})} \qquad (9.48)$$

where λ_s is the wavelength in the soil. Of course, the GPR scans are multifrequency data, and so the step (9.48) should be meant conservatively with respect to the minimum involved wavelength. Thus, in the usual multifrequency case we have

$$s = \frac{\lambda_{smin}}{4\sin(\theta_{emax})} \qquad (9.49)$$

If the investigation domain starts quite close to the air–soil interface, of course Eq. (9.49) reduces to $s = \lambda_{smin}/4$. Indeed, this is the most common case, so that the criterion $s = \lambda_{smin}/4$ (sometime relaxed to $s = \lambda_{sc}/4$, i.e. referred to the central wavelength instead of the minimum one) is probably the most reported one (Daniels, 2004; Jol, 2009). However, by virtue of the loss of resolution versus the depth, if the targets looked for are deep enough, the spatial step can be theoretically reliably relaxed according to Eq. (9.49). This can have a practical relevance in some special cases—for example, in cases where the data have to be gathered in a discrete (noncontinuous) mode (Persico et al., 2010).

Finally, it is also worth noting that, in some cases, the targets looked for might be very large with respect to the wavelength—for example, in the case of the investigation of geological stratifications. In these cases the horizontal spatial band of spectrum of the contrast is possibly quite narrow, and useful results can be achieved even with a spatial step larger than that given in Eq. (9.49).

Equation (9.49) involves some redundancy in the gathered data, because the step is conservatively optimised with respect to the highest involved frequency. This redundancy cannot be avoided, because it is unpractical to think of modifying the spatial step versus the frequency (it is just impossible with a pulsed system and it would be uselessly time-consuming with a stepped frequency system). However, if needed for computational reasons, some decimation of the data might be done before processing them.

As a rule of thumb, it is not wrong to gather data with a spatial step narrower than (9.49), due to the losses and to the parametric uncertainties about the characteristics of the soil, and also in order to average some noise. However, it is pointless to gather data with a spatial step one order of magnitude smaller than that of Eq. (9.49). In Section 15.2 the reader will find some exercises on spatial step and horizontal resolution. Please note that they are also based on the 2D migration exposed in Chapter 10 and on the SVD theory exposed in Chapter 14.

9.8 FREQUENCY STEP

Let us now examine the problem of the needed frequency step. Let us specify that what we are looking for is the needed frequency step in order to process correctly GPR data in the frequency domain, which refers to data gathered in either the time domain or the frequency domain. Instead, the result worked out in Chapter 3 was referred to the frequency step needed in order to gather the data correctly with a stepped frequency system. The two frequency steps can be different because, from the raw data, we might see that the targets of interest are confined within a subset of the entire initially probed depth range, so that we might be interested in focusing only the targets enclosed in a specific depth range smaller than the entire "gathered" depth range.

That said, let us note that, within the DT relationship, a single frequency corresponds to an arch of circle within the retrievable spectral set. The spectral set is therefore somehow sampled by these arches of circle. This involves the fact that the vertical spatial frequency sampling is not independent from the horizontal frequency sampling, due to the nonlinear relationship (9.14). So, to simplify the problem, in a first moment we will neglect the curvature of the arches of circle linked to each frequency, and we will approximate them as horizontal segments. In this way, we can identify a linear relationship between the vertical spatial frequency step and the frequency step. In particular, under this approximation, the vertical spatial frequency sampling step is related to the (temporal) frequency step by the following proportionality relationship:

$$\Delta\varsigma = 2\Delta k_s = 2\frac{2\pi}{c}\Delta f = \frac{4\pi\sqrt{\varepsilon_{sr}\mu_{sr}}}{c_0}\Delta f \tag{9.50}$$

where c is the propagation velocity of the wave in the soil and c_0 is the propagation velocity of the waves in free space, ε_{sr} and μ_{sr} are the relative dielectric permittivity and magnetic permeability of the soil. In order to establish whether the sampling of the spatial frequencies is correct, we resort again to the Nyquist criterion. This time the role of the "band", is played by the vertical extent of the investigation domain, because of the formal similarity between the direct and the inverse Fourier transforms. So, if we label as b the vertical extent of this domain (see Figure 4.1), we have that the Nyquist criterion requires

$$\Delta \varsigma \le \frac{2\pi}{b} \tag{9.51}$$

Substituting Eq. (9.51) into Eq. (9.50), we obtain the condition

$$\frac{\Delta \varsigma}{2\pi} = \frac{2\sqrt{\varepsilon_{sr}\mu_{sr}}}{c_0} \Delta f \le \frac{1}{b} \Rightarrow \Delta f \le \frac{c_0}{2b\sqrt{\varepsilon_{sr}\mu_{sr}}} \tag{9.52}$$

Equation (9.52) reduces to Eq. (3.12) if $b = D$; that is, if the processed depth range coincides with the entire "gathered" depth. However, if we focus on only a subset of the maximum investigated (nonambiguous) depth, then $b < D$ and Eq. (9.52) is a condition relaxed with respect to Eq. (3.12), which can be computationally important. It is important to emphasize, however, that in our model (see Figure 4.1) we suppose apriori that the targets are present only within the investigation domain. In particular, we assume that there is no target shallower than it and no target deeper than it. This assumption is quite strong, and in many cases a simple glance to the raw data is sufficient to prove that it is unacceptable. In these cases, in order to focus a precise range of depth, one should preventively mute the signal in the time domain before and after the depth range of interest. This operation is an anti-aliasing spatial prefiltering.

At this point, let us now consider the curvature of the spectral arches of circle corresponding to each time frequency within the retrievable spectral set. In particular, we can observe that the calculated vertical spatial frequency sampling is based on the distance along the ς-axis of two spectral circles corresponding to two different frequencies. This is to say that we have founded our calculations on $\Delta \varsigma$ for $\eta = 0$. For any $\eta \ne 0$, the vertical frequency sampling between two circles, from Eqs. (9.14), is given by

$$\Delta \varsigma(\eta) = \sqrt{4(k_s + \Delta k_s)^2 - \eta^2} - \sqrt{4k_s^2 - \eta^2} \tag{9.53}$$

However, the minimum versus η of the function (9.53) is achieved for $\eta = 0$, and therefore condition (9.52) is conservative. Similar to the spatial step, the frequency step is to be meant as an order of magnitude. A good rule of thumb is to choose a frequency step for data processing slightly but not much (namely one order of magnitude or more) smaller than that given in Eq. (9.52), in order to mitigate some possible effect of the losses, of the model approximations, and of the parametric uncertainties. In Section 15.3 the reader can find some exercises on frequency step and vertical resolution. Please note that they are also based on the 2D migration exposed in Chapter 10 and on the SVD theory exposed in Chapter 14.

9.9 TIME STEP

As said, GPR data can be processed either in the frequency domain or in the time domain. So, for completeness, now we provide the time step needed in order to process the data in the time domain. The reasoning is dual to the case of processing in frequency domain, and also in this case the time step needed for the processing is not necessarily equal to that needed in order to gather data correctly by means of a pulsed system [see Eq. 3.39)]. In particular, we might be interested in the result relative to a subset of the entire band of the data, because (for example) we have recognized that there are some strong interferences on some frequencies). In this case, after filtering the data [thus restricting the spectrum to a band of interest (let us call it B_{int}) narrower than the band B of the signal initially gathered] and applying the Nyquist criterion, we have that the time step needed for a processing in the time domain is given by

$$\Delta t = \frac{1}{B_{int}} \tag{9.54}$$

It is obvious that condition (9.54) is relaxed with respect to condition (3.39). Actually, in the GPR common praxis the time step is rarely a real problem, because the allowed number of time samples for a given time bottom scale is in general so high to guarantee the absence of any meaningful aliasing without particular computational problems (see also Section 3.4, question 2 in Chapter 3, and answer 2 in Appendix G, Section on Chapter 3).

9.10 THE EFFECT OF A NON-NULL HEIGHT OF THE OBSERVATION LINE

Let us now consider the case of an observation line at height $h \neq 0$ with respect to the soil.[5] This removes hypothesis 2 in Section 9.2. We retain in a first moment hypotheses 1, 3, 4, and 5. In particular, we consider first the ideal case of an infinite observation line, then we will pass to the case of a limited observation line. So, we can start from the relationship

$$E_s(x;\omega) = k_s^2 \iint_D G_e(x,x',h,z';\omega) E_{inc}(x+\Delta,x',h,z';\omega)\chi_e(x',z')\,dx'dz', \quad x \in \Sigma \tag{9.55}$$

Substituting in Eq. (9.55) the expressions of the external Green's function and of the incident field given in Eqs. (4.47)–(4.48), we have

$$E_s(x;\omega) = \frac{j\omega I_0 \mu_0^2 \mu_s k_s^2}{2\pi} \int_{-\infty}^{+\infty}\int_{-\infty}^{+\infty}\int_{-\infty}^{+\infty}\int_{-\infty}^{+\infty} \frac{\exp(-j(k_{zs}(u)+k_{zs}(v))z')\exp(j(u+v)x)}{(\mu_0 k_{zs}(u)+\mu_s k_{z0}(u))(\mu_0 k_{zs}(v)+\mu_s k_{z0}(v))}$$

$$\times \exp(jv,\Delta)\exp(-j(u+v)x')\exp(jh(k_{z0}(u)+k_{z0}(v)))\chi_e(x',z')\,dx'dz'dudv, \quad x \in \Sigma \tag{9.56}$$

---

[5] Let us remind ourselves that, by virtue of the Cartesian reference assumed (see Figure 4.1), h is to be taken as negative.

Equation (9.56) is the same as Eq. (9.2) but for the term $\exp(jh(k_{z0}(u) + k_{z0}(v)))$. This term is identically unitary for $h = 0$ and is almost unitary when the height h of the antennas is small in terms of the minimum involved wavelength in air. Now, we will instead assume the opposite hypothesis—that is, that h is large with respect to the maximum involved wavelength. In this case, putting $p = u + v \Rightarrow u = p - v$ as previously done with regard to Eq. (9.2), we can develop the calculations as follows (Persico, 2006):

$$E_s(x;\omega) = \frac{j\omega I_0 \mu_0^2 \mu_s k_s^2}{2\pi} \int_{-\infty}^{+\infty}\int_{-\infty}^{+\infty}\int_{-\infty}^{+\infty}\int_{-\infty}^{+\infty} \frac{\exp(-j(k_{zs}(p-v) + k_{zs}(v))z')\exp(jpx)}{(\mu_0 k_{zs}(p-v) + \mu_s k_{z0}(p-v))(\mu_0 k_{zs}(v) + \mu_s k_{z0}(v))}$$

$$\times \exp(jh(k_{z0}(p-v) + k_{z0}(v)))\exp(jv\Delta)\exp(-jpx')\chi_e(x',z')\,dx'dz'dpdv, \quad x \in \Sigma$$

$$(9.57)$$

Rearranging the integrals in the same way followed for the calculation carried out in the case $h = 0$, we obtain

$$\hat{E}_s(p;\omega) = j\omega I_0 \mu_0^2 \mu_s k_s^2 \int_{-\infty}^{+\infty} dz' \hat{\chi}_e(p,z')$$

$$\times \int_{-\infty}^{+\infty} \frac{\exp(-j(k_{zs}(p-v) + k_{zs}(v))z')\exp(jh(k_{z0}(p-v) + k_{z0}(v)))\exp(jv\Delta)}{(\mu_0 k_{zs}(p-v) + \mu_s k_{z0}(p-v))(\mu_0 k_{zs}(v) + \mu_s k_{z0}(v))}\,dv, \quad x \in \Sigma$$

$$(9.58)$$

Equation (9.58) is similar to Eq. (9.4), with the only difference constituted by the presence of the term $\exp(jh(k_{z0}(p - v) + k_{z0}(v)))$. Now, in order to analyze the effect of this term, we distinguish two cases: The first one is the case $|h| << z'$, and the second one is $|h| >> z'$.

Let us now focus on the case $|h| << z'$. In this case the term $\exp(jh(k_{z0}(p - v) + k_{z0}(v)))$ can be regarded as slowly varying with respect to the term $\exp(jz'(k_{z0}(p - v) + k_{z0}(v)))$. So, we can apply the same stationary phase method applied in the case $h = 0$, and the result is

$$\hat{E}_s(p;\omega) = f(p;\omega)\exp\left(j2hk_{z0}\left(\frac{p}{2}\right)\right)\hat{\chi}_{e1}(\eta(p,\omega),\varsigma(p,\omega)) \tag{9.59}$$

where $f(p;\omega)$ is given in Eq. (9.12) and $\hat{\chi}_{e1}(\eta(p,\omega),\varsigma(p,\omega))$ is given in Eq. (9.13). Thus, the achieved result is just a modification of the spectral weight times the factor $\exp(j2hk_{z0}(p/2))$. Now, while p ranges in the interval $I_{vh} = (-2k_0, 2k_0)$ (the suffix h refers to the "non-null height"), the quantity $\exp(j2hk_{z0}(p/2))$ is just a phase term that does not affect the modulus of the spectral weight. However, beyond this range, the exponent has a negative real part that attenuates meaningfully the spectral weight. An example is given in Figure 9.13. In particular, Figure 9.13 shows the comparison between the normalized modulus of the spectral weight at the air–soil interface [calculated according to Eq. (9.12)] and at a height equal to one-fifth of the minimum

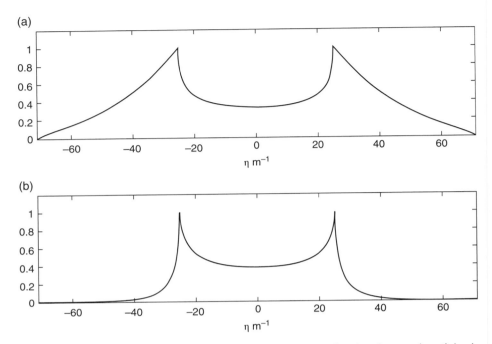

Figure 9.13. Comparison between the spectral weight at the soil and at the wavelength in air. In both panels, we have $\varepsilon_{sr} = 4$, $\mu_{sr} = 2$, and $f = 600$ MHz. **Panel a:** $h = 0$. **Panel b:** $h = 0.1$ m. In the example at hand $2k_s = 71.13$ m^{-1} and $2k_o = 25.15$ m^{-1}

wavelength in air [calculated accordingly to Eq. (9.59)]. This shows that, even for a relatively small electrical height of the observation line, the ideal visible interval reduces rapidly from $I_v = (-2k_s, 2k_s)$ up to $I_{vh} = (-2k_o, 2k_o)$. We can give a physical interpretation of this fact by reminding that the main contribution to the spectrum of the data at the air–soil interface, for any value of p within the visible interval I_v, is given by a plane wave coming from the direction $2k_s \sin(\theta)$. However, as is well known, if this plane wave impinges from the underground beyond the critic angle θ_c, the wave refracted in the air attenuates exponentially versus the height. Due to Snell's law, the critic angle is the angle at which $\sin(\theta_c) = 1/\sqrt{\varepsilon_{sr}\mu_{sr}}$ (Franceschetti, 1997), where ε_{sr} and μ_{sr} are the relative permittivity and the relative magnetic permeability of the soil. Consequently, the visible interval for a high observation line might also be rewritten as $I_{vh} = (-2k_s \sin(\theta_c), 2k_s \sin(\theta_c))$. This means that, physically, the effect of the height of the observation line is that of limiting the maximum and minimum view angle, even in the case of an ideal infinite observation line. From Figure 9.13, we also appreciate the fact that the maxima of the spectral weight for $h = 0$ are close to the extremes of I_{vh}. Actually, this is related to the radiation pattern of a filamentary current on the soil, which tends (for high relative permittivity of permeability of the soil) to show its maxima at the critic angles (Engheta et al., 1982).

Consequently to the reduction of the visible interval, the retrievable spectral set restricts too, as depicted in Figure 9.14. Let us now pass to the case where $|h| >> z'$

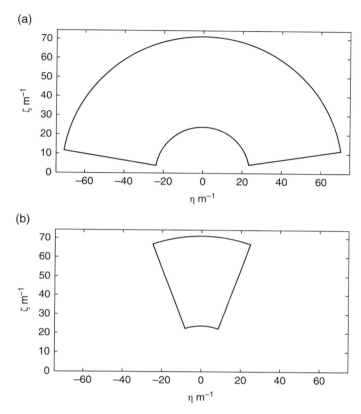

Figure 9.14. Comparison between the spectral set at the soil (panel a) and for a high observation line (panel b). $\varepsilon_{sr} = 4$, $\mu_{sr} = 2$, $f_{min} = 200$ MHz, $f_{max} = 600$ MHz. In panel a, the view angle has been made to range from $\theta_{min} = -\pi/2 + \pi/20$ to $\theta_{max} = \pi/2 - \pi/20$, in order to provide a pictorial effect of the vanishing of the spectral weight toward the extremes of the visible interval (see Figure 9.4).

$\forall z' \in D$. In this case, when applying the stationary phase method to the integral in dv, we will regard $\exp(jh(k_{z0}(p-v) + k_{z0}(v)))$ as the oscillating term, whereas the term $\exp(-j(k_{zs}(p-v) + k_{zs}(v))z')$ is included among the "smooth" ones. Please note that this time we can remove the hypothesis that the targets are necessarily deep in terms of wavelength, because the oscillations of the terms $\exp(jh(k_{z0}(p-v) + k_{z0}(v)))$ are sufficient to enable us to apply the stationary phase method. Incidentally, in the case of an electrically large height of the data, the removing of the hypothesis of deep targets is also physically reasonable, because in this case it is hard to think of detecting electrically deep targets.

That said, under these new assumption, there is again a unique first-order stationary point at $v = p/2$, and the result is

$$\hat{E}_s(p;\omega) = f_h(p;\omega)\hat{\chi}_e(\eta(p), \varsigma(p)) \tag{9.60}$$

where

$$f_h(p;\omega) = \frac{j\omega I_0 \mu_0^2 \mu_s k_s^2 \exp\left(j\Delta\frac{p}{2}\right)\exp\left(j2hk_{z0}\left(\frac{p}{2}\right)\right)\exp\left(j\frac{\pi}{4}\right)}{k_o\sqrt{h}\left(\mu_0 k_{zs}\left(\frac{p}{2}\right)+\mu_s k_{z0}\left(\frac{p}{2}\right)\right)^2}\sqrt{\pi\left(k_{z0}\left(\frac{p}{2}\right)\right)^3} \tag{9.61}$$

where the subscript h stands for "high measurement line." Let us explicitly note that in (9.60) we have the spectrum of the contrast function, and not the spectrum of the contrast normalized to the square root of the depth. This because the stationary phase is applied with respect to the propagation in air, and so the variable factor $1/\sqrt{z'}$ has been replaced by the constant factor $1/\sqrt{h}$.

Apart from the formal differences between Eq. (9.11) and Eq. (9.60), the essential physical difference is the again the presence of the factor $\exp(j2hk_{z0}(p/2))$ in the spectral weighting function. Thus, the same considerations just exposed can be repeated. An example is given in Figure 9.15. In particular, Figure 9.15 shows the comparison between the normalized modulus of the spectral weight at the air–soil interface [calculated according to Eq. (9.12)] and at a height equal to the wavelength in air [calculated accordingly to Eq. (9.61)].

The visible interval and the retrievable spectral set are restricted further on in the case of a limited line, and these reductions can be described again resorting to the effective maximum and minimum view angle. However, the effective view angle should not be based on the maximum abscissa at which we see the last "tail" of the diffraction curve. In fact, due to Snell's law, this leads to an apparent maximum view angle larger than the effective one, which should be calculated at the air–soil interface. Basically, this is the same physical phenomenon that makes a target embedded under a shallow layer of water appear larger than it is to our eye. The phenomenon can be accounted for by means of the quantities $x_1(x_{max})$ and $x_1(x_{min})$ related to x_{max} and x_{min} as described in Chapter 2 (see Section 2.3.4). In particular, the maximum effective view angle should be evaluated as

$$\sin\left(\theta_{emax}\right) = \frac{x_1\left(x_{max}\right)}{\sqrt{\left[x_1\left(x_{max}\right)\right]^2 + d^2}} \tag{9.62}$$

where d is the depth of the target. A geometrical scheme is provided in Figure 9.16. A similar formula holds for the minimum view angle too.

Of course, the case dealt with represents a limit situation, where the observation line and the interface are far from each other with respect to the minimum involved wavelength. In general, intermediate situations are possible, so that at the lower frequencies some meaningful "tail" beyond I_{vh} are still present. In these cases, we have some formal difficulty in describing an intermediate spectral set, because the hypotheses exploited for retrieving the spectral sets do not hold any longer. At any rate, what is physically important is "the trend," which shows how the height of the observation line

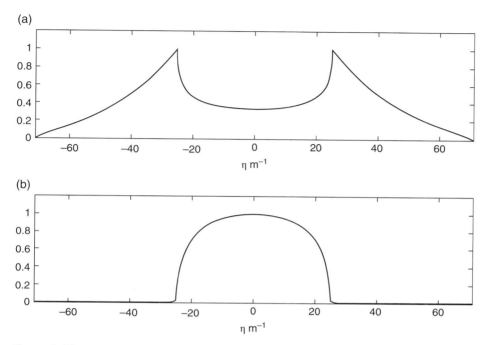

Figure 9.15. Comparison between the spectral weight at the soil and at height equal to the wavelength in air. In both panels, we have $\varepsilon_{sr} = 4$, $\mu_{sr} = 2$, and $f = 600$ MHz. **Panel a:** $h = 0$. **Panel b:** $h = 0.5$ m. In the example at hand $2k_s = 71.13$ m^{-1} and $2k_0 = 25.15$ m^{-1}.

is paid with a loss of information extractable from the data (Persico, 2006). When possible, therefore, the best choice is to gather contact data. However, there are applications where the data are necessarily contactless (e.g., asphalt monitoring or airborne GPR surveys).

The horizontal and vertical resolutions, as well as the spatial step, are straightforwardly derived from the shape of the spectral set. In particular, with regard to the horizontal resolution, we have

$$HR_h = \frac{\lambda_{sec}}{2\sin(\theta_{emax})} \tag{9.63}$$

which is formally identical to Eq. (9.44) but substantially different because of the reduced $\theta_{e\,max}$. In particular, we have now the physical constraint $\theta_{emax} \leq \theta_c$. The same holds for the needed spatial step. Instead, the vertical resolution and the frequency step keep substantially unchanged with respect to the case with data at the air–soil interface, because the view angle does not enter the calculations in this case. In Section 15.6 the reader will find some exercises on the effects of the height of the observation line. Please note that they are also based on the 2D migration exposed in Chapter 10 and on the SVD theory exposed in Chapter 14.

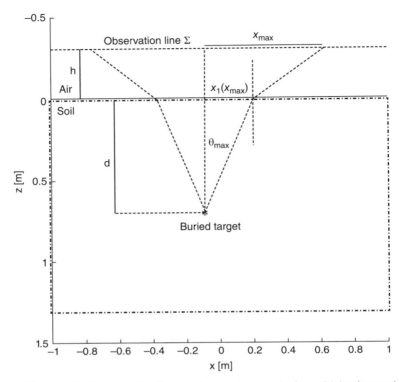

Figure 9.16. Pictorial for the effective maximum view angle for a high observation line. The depth of the target is exaggerated for representation exigencies. The units are arbitrary.

9.11   THE EFFECT OF THE RADIATION CHARACTERISTICS OF THE ANTENNAS

The effect of the radiation characteristics of the antennas can be accounted for by considering the scattering equation with the incident field given by Eq. (4.84) and the external Green's function given by Eq. (4.115). We will consider first the case with antennas at the air–soil interface, with an infinite observation line and with no magnetic target. The passages are fully analogous to those exposed in Section 9.2. Thus, we won't repeat them and will provide directly the DT relationship, which in this case is given by

$$\hat{E}_s(p;\omega) = f_a(p;\omega)\hat{\bar{\chi}}_{e1}(\eta(p),\varsigma(p)) \tag{9.64}$$

where $\chi_{e1}(x',z')$ is the contrast normalized to the square root of the depth as given in Eq. (9.13), and the spectral weight (the suffix stands for "antennas") is given by

$$f_a(p;\omega) = \frac{-jm\mu_0\mu_s\lambda_0 \exp\left(j\Delta\frac{p}{2}\right)\exp\left(j\frac{\pi}{4}\right)\hat{E}_{Tx}\left(\frac{p}{2}\right)\hat{E}_{Rx}\left(\frac{p}{2}\right)k_{z0}^2\left(\frac{p}{2}\right)}{2\pi^3 I_R\varsigma_0\left(\mu_0 k_{zs}\left(\frac{p}{2}\right)+\mu_s k_{z0}\left(\frac{p}{2}\right)\right)^2}\sqrt{\pi\left(k_{zs}\left(\frac{p}{2}\right)\right)^3} \tag{9.65}$$

With respect to the relationship (9.12), we can see that the retrievable spectral set is the same in principle, because the arguments of the Fourier transforms are the same, and also the set where the spectral variables range are the same. The difference is, one more time, in the weighting spectral function, which depends among other things on the radiation characteristics of the antennas. Now, for practical reasons, the GPR antennas are never electrically large; and this, in general, makes their radiation pattern "smooth" (Daniels, 2004). This also implicitly explains the fact that, in most cases, the GPR data processing is done without having a precise quantification of the radiation characteristics of the antennas. Let us also stress the fact that the measure of these characteristics is not a trivial task, and in particular buried antennas would be required (Meincke and Hansen, 2004). Finally, in principle the characteristics of the antennas are site-dependent too.

It is worth outlining that, beyond the fact that a very directive GPR antenna pattern is usually unrealistic, it is not desirable either. In fact, in this limit case, based on the dealing shown in Sections 4.4–4.8, we have that the plane wave spectra of the antennas [and thus the spectral weighting function (9.65)] becomes very peaked near the value $p = 0$. This means that we have a restriction of the effective retrievable spectral set near the line $p = 0$. The effect is therefore, again, a reduction of the value of the effective maximum view angle; and the higher the directivity, the stronger the reduction. Physically, this means that an extremely directive source eliminates the possibility to see the targets from a wide view angle. In fact, in the limit for high directivity, the transmitting antenna emits (and the receiving antenna receives) only TEM broadside plane waves. It is sometimes erroneously thought that very directive antennas radiate just a "thin ray" in front of them. However, the narrowness of the radiation pattern is in angular terms, and not in terms of linear distance, and let us also remind ourselves that the directivity is in turn a limit concept that is defined only in the far-field zone of the antenna, which in turn depends on the size of the antenna too (Collin, 1985; Franceschetti, 1997).

It is clear that, in the limit case of only TEM broadside waves, it would be useless to move the source, because the incident field would remain the same (Soldovieri et al., 2005b). This limit case shows again that the quality of the image improves if the antennas "see" the targets from several viewpoints. Of course, the data processing is a key point with regard to this. In fact, in any fixed position, a couple of not-much-directive antennas illuminate and gather the echoes coming from many directions, and these contributions are superposed on each other. However, when the radiation and reception is performed in several positions and at several frequencies for each position, a suitable data processing is able to distinguish the origin of each contribution.

The passage to a finite observation line is analogous to those exposed with regard to the case of a filamentary current, and so it will be not repeated. Because the antenna pattern is expected (at least in air) to be quite roundish and not varying so much among the practical antennas, the spatial and frequency steps, as well as the horizontal and vertical resolutions, are expected to be of the same order as those provided in Sections 9.5–9.8.

As a last observation, we can note that the spectral weight is composed of (a) a part dependent on the current source and receiver (i.e., the product $\hat{E}_{Tx}(p/2)\hat{E}_{Rx}(p/2)$) and (b) a part independent of it, given by

$$f_1(p;\omega) = \frac{-jm\mu_0\mu_s\lambda_0\exp\left(j\Delta\frac{p}{2}\right)\exp\left(j\frac{\pi}{4}\right)k_{z0}^2\left(\frac{p}{2}\right)}{2\pi^3 I_R\varsigma_0\left(\mu_0 k_{zs}\left(\frac{p}{2}\right)+\mu_s k_{z0}\left(\frac{p}{2}\right)\right)^2}\sqrt{\pi\left(k_{zs}\left(\frac{p}{2}\right)\right)^3} \tag{9.66}$$

The modulus of $f_1(p;\omega)$ is shown in Figure 9.17.

Figure 9.17 shows that, independently from the exploited antennas, the visible interval vanishes at the extreme points $\pm 2k_s$. There are also two further roots in the points $\pm 2k_0$, corresponding to the critic angle, but these can be counteracted by singularities of the plane wave spectrum of the source and receiver, as seen in Eq. (9.12).

9.12   DT RELATIONSHIP IN THE PRESENCE OF MAGNETIC TARGETS

At this point, let us consider the possibility of magnetic targets. To deal with this case, we come back to the case of a filamentary current as source and for the scattered field as datum; moreover, we will consider the case of data gathered at the air–soil interface and will limit our discussion to the ideal case of an infinite observation line.

Thus, the starting equation is the complete linear scattering equation (8.7), with the incident field given by Eq. (4.48) and the external electrical Green's function given by Eq. (4.47). The substitutions of these quantities, together with the calculation of the gradients under the sign of integral, provide

$$E_s(x;\omega) = \frac{j\omega I_0\mu_0^2\mu_s k_s^2}{2\pi}\int_{-\infty}^{+\infty}\int_{-\infty}^{+\infty}\int_{-\infty}^{+\infty}\int_{-\infty}^{+\infty}\frac{\exp(-j(k_{zs}(u)+k_{zs}(v))z')\exp(j(u+v)x)}{(\mu_0 k_{zs}(u)+\mu_s k_{z0}(u))(\mu_0 k_{zs}(v)+\mu_s k_{z0}(v))}$$

$$\times\exp(jv\Delta)\exp(-j(u+v)x')\chi_e(x',z')\,dx'dz'dudv$$

$$-\frac{j\omega I_0\mu_0^2\mu_s}{2\pi}\int_{-\infty}^{+\infty}\int_{-\infty}^{+\infty}\int_{-\infty}^{+\infty}\int_{-\infty}^{+\infty}\frac{\exp(-j(k_{zs}(u)+k_{zs}(v))z')\exp(j(u+v)x)}{(\mu_0 k_{zs}(u)+\mu_s k_{z0}(u))(\mu_0 k_{zs}(v)+\mu_s k_{z0}(v))}$$

$$\times\exp(jv\Delta)\exp(-j(u+v)x')\chi_{m1}(x',z')(uv+k_{zs}(u)k_{zs}(v))\,dx'dz'dudv,\quad x\in\Sigma \tag{9.67}$$

with $\chi_{m1} = \dfrac{\chi_m(x',z')}{1+\chi_m(x',z')}$ as in Section 8.2.

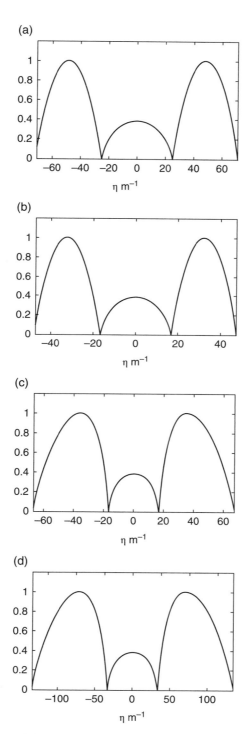

Figure 9.17. The factor of the spectral weight source and receiver independent at the air–soil interface. **Panel a:** $f = 600$ MHz, $\varepsilon_{rs} = 4\mu_{rs} = 2$. **Panel b:** $f = 400$ MHz, $\varepsilon_{rs} = 4\mu_{rs} = 2$. **Panel c:** $f = 400$ MHz, $\varepsilon_{rs} = 4\mu_{rs} = 4$. **Panel d:** $f = 800$ MHz, $\varepsilon_{rs} = 4\mu_{rs} = 4$.

From Eq. (9.67), we can develop calculations fully analogous to those shown before. The details are left as an exercise. The final result is

$$\hat{E}_s(p;\omega) = f(p;\omega)\hat{\tilde{\chi}}_{e1}(\eta(p),\varsigma(p)) - f(p;\omega)\hat{\tilde{\chi}}_{m11}(\eta(p),\varsigma(p)) \tag{9.68}$$

where

$$\chi_{m1}(x',z')\frac{\chi_{m1}(x',z')}{\sqrt{z'}} \tag{9.69}$$

and $f(p;\omega)$ is the electrical spectral weight as provided in Eq. (9.12).

Equation (9.68) states the theoretical impossibility, within the DT approximations, to distinguish the dielectric or the magnetic nature of the buried targets. In fact, the retrieved quantity is the difference of the spectra of the two object functions, electric and magnetic, and "unfortunately" the spectra are calculated in the same points of the plane (η,ς) and with the same spectral weights, so that there is no way to distinguish the two contributions from each other unless some further off-line information is achievable.

9.13   DT RELATIONSHIP FOR A DIFFERENTIAL CONFIGURATION

Let us now deal with the DT relationship for differential configuration. The following hypotheses are assumed: The source is a filamentary current, the observation line is at the air–soil interface and is limited, and no magnetic target is present. The presence of small losses is accounted for perturbatively. Let us remind that in the differential configuration the datum is the difference of the scattered field values gathered in two points at a fixed offset and symmetrical with respect to the source. Thus, the datum is given by

$$\Delta E(x,\omega) = E_s(x+\Delta,\omega) - E_s(x-\Delta,\omega) \tag{9.70}$$

where Δ is the offset between the central receiver and any of the two transmitters (see Section 7.3). By Fourier transforming along the x-axis Eq. (9.70), we achieve, after the same straigthforward passages (Persico and Soldovieri, 2006).

$$\hat{E}_s(p;\omega) = f_d(p;\omega)\hat{\tilde{\chi}}_{e1}(\eta(p),\varsigma(p)) \tag{9.71}$$

where

$$f_d(p;\omega) = \frac{j\omega I_0\mu_0^2\mu_s k_s \exp\left(j\frac{\pi}{4}\right)}{\left(\mu_0 k_{zs}\left(\frac{p}{2}\right) + \mu_s k_{z0}\left(\frac{p}{2}\right)\right)^2}\sqrt{\pi\left(k_{zs}\left(\frac{p}{2}\right)\right)^3}\left(\exp\left(j\Delta\frac{p}{2}\right) - \exp\left(-j\Delta\frac{p}{2}\right)\right)$$

$$= 2j\sin\left(\frac{\Delta p}{2}\right)\frac{j\omega I_0\mu_0^2\mu_s k_s \exp\left(j\frac{\pi}{4}\right)}{\left(\mu_0 k_{zs}\left(\frac{p}{2}\right) + \mu_s k_{z0}\left(\frac{p}{2}\right)\right)^2}\sqrt{\pi\left(k_{zs}\left(\frac{p}{2}\right)\right)^3} \tag{9.72}$$

Equations (9.71)–(9.72) are different from Eqs. (9.11)–(9.12) only for the spectral weighting function; however, the difference is meaningful because the purely phase term $\exp(j\Delta p/2)$ has now been replaced by $2j\sin(\Delta p/2) = 2j\sin(\Delta\eta/2)$, which involves an amplitude weighting of the spatial frequencies within the spectral retrievable set. In particular, the retrievable spectral set is reduced, because this term erases the spatial frequencies corresponding to the zeroes of this sine function, given by

$$\eta_0 = \frac{2n\pi}{\Delta} \tag{9.73}$$

where n is any integer. In particular, this erases a vertical belt about the zero horizontal frequency. The offset Δ affects the wideness of the belt to be erased. The erased spectral belt is actually case-dependent (in particular, noise-dependent), because it is the belt within which we "can't trust" the diffraction tomography result. To provide a graphical representation, let us now assume that the reconstruction is not reliable in the belt corresponding to the −20 dB level of the sine function with respect to the maximum, and let us focus around $\eta = 0$. The wideness of the erased belt is provided by the equation

$$\left|\sin\left(\frac{\eta\Delta}{2}\right)\right| = 0.1 \Rightarrow \frac{\eta\Delta}{2} \simeq \pm 0.1 \Rightarrow \eta \simeq \pm \frac{0.2}{\Delta} \tag{9.74}$$

If the data are noisier, the erasing band enlarges; and if the data are cleaner, the belt restricts; but it is, in any case, inversely proportional to the offset. The wideness EB of the erased belt is given the distance between the two solutions of Eq. (9.74), and so it is given by

$$EB = \frac{0.4}{\Delta} \tag{9.75}$$

Physically, this means that, when the two source points are too close to each other, the two subtracted values of the scattered field are substantially the same, because (whatever the buried targets) the field does not have "enough space" to vary meaningfully. Therefore, for short offsets, the scattering operator under differential configuration will some way tend to an identically null operator. Moreover, the fact that the zero horizontal spatial frequency is erased modifies the properties of the inverse scattering operator, making the retrievable reconstruction no longer horizontally low-pass and vertically band-pass, but rather band-pass along both the horizontal and the vertical directions. As a consequence, we will tend to reconstruct essentially the edges of the target, both along the abscissa and along the depth. This is physically easily explained. In fact, let us think of a long horizontal target; in this case, apart from a brief path over the left- and right-hand edges, when the system of antennas flies over this target, the scenario becomes symmetrical (see Figure 9.18), and therefore the gathered signal is null over most of the crossed target. In the end, this is a drawback coherent with the main purpose

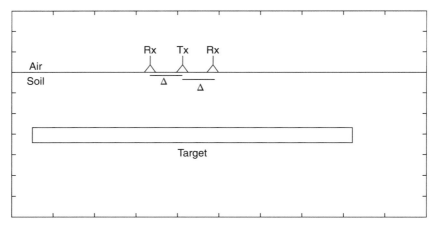

Figure 9.18. Pictorial of a prospecting in differential configuration on a long target (x and z are in arbitrary units). If the edges are far enough, the buried scenario is symmetric and the datum vanishes.

of the differential configuration—that is, to erase the contribution of the air–soil interface and of any further quasi-horizontal layers. In particular, these interfaces can be assimilated to long quasi-horizontal targets, and we can expect that, if they are filtered out by the differential configuration, then also long horizontal target will be filtered out.

The choice of the offset Δ is therefore crucial, because it influences the class of the retrievable or at least detectable targets. In particular, this choice should be done, as far as possible, also related to the size of the targets looked for. In particular, a rough quantification of the class of targets specifically erased by the differential configuration can be identified by considering the Fourier transform of a centered rectangular homogeneous target large L. The vertical size and the depth of the center are at a first approximation unessential, but let us label them as vs and z_c. The spectrum of such a dielectric contrast[6] is given by

$$\hat{\tilde{\chi}}_e = \chi_o Lh \exp(-j\varsigma z_c)\,\mathrm{sinc}\left(\frac{\varsigma vs}{2}\right)\mathrm{sinc}\left(\frac{\eta L}{2}\right) \tag{9.76}$$

The main lobe of the sinc function along the horizontal frequencies is large $2\pi/L$; and so, if we desire a "good" reconstruction of a target large L, we should make this main lobe at least larger than the erased belt EB. This means that, due to Eq. (9.75), we should have at least

$$\frac{0.4}{\Delta} \le \frac{2\pi}{L} \Leftrightarrow \Delta \ge \frac{0.2L}{\pi} = 0.064L \tag{9.77}$$

---

[6] Actually we should consider the spectrum of the target normalized to the square root of the depth; but for a qualitative evaluation we can neglect this, especially if we consider a vertically thin target.

On the other hand, to increase the offset indeterminately is not a good idea. In fact, apart some possible degradation of the signal-to-noise ratio, we have to consider the problem of the periodicity of the sine function. In particular, the zeroes of the sine are at a $2\pi/\Delta$ distance from each other; and this means that, in order to avoid that further erasing belts enter the retrievable spectral set, we should also guarantee the condition

$$\frac{2\pi}{\Delta} - \frac{0.2}{\Delta} = \frac{6.08}{\Delta} \geq 2k_{smax} \sin\left(\theta_{emax}\right) \Leftrightarrow \Delta \leq \frac{3.04}{k_{smax}\sin\left(\theta_{emax}\right)} = \frac{0.48\lambda_{smin}}{\sin\left(\theta_{emax}\right)} \qquad (9.78)$$

It is not an easy task to establish a priori an optimal choice for Δ, but Eqs. (9.77) and (9.78) provide two relevant trade-offs.

Let us also note that the two conditions (9.77) and (9.78) are incompatible with each other if $L > 0.48\lambda_{smin}/0.064 \sin\left(\theta_{emax}\right) = 7.5\lambda_{smin}/\sin\left(\theta_{emax}\right)$. The coefficient 7.5 becomes larger if the criterion for the choice of the erase band is relaxed (e.g. -30 dB instead of -20 dB). However, the general sense of Eq. (9.78) is that the targets horizontally large in terms of wavelength will be poorly reconstructed under a differential configuration, especially if buried at a shallow depth.

Four examples are shown in Figures 9.19 and 9.20. In particular, in Figure 9.19 the erasing effect is shown at several values of the offset, and the resulting spectral sets are shown in Figure 9.20.

9.14   DT RELATIONSHIP IN THE PRESENCE OF BACKGROUND REMOVAL

Let us now consider the DT in the presence of background removal (BKGR). The hypotheses adopted are the same of the previous section. The BKGR is a procedure widely known and exploited (Conyers, 2004), and it is also widely known that the BKGR constitutes a spatial filtering. A quantitative evaluation of the spatial filtering effects of the BKGR has been provided, in recent years, in Persico and Soldovieri (2008) and in Persico and Soldovieri (2010).

So, let us consider relationship (7.11), that provides the datum under background removal in frequency domain. The Fourier transform of Eq. (7.11) with respect to the abscissa provides

$$\hat{E}_b(p,\omega) = \hat{E}_s(p,\omega)\left[1 - \frac{1}{2N+1}\sum_{n=-N}^{N}\exp(-jpns)\right] \qquad (9.79)$$

By following the same formal passages shown in Chapter 3 [see Eq. (3.18)], Eq. (9.79) can be still rewritten as

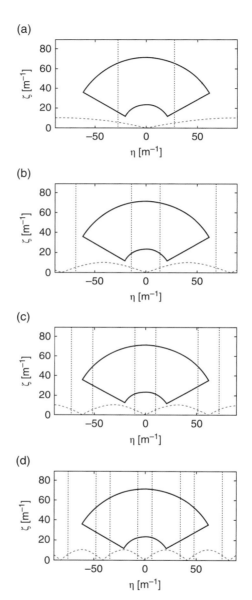

Figure 9.19. Red dashed lines: the modulus of the erasing function (emphasized by a factor 5 for graphic reasons). Blue dot lines: the erasing belts deriving from the erasing function. Solid black line: the spectral set without the differential effect. The parameters are: $f_{min} = 200$ MHz, $f_{max} = 600$ MHz, $\varepsilon_{sr} = 4$, $\mu_{sr} = 2$. **Panel a**: $\Delta = 3.83$ cm. **Panel b**: $\Delta = 7.65$ cm. **Panel c**: $\Delta = 10.21$ cm. **Panel d**: $\Delta = 15.31$ cm. The erased belt is calculated according to Eq. (9.75). For color detail, please see color plate section.

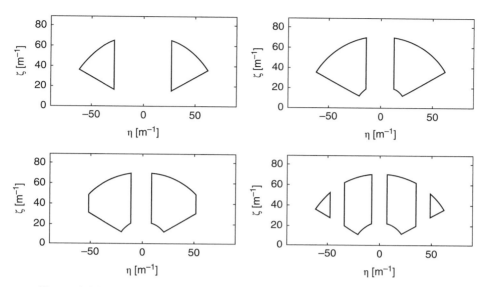

Figure 9.20. The resulting retrievable spectral sets in relationship to Figure 9.19.

$$\hat{E}_b(p,\omega) = f_1(p)\hat{E}_s(p,\omega) = \left[1 - \frac{1}{2N+1} \frac{\sin\left((2N+1)\dfrac{ps}{2}\right)}{\sin\left(\dfrac{ps}{2}\right)} \right] \hat{E}_s(p,\omega) \tag{9.80}$$

Substituting $\hat{E}_s(p,\omega)$ from Eq. (9.11), we have

$$\hat{E}_b(p,\omega) = f(p,\omega)f_1(p)\hat{\chi}_{e1}(\eta(p),\varsigma(p)) \tag{9.81}$$

where the transformation of coordinates is given by Eq. (9.14), the spectral weight $f(p,\omega)$ is given in Eq. (9.12), and the further spectral weight $f_1(p)$ is specified in Eq. (9.80) and is the complement to 1 of a normalized Dirchlet sine function. With respect to Eq. (9.11), Eq. (9.80) just shows the extra-spectral weight $f_1(p)$. The dealing reveals noticeable similarities to that relative to a differential configuration, where the substantial effect of the displacement of source and receivers is to provide a spectral extra-weight. We might even say that the differential configuration is an hardware way to implement a simple kind of background removal.

The extra-weight $f_1(p)$ is quantitatively represented in four cases in Figure 9.21. The effect of the weighting function on the retrievable spectral set is in some aspect analogous to that illustrated in Figures 9.19 and 9.20 with regard to a differential configuration.

In particular, similar to the case of a differential configuration, the extra-weight is periodical and has a root at $p = \eta = 0$. So, also the background removal necessarily involves a filtering of the lowest horizontal spatial frequencies, with periodical erasing belts of the same thickness in the plane (η,ς). The wideness of the main lobe of the Dirichlet sine at hand is given by

$$\frac{4\pi}{(2N+1)s} \tag{9.82}$$

and the function is periodic with period

$$P = \frac{2\pi}{s} \tag{9.83}$$

Consequently, in order to guarantee that there is only the central "erasing belt" within the retrievable spectral set, we have to guarantee that

$$\frac{2\pi}{s} - \frac{2\pi}{(2N+1)s} = \frac{4\pi N}{(2N+1)s} \geq 2k_{slmax} \sin\left(\theta_{eslmax}\right) \Leftrightarrow s \leq \frac{N\lambda_{slmin}}{(2N+1)\sin\left(\theta_{eslmax}\right)} \tag{9.84}$$

Also in this case, we have to minimize the thickness of the central horizontal belt. However, this time we have at our disposal two parameters to handle, namely the number of averaged traces and the spatial step s of the data.[7] Actually, the product $(2N+1)s$ in Eq. (9.82) has the physical meaning of the "averaged distance" along the B-scan, starting at a distance $s/2$ before the first averaged trace and ending at a distance $s/2$ after the last one. The importance of having two parameters at our disposal is that we can reduce the thickness of the central erasing belt without "pulling in" further erasing belts within the spectral retrievable domain. To do this, we have to increase the number of averaged traces without enlarging the spatial step. On the other hand, it is also wrong to choose a spatial step too narrow. In this way, in fact, the background removal makes the scattering operator tend toward a null operator unless the number of averaged traces is correspondingly increased. In other words, if the averaged traces are too close to each other, their average value is substantially the same as the central trace, and so the difference between the two quantities vanishes. The averaged length should also account for the kind of targets looked for. In particular, we can define a critical length equal to

$$L = (2N+1)s \tag{9.85}$$

The critical length represents the maximum horizontal size that a target should have in order not to be strongly filtered out by the background removal. The proof is straightforward on the basis of the same reasoning of the previous section. With regard to the choice of an optimal N, due to Eq. (9.82), we have that the maximum amount of information is achieved when all the gathered traces are averaged, because this choice makes the critical length substantially as long as the entire B-scan.

Practically, some further issues should be accounted for. One of them is the smoothness of the possible layering—that is, the flatness of the interface layers. Another issue is the possibility that a strong scattering object modifies the retrieved average value of the traces at a fixed time depth. Both these phenomena are not accounted for within a BA

---

[7] The offset between the antennas, unlike the differential configuration, represents at a first approximation just a phase term.

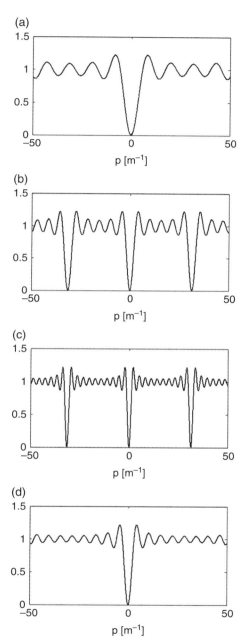

Figure 9.21. The masking function $f_1(p)$ under background removal. **Panel a**: $N = 5$, $s = 10$ cm. **Panel b**: $N = 5$, $s = 20$ cm. **Panel c**: $N = 10$, $s = 20$ cm. **Panel d**: $N = 10$, $s = 10$ cm.

model funded on a couple of homogeneous half-spaces. It is also important to emphasize the fact that, usually, commercial codes do not account for the performed background removal when performing the migration (dealt with in the next chapter) of the data, which instead should theoretically be done. All these factors can make a moving average BKGR preferable. The evaluation of the best choice, in many cases, can be only heuristic—that is, performed on the basis of the results achieved from several applied BKGRs.

To sum up, it has been shown that the BKGR, as any spatial filtering, has some drawbacks. In particular, it unavoidably involves some filtering effect on the targets of interest too. In many cases the background removal is needed, and this drawback is just a price to be paid. However, sometimes the targets are deep enough and the scenario is quite "clean." In these cases, it is not well-advised to apply the BKGR.

In Section 15.8 the reader can find some exercises on the effects of the background removal. Please note that they are also based on the 2D migration exposed in Chapter 10 and on the SVD theory exposed in Chapter 14.

QUESTIONS

1. On which parameters does the visible interval depend?
2. Do the spatial and frequency steps affect the retrievable spectral set?
3. Do the time step and the time bottom scale influence the spectral retrievable set?
4. Does a narrowing of the spatial (frequency) step essentially improve the available horizontal (vertical) resolution in any case?
5. Does the resolution essentially improve if the observation line becomes longer? And is this improvement progressive?
6. Does a more directive couple of antennas improve the horizontal resolution?
7. Does a more directive couple of antennas influence the vertical resolution?
8. Could the frequency step needed for data processing be relaxed with respect to that needed for gathering the data with a stepped frequency system?
9. Can the time step needed for data processing be relaxed with respect to that needed in order to properly gather the data with a pulsed system?
10. Can an enlargement of the available band improve the reconstruction? Might it also cause problems?
11. Does a background removal on an increased number of traces increase the theoretical amount of information indefinitely?
12. Does an increased amount of information improve the achieved image?

TWO-DIMENSIONAL MIGRATION ALGORITHMS

10.1 MIGRATION IN THE FREQUENCY DOMAIN

In this section we will focus on data in the frequency domain, and in the next one we will focus on data in the time domain. In both cases, we will limit ourselves to the case of common offset data gathered at the air–soil interface, as well as to the case of only dielectric soils and dielectric targets.

Let us start from Eq. (9.11), reported here for sake of easy readability;

$$\hat{E}_s(p;\omega) = f(p;\omega)\hat{\hat{\chi}}_{e1}(\eta(p,\omega),\varsigma(p,\omega)) \tag{10.1}$$

where (let us remind ourselves)

Introduction to Ground Penetrating Radar: Inverse Scattering and Data Processing,
First Edition. Raffaele Persico.
© 2014 The Institute of Electrical and Electronics Engineers, Inc. Published 2014 by John Wiley & Sons, Inc.

$$f(p;\omega) = -\frac{\omega^2 I_0 \mu_0^2 \mu_s \exp\left(j\Delta\frac{p}{2}\right) \exp\left(j\frac{\pi}{4}\right)}{c\left(\mu_0 k_{zs}\left(\frac{p}{2}\right) + \mu_s k_{z0}\left(\frac{p}{2}\right)\right)^2} \sqrt{\pi\left(k_{zs}\left(\frac{p}{2}\right)\right)^3}$$

$$= -\frac{\omega^2 \sqrt{\pi} I_0 \mu_0^2 \mu_s \exp\left(j\Delta\frac{p}{2}\right) \exp\left(j\frac{\pi}{4}\right) \left(\frac{\omega^2}{c^2} - \frac{p^2}{4}\right)^{3/4}}{c\left(\mu_0 \sqrt{\frac{\omega^2}{c^2} - \frac{p^2}{4}} + \mu_s \sqrt{\frac{\omega^2}{c_0^2} - \frac{p^2}{4}}\right)^2} \tag{10.2}$$

$$\chi_{e1}(x',z') = \frac{\chi_e(x',z')}{\sqrt{z'}} \tag{10.3}$$

$$\eta(p,\omega) = p, \quad \varsigma(p,\omega) = 2k_{zs}\left(\frac{p}{2}\right) = 2\sqrt{k_s^2 - \left(\frac{p}{2}\right)^2} = 2\sqrt{\left(\frac{\omega}{c}\right)^2 - \left(\frac{p}{2}\right)^2} = \sqrt{\frac{4\omega^2}{c^2} - p^2} \tag{10.4}$$

The transformation (10.1) is algebraically inverted as

$$\hat{\chi}_{e1}(\eta(p,\omega),\varsigma(p,\omega)) = g(p,\omega)\hat{E}_s(p;\omega) \tag{10.5}$$

where

$$g(p,\omega) = \begin{cases} \dfrac{1}{f(p;\omega)} = -\dfrac{c\left(\mu_0\sqrt{\dfrac{\omega^2}{c^2} - \dfrac{p^2}{4}} + \mu_s\sqrt{\dfrac{\omega^2}{c_0^2} - \dfrac{p^2}{4}}\right)^2 \exp\left(-j\left(\Delta\dfrac{p}{2} + \dfrac{\pi}{4}\right)\right)}{\omega^2 I_0 \mu_0^2 \mu_s \sqrt{\pi}\left(\dfrac{\omega^2}{c^2} - \dfrac{p^2}{4}\right)^{3/4}}, & (p,\omega) \in \Omega \\ \\ 0 \quad (p,\omega) \notin \Omega \end{cases} \tag{10.6}$$

where Ω is meant as the visible 2D multifrequency set, that ranges in the interval $[\omega_{min}, \omega_{max}]$ along the ω-axis and ranges in the interval $[-2k_s\sin(\theta_{emax}),$ $2k_s\sin(\theta_{emax})] = \left[-2\dfrac{\omega}{c}\sin(\theta_{emax}), 2\dfrac{\omega}{c}\sin(\theta_{emax})\right]$ along the p-axis at any fixed ω. Ω is of course a trapezium in the plane (p,ω).

At this point, we can solve for the spectrum of the object function inverting relationship (10.4) between the coordinatives, thereby obtaining a relationship of the form

$$\hat{\chi}_{e1}(\eta,\varsigma) = g(p(\eta,\varsigma),\omega(\eta,\varsigma))\hat{E}_s(p(\eta,\varsigma),\omega(\eta,\varsigma)) \tag{10.7}$$

Equation (10.7) is a nonambiguous relationship because the correspondence (10.4) between the point (p,ω) (with p belonging to the visible interval) and the point (η,ς) is bi-univocal.[1] Finally, the inverse Fourier transform of Eq. (10.7) in the spatial domain provides the solution of the problem, given by

$$\chi_{e1}(x',z') = \frac{1}{4\pi^2} \int\limits_{-\infty}^{+\infty} \int\limits_{-\infty}^{+\infty} g(p(\eta,\varsigma),\omega(\eta,\varsigma)) \hat{E}_s(p(\eta,\varsigma),\omega(\eta,\varsigma)) \exp(j\eta x') \exp(j\varsigma z') \, d\eta d\varsigma$$

(10.8)

Equation (10.8) can be explicated inverting Eq. (10.4), thereby obtaining[2]

$$p(\eta,\varsigma) = \eta, \quad \omega(\eta,\varsigma) = \frac{c\sqrt{\eta^2 + \varsigma^2}}{2}$$

(10.9)

which provides

$$\chi_{e1}(x',z') = \frac{1}{4\pi^2} \int\limits_{-\infty}^{+\infty} \int\limits_{-\infty}^{+\infty} g_1(\eta,\varsigma) \hat{E}_s\left(\eta, \frac{c\sqrt{\eta^2 + \varsigma^2}}{2}\right) \exp(j\eta x') \exp(j\varsigma z') \, d\eta d\varsigma$$

(10.10)

with

$$g_1(\eta,\varsigma) = g\left(\eta, \frac{c\sqrt{\eta^2 + \varsigma^2}}{2}\right)$$

$$= \begin{cases} -\dfrac{2\sqrt{2}}{c I_0 \mu_0^2 \mu_s \sqrt{\pi}} \dfrac{\left(\mu_0\varsigma + \mu_s\left(\sqrt{\eta^2\left(\left(\frac{c}{c_0}\right)^2 - 1\right) + \varsigma^2}\right)\right)^2 \exp\left(-j\left(\Delta\frac{\eta}{2} + \frac{\pi}{4}\right)\right)}{\left(\eta^2 + \varsigma^2\right)^{3/2}\varsigma}, & (\eta,\varsigma) \in \Omega_1 \\ 0, & (\eta,\varsigma) \notin \Omega_1 \end{cases}$$

(10.11)

where Ω_1 is the image in the plane (η,ς) of the set Ω through the transformation (10.4), which is just the retrievable spectral set. Equation (10.10) constitutes a migration formula in frequency domain. It essentially consists in an interpolation of the spectrum of the data multiplied times a known function and then back Fourier transformed in the spatial domain. Equation (10.10) has the drawback that it requires some interpolation of the spectrum of the data, but has the computational advantage that it is a two-dimensional

---

[1] Let us stress that this depends on the fact that we are considering the multibistatic (common offset) configuration. For example, if the source and the receiver can be moved independently from each other, this does not happen (Soldovieri et al., 2005a).

[2] We maintain the formal difference between p and η, even if they are numerically the same value, because the two quantities are semantically different: The first one is the conjugate variable of the measurement abscissa, and the second one is the conjugate variable of the abscissa in the investigation domain.

inverse Fourier transform and therefore can be implemented by means of computationally effective IFFT algorithms.

Alternatively, we can substitute the following in Eq. (10.10):

$$\frac{c}{2}\sqrt{\eta^2 + \varsigma^2} = \omega \Rightarrow \varsigma = \sqrt{\frac{4\omega^2}{c^2} - \eta^2} \Rightarrow d\varsigma = \frac{4\omega d\omega}{c^2\sqrt{\frac{4\omega^2}{c^2} - \eta^2}} \tag{10.12}$$

thereby obtaining

$$\chi_{e1}(x',z') = \frac{1}{c^2\pi^2} \int_{-\infty}^{+\infty} \int_{-\infty}^{+\infty} g_2(\eta,\omega)\hat{E}_s(\eta,\omega)\exp(j\eta x')\exp\left(j\sqrt{\frac{4\omega^2}{c^2} - \eta^2}z'\right) d\eta d\omega \tag{10.13}$$

where

$$g_2(\eta,\omega)$$

$$= \frac{4\omega g_1\left(\eta, \sqrt{\frac{4\omega^2}{c^2} - \eta^2}\right)}{c^2\sqrt{\frac{4\omega^2}{c^2} - \eta^2}}$$

$$= \begin{cases} -\dfrac{32\sqrt{2}}{cI_0\mu_0^2\mu_s\sqrt{\pi}} \dfrac{\left(\mu_0\sqrt{\frac{4\omega^2}{c^2} - \eta^2} + \mu_s\left(\sqrt{\frac{4\omega^2}{c^2} + \eta^2\left(\left(\frac{c}{c_0}\right)^2 - 2\right)}\right)\right)^2 \exp\left(-j\left(\Delta\frac{\eta}{2} + \frac{\pi}{4}\right)\right)}{\omega\left(\frac{4\omega^2}{c^2} - \eta^2\right)^{5/4}} & (\eta,\omega) \in \Omega, \\ 0, & (\eta,\omega) \notin \Omega \end{cases}$$

$$\tag{10.14}$$

where Ω is the same set defined after Eq. (10.6), this time formally represented in the plane (η,ω) instead of (p,ω)

Equation (10.13) is an alternative form of the migration in the frequency domain, where no interpolation of the data is required but, on the other hand, the achieved integral is not an inverse Fourier transform and thus cannot be calculated by means of a fast IFFT algorithm.

Equations (10.10) and (10.13) express the so-called Stolt's migration or $f-k$ (or $\omega - k$) migration (Stolt, 1978). The name derives from the fact that the integration can be performed either making use of the variable ω or of the variable ς, labeled elsewhere as k.[3] Actually, we can also adopt a more simple migration formula that does not include the factor $g_2(\eta,\omega)$.

---

[3] We have preferred the symbol ς in order to avoid a possible confusion between the vertical spatial frequencies of the spectrum of the object function and the wavenumber.

$$\chi_{e1}(x',z') = \frac{F}{c^2\pi^2} \int\limits_{-\infty}^{+\infty} \int\limits_{-\infty}^{+\infty} \hat{E}_s(\eta,\omega) \exp(j\eta x') \exp\left(j\sqrt{\frac{4\omega^2}{c^2}-\eta^2}z'\right) d\eta d\omega \qquad (10.15)$$

where F is a unitary but not dimensionless factor, included just in order to preserve the dimensional coherence of the equation (the factor $g_2(\eta,\omega)$, in fact, was not dimensionless). In particular, the integration ranges on the entire 2D plane relying on the fact that the integrand $\hat{E}_s(\eta,\omega)$ attenuates outside the multifrequency visible spectrum. Equation (10.13) is obviously mathematically more rigorous than Eq. (10.15.) However, Eq. (10.15) is somehow more robust against uncertainties on the sources and the receiver and against the singular behavior of the spectral weight $g_2(\eta,\omega)$ for $\eta \to \pm 2k_s$.[4] It is also much more similar to the classical migration formulas derived from the seismic (Stolt, 1978).

Assuming the model of Eq. (10.15) for the integration in $d\eta d\omega$, the integration in $d\eta d\varsigma$ can be performed according to the substitution

$$\sqrt{\frac{4\omega^2}{c^2}-\eta^2} = \varsigma \Rightarrow \omega = \frac{c\sqrt{\eta^2+\varsigma^2}}{2} \Rightarrow d\omega = \frac{c\varsigma\,d\varsigma}{2\sqrt{\eta^2+\varsigma^2}} = \frac{c\,d\varsigma}{2\sqrt{1+\frac{\eta^2}{\varsigma^2}}} \qquad (10.16)$$

And consequently (Stolt, 1978; Oden et al., 2007) Eq. (10.15) can be also written as

$$\chi_{e1}(x',z') = \frac{F}{2c\pi^2} \int\limits_{-\infty}^{+\infty} \int\limits_{-\infty}^{+\infty} \frac{\hat{E}_s\left(\eta,\frac{c\sqrt{\eta^2+\varsigma^2}}{2}\right)}{\sqrt{1+\frac{\eta^2}{\varsigma^2}}} \exp(j\eta x') \exp(j\varsigma z')\, d\eta d\varsigma \qquad (10.17)$$

The two formulas (10.15) and (10.17) show the same trade-offs as the formally more rigorous formulas (10.10) and (10.13), namely one of them (10.17) is an inverse Fourier transform but requires some interpolation of the data and the other one (10.15) does not require any interpolation but is not an inverse Fourier transform. There is also a third possible form for the migration integral, that can be worked out expressing the integral (10.10) in polar coordinatives in the plane (η,ς). In particular, let us put

$$\eta = k\cos(\Phi), \quad \varsigma = k\sin(\Phi) \qquad (10.18)$$

---

[4] Theoretically, this diverging behavior is counteracted by the fact that the spectrum of the scattered field should be a null quantity at the edge of the visible interval, but the parametric uncertainties in general affect this equilibrium.

Substituting into Eq. (10.10), we have

$$\chi_{el}(x',z') = \frac{1}{4\pi^2}\int_0^\pi d\phi \int_0^{+\infty} kg_1(k\cos(\phi), k\sin(\phi))\hat{E}_s\left(k\cos(\phi), \frac{ck}{2}\right)$$

$$\times \exp(jk\cos(\phi)x')\exp(jk\sin(\phi)z')\,dk \qquad (10.19)$$

and, after the further change of variable

$$\theta = \phi - \frac{\pi}{2} \qquad (10.20)$$

we obtain

$$\chi_{el}(x',z') = \frac{1}{4\pi^2}\int_{-\pi/2}^{\pi/2} d\theta \int_0^{+\infty} kg_1(-k\sin(\theta), k\cos(\theta))\hat{E}_s\left(-k\sin(\theta), \frac{ck}{2}\right)$$

$$\times \exp(-jk\sin(\theta)x')\exp(jk\cos(\theta)z')\,dk \qquad (10.21)$$

Equation (10.21) requires an interpolation of the data and is not a double inverse Fourier transform. Therefore, it is not commonly exploited. However, it can be instructive because it shows directly the essential role of the view angle and of the band. In particular, in Eq. (10.21), θ is easily recognized to be just the view angle, whereas k is easily recognized to be the wavenumber. This also allows us to express explicitly (instead of implicitly by means of the essential support of g_1) the extremes of integration—that is, to express a regularized migration integral formula. In particular, supposing that the maximum view angle is symmetric, Eq. (10.21) can be rewritten as

$$\chi_{el}(x',z') = \frac{1}{4\pi^2}\int_{-\theta_{emax}}^{\theta_{emax}} d\theta \int_{2k_{smin}}^{2k_{smax}} kg_1(-k\sin(\theta), k\cos(\theta))\hat{E}_s\left(-k\sin(\theta), \frac{ck}{2}\right)$$

$$\times \exp(-jk\sin(\theta)x')\exp(jk\cos(\theta)z')\,dk \qquad (10.22)$$

Of course, it is possible to provide a "polar" migration formula also neglecting the spectral weight, within a nonrigorous but somehow more robust formulation. This can be derived e.g., from Eq. (10.17) after some straightforward manipulations, so to obtain

$$\chi_{el}(x',z') = \frac{F}{2c\pi^2}\int_{-\theta_{emax}}^{\theta_{emax}} \cos(\theta)d\theta \int_{2k_{smin}}^{2k_{smax}} k\hat{E}_s\left(-k\sin(\theta), k\frac{c}{2}\right)$$

$$\times \exp(-jk\sin(\theta)x')\exp(jk\cos(\theta)z')\,dk \qquad (10.23)$$

10.2 MIGRATION IN THE TIME DOMAIN

Raffaele Persico and Raffaele Solimene

In GPR prospecting, it is often found comfortable to migrate the data in the time domain. The formula for the migration in the time domain can be worked out starting from Eq. (10.13) and observing that, by definition,

$$\hat{E}_s(\eta,\omega) = \int\limits_{-\infty}^{+\infty} \int\limits_{0}^{+\infty} E_s(x,t)\exp(-j\eta x)\exp(-j\omega t)\, dx dt \qquad (10.24)$$

where the extremes of the integral account for the fact that the signal starts at the time instant $t=0$.

Substituting Eq. (10.24) into Eq. (10.13), we obtain

$$\chi_{e1}(x',z') = \frac{1}{c^2\pi^2} \int\limits_{-\infty}^{+\infty} dx \int\limits_{0}^{+\infty} dt E_s(x,t) \int\limits_{-\infty}^{+\infty} \exp(-j\omega t) \left[\int\limits_{-\infty}^{+\infty} g_2(\eta,\omega) \right.$$

$$\left. \times \exp(-j\eta(x-x'))\exp\left(j\sqrt{\frac{4\omega^2}{c^2}-\eta^2}z'\right) d\eta \right] d\omega$$

$$= \frac{1}{c^2\pi^2} \int\limits_{-\infty}^{+\infty} dx \int\limits_{0}^{+\infty} g_3(x-x',z',t)E_s(x,t)\, dt \qquad (10.25)$$

Equation (10.25) expresses the most general form of the migration in the time domain, also called Kirchhoff's migration, where the convolutional weight in the time domain is given by

$$g_3(x-x',z',t) = \int\limits_{-\infty}^{+\infty} \exp(-j\omega t) \left[\int\limits_{-\infty}^{+\infty} g_2(\eta,\omega) \times \exp(-j\eta(x-x'))\exp\left(j\sqrt{\frac{4\omega^2}{c^2}-\eta^2}z'\right) d\eta \right] d\omega \qquad (10.26)$$

The integral (10.26) cannot be solved in a closed form, but can be effectively calculated because, for any fixed z', it is the double Fourier transform of the quantity $g_2(\eta,\omega)$ $\exp\left(j\sqrt{\frac{4\omega^2}{c^2}-\eta^2}z'\right)$ calculated in the points $x-x'$ and t. However, as said in the previous section, the spectral weighting is often neglected, following seismic models. In the case at hand, this provides the following expression for the migration in the time domain:

$$
\chi_{e1}(x',z') = \frac{F}{c^2\pi^2} \int_{-\infty}^{+\infty} dx \int_{0}^{+\infty} dt E_s(x,t) \int_{-\infty}^{+\infty} \exp(-j\omega t) \left[\int_{-\infty}^{+\infty} \exp\left(j\sqrt{\frac{4\omega^2}{c^2} - \eta^2} z' \right) \right.
$$

$$
\left. \times \exp(-j\eta(x-x')) d\eta \right] d\omega \tag{10.27}
$$

where again F is meant as unitary but not dimensionless. Under the approximated model of Eq. (10.27), the double integral in square brackets can be calculated in a closed form. To show this, let us preliminarily write

$$
g_4(x-x',z',t) = \int_{-\infty}^{+\infty} \int_{-\infty}^{+\infty} \frac{\exp\left(j\sqrt{\frac{4\omega^2}{c^2} - \eta^2} z' \right)}{j\sqrt{\frac{4\omega^2}{c^2} - \eta^2}} \exp(-j\eta(x-x')) \exp(-j\omega t) \, d\eta d\omega \tag{10.28}
$$

The integral in $d\eta$ in Eq. (10.27) is proportional to the Hankel function of zeroth order and of second kind (Harrington, 1961). Physically, this means that by neglecting the spectral weight we essentially neglect the interface and assimilate the air–soil space to a homogeneous medium entirely composed of "soil."

It is immediate that the integral (10.28) can be rewritten as

$$
\chi_{e1}(x',z') = \frac{F}{c^2\pi^2} \frac{\partial}{\partial z'} \int_{-\infty}^{+\infty} dx \int_{0}^{+\infty} g_4(x-x',z',t) E_s(x,t) \, dt \tag{10.29}
$$

At this point, the integral expansion of Eq. (10.28) can be also rewritten as follows (Heidary, 2003):

$$
g_4(x-x',z',t) = \int_{-\infty}^{+\infty} \int_{-\infty}^{+\infty} \frac{\exp\left(j\frac{2\omega}{c}\sqrt{(x-x')^2 + z'^2 + y^2} \right)}{\sqrt{(x-x')^2 + z'^2 + y^2}} \exp(-j\omega t) \, dy d\omega
$$

$$
= 2 \int_{-\infty}^{+\infty} \exp(-j\omega t) d\omega \int_{0}^{+\infty} \frac{\exp\left(j\frac{2\omega}{c}\sqrt{(x-x')^2 + z'^2 + y^2} \right)}{\sqrt{(x-x')^2 + z'^2 + y^2}} \, dy \tag{10.30}
$$

The passages that work out Eq. (10.30) from Eq. (10.28) are shown in Appendix A. So, substituting Eq. (10.30) into Eq. (10.29), we obtain

$$\chi_{e1}(x',z') =$$

$$= \frac{2F}{c^2\pi^2} \frac{\partial}{\partial z'} \int\limits_{-\infty}^{+\infty} dx \int\limits_{0}^{+\infty} dt E_s(x,t) \left[\int\limits_{-\infty}^{+\infty} \exp(-j\omega t) \int\limits_{0}^{+\infty} \frac{\exp\left(j\dfrac{2\omega}{c}\sqrt{(x-x')^2+z'^2+y^2}\right)}{\sqrt{(x-x')^2+z'^2+y^2}} dy \right] d\omega$$

(10.31)

At this point, let us write

$$\tau = \frac{2}{c}\sqrt{(x-x')^2+z'^2+y^2} \Rightarrow y^2 = \left(\frac{\tau c}{2}\right)^2 - \|\mathbf{r}-\mathbf{r}'\|^2$$

$$\Rightarrow y = \sqrt{\left(\frac{\tau c}{2}\right)^2 - \|\mathbf{r}-\mathbf{r}'\|^2} = \frac{c}{2}\sqrt{\tau^2 - \frac{4\|\mathbf{r}-\mathbf{r}'\|^2}{c^2}}$$

$$\Rightarrow dy = \frac{c}{2}\frac{\tau\,d\tau}{\sqrt{\tau^2 - \dfrac{4\|\mathbf{r}-\mathbf{r}'\|^2}{c^2}}}$$

(10.32)

where of course $\|\mathbf{r}-\mathbf{r}'\| = \sqrt{(x-x')^2+z'^2}$. Substituting in Eq. (10.31), we have, with straightforward manipulations, the following:

$$\chi_{e1}(x',z') =$$

$$= \frac{2F}{c^2\pi^2} \frac{\partial}{\partial z'} \int\limits_{-\infty}^{+\infty} dx \int\limits_{0}^{+\infty} dt E_s(x,t) \left[\int\limits_{-\infty}^{+\infty} \exp(-j\omega t) \times \int\limits_{\frac{2\|\mathbf{r}-\mathbf{r}'\|}{c}}^{+\infty} \frac{\exp(j\omega\tau)}{\frac{c\tau}{2}}\frac{c\tau}{2}\frac{1}{\sqrt{\tau^2 - \dfrac{4\|\mathbf{r}-\mathbf{r}'\|^2}{c^2}}} d\tau \right] d\omega$$

$$= \frac{F}{c^2\pi^2} \frac{\partial}{\partial z'} \int\limits_{-\infty}^{+\infty} dx \int\limits_{0}^{+\infty} dt E_s(x,t) \int\limits_{\frac{2\|\mathbf{r}-\mathbf{r}'\|}{c}}^{+\infty} \frac{d\tau}{\sqrt{\tau^2 - \dfrac{4\|\mathbf{r}-\mathbf{r}'\|^2}{c^2}}} \int\limits_{-\infty}^{+\infty} \exp(j\omega(\tau-t))\,d\omega$$

$$= \frac{2F}{c^2\pi} \frac{\partial}{\partial z'} \int\limits_{-\infty}^{+\infty} dx \int\limits_{0}^{+\infty} dt E_s(x,t) \int\limits_{\frac{2\|\mathbf{r}-\mathbf{r}'\|}{c}}^{+\infty} \frac{\delta(\tau-t)\,d\tau}{\sqrt{\tau^2 - \dfrac{4\|\mathbf{r}-\mathbf{r}'\|^2}{c^2}}}$$

$$= \frac{2F}{c^2\pi} \frac{\partial}{\partial z'} \int\limits_{-\infty}^{+\infty} dx \int\limits_{0}^{+\infty} dt E_s(x,t) \int\limits_{\frac{2\|\mathbf{r}-\mathbf{r}'\|}{c}-t}^{+\infty} \frac{\delta(u)\,du}{\sqrt{(u+t)^2 - \dfrac{4\|\mathbf{r}-\mathbf{r}'\|^2}{c^2}}}$$

(10.33)

At this point, we have

$$
\int\limits_{\frac{2\|\mathbf{r}-\mathbf{r}'\|}{c}-t}^{+\infty} \frac{\delta(u)\,du}{\sqrt{(t+u)^2 - \frac{4\|\mathbf{r}-\mathbf{r}'\|^2}{c^2}}}
$$

$$
= \begin{cases}
0 & \text{if } \dfrac{2\|\mathbf{r}-\mathbf{r}'\|}{c}-t>0 \\[3ex]
\dfrac{1}{\sqrt{t^2 - \dfrac{4\|\mathbf{r}-\mathbf{r}'\|^2}{c^2}}} & \text{if } \dfrac{2\|\mathbf{r}-\mathbf{r}'\|}{c}-t<0
\end{cases}
\Rightarrow
\int\limits_{\frac{2\|\mathbf{r}-\mathbf{r}'\|}{c}-t}^{+\infty} \frac{\delta(u)\,du}{\sqrt{(t+u)^2 - \frac{4\|\mathbf{r}-\mathbf{r}'\|^2}{c^2}}}
=
\frac{H\left(t-\dfrac{2\|\mathbf{r}-\mathbf{r}'\|}{c}\right)}{\sqrt{t^2 - \dfrac{4\|\mathbf{r}-\mathbf{r}'\|^2}{c^2}}}
$$

$$(10.34)$$

where H is the Heaviside's function, equal to 1 if its argument is greater than 0 and equal to 0 if its argument is negative. Substituting Eq. (10.34) into Eq. (10.33), we achieve

$$
\chi_{el}(x',z') = \frac{2F}{c^2\pi}\frac{\partial}{\partial z'}\int\limits_{-\infty}^{+\infty} dx \int\limits_{0}^{+\infty} dt E_s(x,t)\frac{H\left(t-\dfrac{2\|\mathbf{r}-\mathbf{r}'\|}{c}\right)}{\sqrt{t^2 - \dfrac{4\|\mathbf{r}-\mathbf{r}'\|^2}{c^2}}}
$$

$$
= \frac{2F}{c^2\pi}\frac{\partial}{\partial z'}\int\limits_{-\infty}^{+\infty} dx \int\limits_{\frac{2\|\mathbf{r}-\mathbf{r}'\|}{c}}^{+\infty} \frac{E_s(x,t)}{\sqrt{t^2 - \dfrac{4\|\mathbf{r}-\mathbf{r}'\|^2}{c^2}}}\,dt \tag{10.35}
$$

Equation (10.35) expresses the migration in time domain, or Kirchhoff's migration, in a way similar to the classical seismic formulation as given in (Schneider, 1978). In particular, having neglected the spectral weight $g_2(\eta,\omega)$ the convolutional weight $g_3(x-x',z',t)$ has been calculated as $H\left(t-\frac{2\|r-r'\|}{c}\right)\Big/\sqrt{t^2 - \frac{4\|r-r'\|^2}{c^2}}$.

Performing the derivative $\partial/\partial z'$ under the sign of integral (and accounting for the fact that $\partial\|\mathbf{r}-\mathbf{r}'\|/\partial z' = z'/\|\mathbf{r}-\mathbf{r}'\| = \cos(\theta)$ is just the cosine of the view angle), we obtain another expression for the Kirchhoff's migration formula:

$$\chi_{e1}(x',z') = \frac{2F}{c^2\pi} \frac{\partial}{\partial z'} \int_{-\infty}^{+\infty} dx \int_{\frac{2\|\mathbf{r}-\mathbf{r}'\|}{c}}^{+\infty} \frac{E_s(x,t)}{\sqrt{t^2 - \frac{4\|\mathbf{r}-\mathbf{r}'\|^2}{c^2}}} dt$$

$$= \frac{2F}{c^2\pi} \frac{\partial}{\partial z'} \int_{-\infty}^{+\infty} dx \int_{0}^{+\infty} \frac{E_s\left(x, u + \frac{2\|\mathbf{r}-\mathbf{r}'\|}{c}\right)}{\sqrt{\left(u + \frac{2\|\mathbf{r}-\mathbf{r}'\|}{c}\right)^2 - \frac{4\|\mathbf{r}-\mathbf{r}'\|^2}{c^2}}} du$$

$$= \frac{2F}{c^2\pi} \frac{\partial}{\partial z'} \int_{-\infty}^{+\infty} dx \int_{0}^{+\infty} \frac{E_s\left(x, u + \frac{2\|\mathbf{r}-\mathbf{r}'\|}{c}\right)}{\sqrt{u\left(u + \frac{4\|\mathbf{r}-\mathbf{r}'\|}{c}\right)}} du = \frac{2F}{c^2\pi} \frac{\partial}{\partial z'} \int_{-\infty}^{+\infty} dx \int_{0}^{+\infty} \frac{1}{\sqrt{t}} \frac{E_s\left(x, t + \frac{2\|\mathbf{r}-\mathbf{r}'\|}{c}\right)}{\sqrt{\left(t + \frac{4\|\mathbf{r}-\mathbf{r}'\|}{c}\right)}} dt$$

$$= \frac{2F}{c^2\pi} \int_{-\infty}^{+\infty} dx \int_{0}^{+\infty} \frac{1}{\sqrt{t}} \frac{\partial}{\partial z'} \frac{E_s\left(x, t + \frac{2\|\mathbf{r}-\mathbf{r}'\|}{c}\right)}{\sqrt{\left(t + \frac{4\|\mathbf{r}-\mathbf{r}'\|}{c}\right)}} dt$$

$$= \frac{4F}{c^3\pi} \int_{-\infty}^{+\infty} dx\cos(\theta) \int_{0}^{+\infty} \frac{1}{\sqrt{t}} \left(\frac{\frac{\partial}{\partial t}E_s\left(x, t + \frac{2\|\mathbf{r}-\mathbf{r}'\|}{c}\right)}{\sqrt{\left(t + \frac{4\|\mathbf{r}-\mathbf{r}'\|}{c}\right)}} - \frac{E_s\left(x, t + \frac{2\|\mathbf{r}-\mathbf{r}'\|}{c}\right)}{\left(t + \frac{4\|\mathbf{r}-\mathbf{r}'\|}{c}\right)^{3/2}} \right) dt \qquad (10.36)$$

sometimes (indeed more often in seismic than in GPR applications) it is used a far-field approximation of Eq. (10.36) that retains only the first term in the parentheses, which decays more slowly than the other one versus the distance $\|\mathbf{r} - \mathbf{r}'\|$. Making use of Heaviside's function, this approximated far-field formula can be written as

$$\chi_{e1}(x',z') = \frac{4F}{c^3\pi} \int_{-\infty}^{+\infty} dx\cos(\theta) \int_{0}^{+\infty} \frac{1}{\sqrt{t}} \frac{\frac{\partial}{\partial t}E_s\left(x, t + \frac{2\|\mathbf{r}-\mathbf{r}'\|}{c}\right)}{\sqrt{\left(t + \frac{4\|\mathbf{r}-\mathbf{r}'\|}{c}\right)}} dt$$

$$= \frac{4F}{c^3\pi} \int_{-\infty}^{+\infty} dx\cos(\theta) \int_{-\infty}^{+\infty} \frac{H(t)}{\sqrt{t}} \frac{\frac{\partial}{\partial t}E_s\left(x, t + \frac{2\|\mathbf{r}-\mathbf{r}'\|}{c}\right)}{\sqrt{\left(t + \frac{4\|\mathbf{r}-\mathbf{r}'\|}{c}\right)}} dt$$

$$= \frac{4F}{c^3\pi} \int_{-\infty}^{+\infty} dx\cos(\theta) \int_{-\infty}^{+\infty} \frac{H\left(\tau - \frac{2\|\mathbf{r}-\mathbf{r}'\|}{c}\right)}{\sqrt{\tau - \frac{2\|\mathbf{r}-\mathbf{r}'\|}{c}}} \frac{\frac{\partial}{\partial t}E_s(x, t = \tau)}{\sqrt{\left(\tau + \frac{2\|\mathbf{r}-\mathbf{r}'\|}{c}\right)}} d\tau \qquad (10.37)$$

At this point, let us note that the integrand is null up to the time instant $\tau = \frac{2\|\mathbf{r}-\mathbf{r}'\|^-}{c}$ whereas, afterward, we have that the piece with $\sqrt{\tau - \frac{2\|\mathbf{r}-\mathbf{r}'\|}{c}}$ is singular at $\tau = \frac{2\|\mathbf{r}-\mathbf{r}'\|^+}{c}$, whereas the piece with $\sqrt{\tau + \frac{2\|\mathbf{r}-\mathbf{r}'\|}{c}}$ is instead quite flat because of the far field hypothesis. So, there is an error vanishing versus $\|\mathbf{r}-\mathbf{r}'\|$ if we introduce the further approximation

$$\sqrt{\tau + \frac{2\|\mathbf{r}-\mathbf{r}'\|}{c}} \approx \left. \sqrt{\tau + \frac{2\|\mathbf{r}-\mathbf{r}'\|}{c}} \right|_{\tau = \frac{2\|\mathbf{r}-\mathbf{r}'\|}{c}} = \sqrt{\frac{4\|\mathbf{r}-\mathbf{r}'\|}{c}} = \frac{2\sqrt{\|\mathbf{r}-\mathbf{r}'\|}}{\sqrt{c}} \qquad (10.38)$$

Under this approximation, we have

$$\chi_{el}(x',z') = \frac{2F}{c^{2.5}\pi} \int\limits_{-\infty}^{+\infty} dx \frac{\cos(\theta)}{\sqrt{\|\mathbf{r}-\mathbf{r}'\|}} \int\limits_{-\infty}^{+\infty} \frac{H\left(\tau - \frac{2\|\mathbf{r}-\mathbf{r}'\|}{c}\right)}{\sqrt{\tau - \frac{2\|\mathbf{r}-\mathbf{r}'\|}{c}}} \frac{\partial}{\partial t} E_s(x,t=\tau)\, d\tau \qquad (10.39)$$

The integral in $d\tau$ is the convolution product between the two functions $\frac{H(t)}{\sqrt{t}}$ and $\frac{\partial}{\partial t}E_s(x,t)$, calculated in the point $t = 2\|\mathbf{r}-\mathbf{r}'\|/c$. This convolution is equal to the so-called "half-derivative" of the electric field (see Appendix B), so that Eq. (10.37) can be also written with a more compact notation as (Gazdag and Squazzero, 1984)

$$\chi_{el}(x',z') = \frac{2F}{c^{2.5}\pi} \int\limits_{-\infty}^{+\infty} dx \frac{\cos(\theta)}{\sqrt{\|\mathbf{r}-\mathbf{r}'\|}} \frac{\partial^{0.5}}{\partial^{0.5}t} E_s\left(x, t = \frac{2\|\mathbf{r}-\mathbf{r}'\|}{c}\right) \qquad (10.40)$$

In order to compare the obtained formulas (both in frequency and in time domain) with the "classical" formulas available in the literature, a brief aside is still needed. In particular, in the classical migration algorithms derived from the seismic processing (Schneider, 1978), the unknown looked for is not the contrast but the scattered field in the time domain in the buried investigated points $(E_s(x',z',0))$ starting from the scattered field in the observation line in air $(E_s(x',0,t))$. This involves the fixing of a conventional zero time at the instant when the buried point begins its re-scattering of the energy impinging on it. Of course, the buried targets are not illuminated at the same instant, because of the finite velocity of propagation of the incident field radiated by the primary sources, and this is "amortized" by considering waves that "start" from the buried points within the investigation domain at the same instant but propagate at one-half the actual propagation velocity of the electromagnetic waves in the soil. Such a formulation is not rigorous in the framework of Maxwell's equations, because the propagation velocity is univocally determined by the same Maxwell's equations. So, we have recast (as often is done in the literature on inverse problems) the problem in terms of contrast functions. In particular, the contrast does not represent the "temporal origin" of the scattered field, but rather its "causal origin." That said, the interested reader can compare the retrieved migration formulas in frequency domain with Eqs. (51) and (52) in (Stolt, 1978) and the retrieved migration formulas in time domain with Eq. (12) in Schneider (1978): He/she will see that, if the spectral weight is neglected, then the integrals are the same.

The Stolt's and the Kirchhoff's migrations are equivalent in theory, but the Kirchhoff's formulation allows us to impose more immediately some regularization. In particular, when performing the numerical integration of the data along the x-axis, we can limit the integration interval on the basis of the wideness of the visible diffraction curves. This implicitly means to choose the extension of the spectral retrievable set in relationship to the maximum effective view angle. Several commercial codes for GPR data processing allow us to choose the number of traces to be taken into account when performing the Kirchhoff's migration (but usually not when performing the Stolt's migration), which is equivalent to limiting the extremes for the integration in dx.

QUESTIONS

1. Do 2D migration algorithms account for all the aspects of the 2D electromagnetic scattering?
2. Are 2D migration algorithms directly derived by the Born approximation?
3. Is the air–soil interface accounted for in migration algorithms?
4. Can the radiation pattern of the antennas be partially (roughly) accounted for within a commercial migration code for migration, even if not explicitly enclosed in the available menus?
5. Is it always true that to account for the spectral weight improves the result?
6. Suppose we perform a GPR prospecting crossing a buried empty crypt. The crypt is at the depth of 1 m, and its cross section is about 2 m long and 3 m deep. The soil shows a relative permittivity equal to 9, no magnetic properties are present, and the losses are negligible. The B-scan is centered on the crypt and is 9 m long. The central frequency of the antennas is 500 MHz. What apparent size would show the room after being focused by means of a standard migration algorithm accounting for the propagation velocity of the surrounding soil? Suppose now that the room is filled with fresh water and that the water is so pure (low concentrations of salts) that it allows the penetration of the GPR signal until the bottom and then back until the antennas. What apparent size would provide the standard migration algorithm in this case?

11

THREE-DIMENSIONAL SCATTERING EQUATIONS

Lorenzo Lo Monte, Raffaele Persico, and Raffaele Solimene

11.1 SCATTERING IN THREE DIMENSIONS: REDEFINITION OF THE MAIN SYMBOLS

So far, the discussion has been focused on 2D inverse scattering. This is the most common model in GPR data processing. However, recent advances in distributed GPR, coherent GPR, HF GPR, and RF tomography led to an extension of classical 2D work in a more proper 3D scenario. The basic principles hold the same, but the formulas become more complicated.

To avoid confusion, we will differentiate between 2D and 3D geometries by renaming our previous variables with different symbols. In this chapter:

- $\mathbf{E}$ is a 3D complex vector representing the electric field; similarly, $\mathbf{E}_{inc}$ will be the incident field and $\mathbf{E}_s$ will be the scattered field.
- $\mathbf{r}_s = x_s i_x + y_s i_y + z_s i_z$ is the source point. Since we are dealing with 3D space, the source must also have an orientation vector, which is the direction of the effective length of the transmitting antenna; this vector is labeled $\hat{\mathbf{a}}$ (see Figure 11.1).

Introduction to Ground Penetrating Radar: Inverse Scattering and Data Processing,
First Edition. Raffaele Persico.

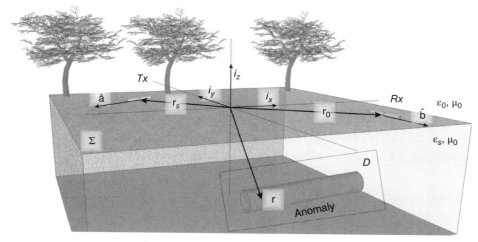

Figure 11.1. Geometry of the problem.

- $\mathbf{r}_0 = x_0 i_x + y_0 i_y + z_0 i_z$ is the receiver point. Since we are dealing with 3D space, the receiver must also have an orientation vector, which is the direction of the effective length of the receiving antenna; this vector is labeled as $\hat{\mathbf{b}}$ (see Figure 11.1).

- $\mathbf{r} = x i_x + y i_y + z i_z$ is a generic point within the investigation domain, beneath the air–soil interface.

Capitalized, bold, and underlined quantities describe dyadics, which are represented by 3×3 matrices as described in Tai, (1991). Bold variables represent vectors, and circumflexed vectors are real-valued with unitary norm, generally representing directions.

Within the proposed 3D discussion, we will not consider the case of magnetic anomalies, nor will we consider the case of magnetic soils.

In general, a set of $n = 1, \ldots, N$ field measurements are collected: A distinct nth measurement is obtained by varying one of the following:

1. transmitter position $\mathbf{r}_s$
2. transmitter direction $\hat{\mathbf{a}}$ (unit norm)
3. receiver position $\mathbf{r}_0$
4. receiver direction $\hat{\mathbf{b}}$ (unit norm)
5. angular frequency $\omega = 2\pi f$

Targets reside only within a volumetric region D named *investigation domain*. The electric field $\mathbf{E}$ [V/m], the magnetic field $\mathbf{H}$ [A/m], the electric current density $\mathbf{J}$ [A/m^2], and the magnetic current density $\mathbf{M}$ [V/m^2] are all expressed in phasor form using the exp $(j\omega t)$ convention (Lo Monte et al., 2010).

Please note that in Figure 11.1 the z-axis is oriented versus the air, unlike the convention followed for the 2D dealing (see Figure 4.1).

11.2 THE SCATTERING EQUATIONS IN 3D

Due to the linearity of Maxwell's equations, we can state that the 2D scalar scattering equations (4.44) and (4.45) evolve into a more general formulation given by

$$\mathbf{E}(\mathbf{r},\mathbf{r}_s) = \mathbf{E}_{inc}(\mathbf{r},\mathbf{r}_s) + k_s^2 \iint_D \chi_e(\mathbf{r}')\mathbf{G}_i(\mathbf{r},\mathbf{r}')\mathbf{E}(\mathbf{r}',\mathbf{r}_s)\, d\mathbf{r}', \qquad \mathbf{r} \in D, \mathbf{r}_s \in \Sigma \qquad (11.1)$$

$$\mathbf{E}_s(\mathbf{r}_0,\mathbf{r}_s) = k_s^2 \iint_D \chi_e(\mathbf{r}')\mathbf{G}_e(\mathbf{r}_0,\mathbf{r}')\mathbf{E}(\mathbf{r}',\mathbf{r}_s)\, d\mathbf{r}', \qquad \mathbf{r}_0, \mathbf{r}_s \in \Sigma \qquad (11.2)$$

where, D is the investigation domain embedded in the soil and Σ is the observation domain in air, as depicted in Figure 11.1. The product between the dyadic Green's functions and the field is implemented as the usual product between a 3×3 matrix times a column vector with three elements. The result is a column vector of three elements that are the components of the vector result along the axes x, y, and z.

So, to specify the 3D scattering equations (11.1) and (11.2) means to find the expressions of Green's functions and (for a given and characterized source) of the incident field.

11.3 THREE-DIMENSIONAL GREEN'S FUNCTIONS

Due to the increased complexity of the 3D dealing, it is useful to generalize the definition of Green's function. First of all, let us remind that the Green's function is the response to a spatially impulsive source as has been shown in 2D. In 3D, any Green's function is dyadic and thus can be expressed by means of a general matrix scheme of the following kind:

$$\mathbf{G} = \begin{pmatrix} G_{xx} & G_{xy} & G_{xz} \\ G_{yx} & G_{yy} & G_{yz} \\ G_{zx} & G_{zy} & G_{zz} \end{pmatrix} \qquad (11.3)$$

Each element of the matrix is a function of the source point, the observation point, and the frequency. The dyadic nature of Green's function corresponds to the fact that the source and the resulting electric field in general are not parallel to each other, unlike in the scalar 2D case. It is easily recognized that the generic element of the matrix G_{hk} ($h, k = x, y, z$) represents the component along i_h of the response to a spatially impulsive source directed along i_k.

It may be useful to warn the reader about the fact that in Eqs. (11.1) and (11.2) the Green's functions have dimensions of m^{-1}, whereas in the scalar 2D equations (4.44) and (4.45) the Green's functions were dimensionless. Of course, this results from the fact that in the 3D case the scattering equations are based on volume integrals instead of surface integrals.

Let us also remind that the internal Green's function represents the impulsive response of the system with both source and observation points embedded in the soil, whereas the external Green's function is the impulsive response of the system with source point embedded in the soil and observation point in air (see Chapter 4). This can be better specified by using the subscripts "a" (which stands for "air") and "s" (which stands for "soil") as follows:

$$\mathbf{G}_{ss}(\mathbf{r},\mathbf{r}') = \mathbf{G}_i(\mathbf{r},\mathbf{r}') \qquad (z,z'<0),$$
$$\mathbf{G}_{sa}(\mathbf{r},\mathbf{r}') = \mathbf{G}_e(\mathbf{r},\mathbf{r}') \qquad (z>0,z'<0) \tag{11.4}$$

It is useful to consider also the two additional quantities

$$\mathbf{G}_{aa}(\mathbf{r},\mathbf{r}') = \text{internal Green's function in air} \qquad (z,z'>0),$$
$$\mathbf{G}_{as}(\mathbf{r},\mathbf{r}') = \text{external air–soil Green's function} \qquad (z<0,z'>0) \tag{11.5}$$

In particular, the dual quantities of Eq. (11.5) represent the Green's function of a formally "mirrored" inverse scattering problem with virtual targets embedded in air and a virtual GPR moving underground. Moreover, it is also useful to consider the degenerated case where the two involved half-spaces have the same electromagnetic characteristics, so that one is actually considering the targets and the GPR embedded in a homogeneous space. In this inverse problem, we cannot distinguish any longer an internal and an external Green's function, but there is just a unique Green's function of the medium. Thus, from now on we will label

$$\mathbf{G}_a(\mathbf{r},\mathbf{r}') = \text{Green's function of a homogeneous space made up of air},$$
$$\mathbf{G}_s(\mathbf{r},\mathbf{r}') = \text{Green's function of a homogeneous space made up of soil} \tag{11.6}$$

Later on in this chapter, we will see that the half-space Green's functions (11.4)–(11.5) can be expressed versus the homogeneous Green's functions of Eq. (11.6). Actually, this happens also in 2D, even if we didn't need to specify the homogeneous Green's function in that case.

11.4 THE INCIDENT FIELD

For the three-dimensional case, we will assume a spatially impulsive source directed along the unitary vector $\hat{\mathbf{a}}$, which means that the primary source is described as

$$\mathbf{J}(\mathbf{r}') = I_0 l \delta(\mathbf{r}' - \mathbf{r}_s)\hat{\mathbf{a}} \tag{11.7}$$

where I_0 is the current flowing into the dipole and l is its length. Please note that, unlike Chapter 4, we will now make use of the common synthetic notation for the delta function with respect to vector variables, so that, by definition, we mean

$$\delta(\mathbf{r}' - \mathbf{r}_s) \equiv \delta(x' - x_s)\delta(y' - y_s)\delta(z' - z_s) \tag{11.8}$$

In order to calculate the incident field due to an impulsive primary source, we can start from Eq. (11.2). In particular, based on Eq. (4.20), on Eq. (11.4), and on the definition of the dielectric contrast [see Eq. (4.1)], we can rewrite Eq. (11.2) as

$$
\begin{aligned}
\mathbf{E}_s(\mathbf{r}_0,\mathbf{r}_s) &= \frac{k_s^2}{j\omega}\iiint_D \mathbf{G}_e(\mathbf{r}_0,\mathbf{r}')j\omega\chi_e(\mathbf{r}')\mathbf{E}(\mathbf{r}',\mathbf{r}_s)\, d\mathbf{r}' \\
&= \frac{k_s^2}{j\omega\varepsilon_0\varepsilon_{sr}}\iiint_D \mathbf{G}_{sa}(\mathbf{r}_0,\mathbf{r}')j\omega\Delta\varepsilon(\mathbf{r}')\mathbf{E}(\mathbf{r}',\mathbf{r}_s)\, d\mathbf{r}' \\
&= \frac{k_s^2}{j\omega\varepsilon_0\varepsilon_{sr}}\iiint_D \mathbf{G}_{sa}(\mathbf{r}_0,\mathbf{r}')\mathbf{J}_{eq}(\mathbf{r}')\, d\mathbf{r}' = -j\omega\mu_0\iiint_D \mathbf{G}_{sa}(\mathbf{r}_0,\mathbf{r}')\mathbf{J}_{eq}(\mathbf{r}')\, d\mathbf{r}', \quad \mathbf{r}_0,\mathbf{r}_s \in \Sigma
\end{aligned}
\tag{11.9}
$$

At this point, let us note that Maxwell's equations are the same for both equivalent (secondary) and actual (primary) sources. Consequently, Eq. (11.9) also provides the electric field evaluated in air due to a generic current distribution embedded in the soil, which can be rewritten as

$$\mathbf{E}(\mathbf{r}_0) = -j\omega\mu_0\iiint_D \mathbf{G}_{sa}(\mathbf{r}_0,\mathbf{r}')\mathbf{J}(\mathbf{r}')d\mathbf{r}', \qquad \mathbf{r}_0 \text{ in air}, \mathbf{r}' \in D \text{ embedded in the soil} \tag{11.10}$$

For the sake of symmetry, the electric field in the soil due to a generic current distribution embedded in air is given by

$$\mathbf{E}(\mathbf{r}_0) = -j\omega\mu_0\iiint_D \mathbf{G}_{as}(\mathbf{r}_0,\mathbf{r}')\mathbf{J}(\mathbf{r}')d\mathbf{r}', \qquad \mathbf{r}_0 \text{ in the soil}, \mathbf{r}' \in D \text{ embedded in air} \tag{11.11}$$

In the special case of the spatially impulsive source given by Eq. (11.7), Eq. (11.11) provides the incident field relative to an impulsive source, that is therefore achieved substituting Eq. (11.7) into Eq. (11.11). The result is

$$\mathbf{E}_{inc}(\mathbf{r}) = -j\mu_0\omega I_0 l \mathbf{G}_{as}(\mathbf{r},\mathbf{r}_s)\cdot\hat{\mathbf{a}}, \qquad z < 0 \tag{11.12}$$

A similar relationship holds also in 2D, and the reader can retrieve it as an exercise starting from Eq. (4.48).

Equation (11.12) provides the incident field to be put into the scattering equation. By a similar derivation, the incident field in air is provided by

$$\mathbf{E}_{inc}(\mathbf{r}) = -j\mu_0\omega I_0 l\mathbf{G}_{aa}(\mathbf{r},\mathbf{r}_s)\cdot\hat{\mathbf{a}}, \qquad z>0 \tag{11.13}$$

Often, in 3D cases, rather than the vector incident field, it is important to consider the component along the effective length of the receiving antenna of this incident field. In this text, we will suppose that the receiving antenna is a Hertzian dipole as the transmitting one. Moreover, the derivation will be simplified in the fact that we will not consider the effective length of the receiving dipole but just its direction $\hat{\mathbf{b}}$. This means that the formal expression of the recorded incident field datum in the observation point (i.e., the direct coupling between the antennas) is given by

$$E_{inc}(\mathbf{r}_0) = -j\omega\mu_0 I_0 l\hat{\mathbf{b}}\cdot\mathbf{G}_{aa}\cdot\hat{\mathbf{a}}, \qquad z>0 \tag{11.14}$$

In Eq. (11.14) we have used the scalar symbol E_{inc} to indicate the projection of the vector $\mathbf{E}_{inc}$ along the unitary vector $\hat{\mathbf{b}}$. In terms of vector–matrix product, the product in Eq. (11.14) is to be meant as a raw vector (1×3) times a matrix (3×3) times a column vector (3×1). The result is clearly a scalar quantity.

The provided dealing shows that, in order to specify the scattering equations, we just need to find Green's functions $\mathbf{G}_{as}$, $\mathbf{G}_{sa}$ and $\mathbf{G}_{ss}$, and in order to specify the incident field everywhere (which means to retrieve the theoretical expression of the direct coupling between the antennas) we have to find $\mathbf{G}_{aa}$. Let us emphasize that this is because we have chosen to model both the transmitting and receiving antennas as two ideal dipole-like ones. If more complicated antennas are considered, the problem of the behavior of the currents on the antennas has to be considered too. In 2D we have seen that this can be done by means of the plane wave spectra (or equivalently the effective lengths) of the antennas, but in the 3D discussion we will avoid this complication, limiting to the case of Hertzian dipoles both in transmission and in reception.

11.5 HOMOGENEOUS 3D GREEN'S FUNCTIONS

As said, the calculation of the half-space Green's functions requires the preliminary calculation of the Green's function of a homogeneous space. So, in this section the expressions of the homogeneous Green's functions $\mathbf{G}_a$ and $\mathbf{G}_s$ are retrieved. Of course, the calculations are the same in the two cases, with just the wavenumbers being different. So, let us focus on the homogeneous Green's function in air $\mathbf{G}_a$.

In the same way followed for a half-space geometry, the dyadic Green's function under study $\mathbf{G}_a(\mathbf{r},\mathbf{r}')$ relates the electric field $\mathbf{E}(\mathbf{r})$ to the electric current density $\mathbf{J}(\mathbf{r}')$ that generated it (either real or equivalent) in free space via the integral relationship:

$$\mathbf{E}(\mathbf{r}) = -j\omega\mu_0\iiint_D \mathbf{G}_a(\mathbf{r},\mathbf{r}')\cdot\mathbf{J}(\mathbf{r}') \, d\mathbf{r}' \tag{11.15}$$

Manipulating the Maxwell's equations as done in Chapter 4 [see, in particular, Section 4.2 and Eq. (4.23)], we have that free space $\mathbf{E}(\mathbf{r})$ is a solution of the vector equation:

$$\nabla \times \nabla \times \mathbf{E}(\mathbf{r}) - k_0^2 \mathbf{E}(\mathbf{r}) = -j\omega\mu_0 \mathbf{J}(\mathbf{r}) \tag{11.16}$$

Substituting Eq. (11.15) into Eq. (11.16), deriving under the integral, and accounting for the fact that the achieved equality holds for any current density (in particular for spatially impulsive currents) we obtain the following differential equation for Green's function[1]:

$$\nabla \times \nabla \times \mathbf{G}_a(\mathbf{r},\mathbf{r}') - k_0^2 \mathbf{G}_a(\mathbf{r},\mathbf{r}') = \mathbf{I}\delta(\mathbf{r}-\mathbf{r}') \tag{11.17}$$

where $\mathbf{I}$ is the 3×3 identity matrix, whose elements are equal to 1 along the diagonal and zero elsewhere. The curl operator on a matrix is meant to be performed on the columns; that is, $\nabla \times \nabla \times \mathbf{G}(\mathbf{r},\mathbf{r}')$ is a 3×3 matrix where each column is given by the curl of the curl of the homologous column of $\mathbf{G}(\mathbf{r},\mathbf{r}')$. The spatial derivations are meant with regard to $\mathbf{r}$.

To find the Green's function $\mathbf{G}_a(\mathbf{r},\mathbf{r}')$, we can use the vector potential theory, in a way formally similar to that followed in Chapter 5 with regard to the magnetic anomalies. Actually, the vector potential theory is fundamental in electromagnetism; a more complete description of this theory can be found in Chew (1995), Balanis (1989), Van Bladel (2007), Jackson (1998), and Tai (1994).

Here we can start from the Maxwell's equations given in Eq. (4.5). Since no magnetic source (neither magnetic anomaly) is involved, Eq. (4.5) can be rewritten as follows:

$$\begin{aligned}
\nabla \times \mathbf{E} &= -j\omega\mu_0 \mathbf{H}, \\
\nabla \times \mathbf{H} &= j\omega\varepsilon_0 \mathbf{E} + \mathbf{J}, \\
\nabla \cdot \varepsilon_0 \mathbf{E} &= \rho, \\
\nabla \cdot \mu_0 \mathbf{H} &= 0
\end{aligned} \tag{11.18}$$

The fourth Maxwell's equation $\nabla \cdot \mu_0 \mathbf{H} = 0$ implies that $\mu_0 \mathbf{H}$ (and thus $\mathbf{H}$, being μ_0 a constant quantity) is solenoidal; therefore, it can be represented as the curl of another vector function, called "potential vector" (let us recall the vector identity $\nabla \times \nabla \cdot \mathbf{A} = 0$). Hence, we can assume that

$$\mu_0 \mathbf{H} = \nabla \times \mathbf{A}(\mathbf{r}) \tag{11.19}$$

where $\mathbf{A}(\mathbf{r})$ is the vector potential. Consequently, the first Maxwell's equations in (11.18) becomes

$$\nabla \times \mathbf{E} = -j\omega\nabla \times \mathbf{A}(\mathbf{r}) \Leftrightarrow \nabla \times (\mathbf{E} + j\omega\mathbf{A}(\mathbf{r})) = 0 \tag{11.20}$$

---

[1] Actually Eq. (11.17) is a general condition that regards Green's function in any geometry, not only in free space.

As is well known (Franceschetti, 1997), Eq. (11.20) ensures the presence of a scalar potential function $\phi(\mathbf{r})$ such as

$$\mathbf{E} = -j\omega\mathbf{A}(\mathbf{r}) - \nabla\phi(\mathbf{r}) \tag{11.21}$$

Substituting Eq. (11.21) in the second of the Maxwell's equations (11.18), we obtain

$$
\begin{aligned}
\nabla \times \nabla \times \mathbf{A}(\mathbf{r}) &= \omega^2 \varepsilon_0 \mu_0 \mathbf{A}(\mathbf{r}) - j\omega\varepsilon_0\mu_0 \nabla\phi(\mathbf{r}) + \mu_0 \mathbf{J}(\mathbf{r}) \\
&= k_0^2 \mathbf{A}(\mathbf{r}) - j\omega\varepsilon_0\mu_0 \nabla\phi(\mathbf{r}) + \mu_0 \mathbf{J}(\mathbf{r})
\end{aligned}
\tag{11.22}
$$

Similar to what is done in Chapter 4 (see Eqs. (4.23) and (4.24)), at this point we exploit the definition of the Laplacian vector, namely $\nabla^2 \vec{v} = \nabla(\nabla \cdot \vec{v}) - \nabla \times \nabla \times \vec{v}$. Applied in Eq. (11.22), this leads to

$$\nabla^2 \mathbf{A}(\mathbf{r}) + k_0^2 \mathbf{A}(\mathbf{r}) = \nabla(\nabla \cdot \mathbf{A} + j\omega\varepsilon_0\mu_0\phi(\mathbf{r})) - \mu_0 \mathbf{J}(\mathbf{r}) \tag{11.23}$$

In the same way shown in Chapter 4 with regard to the Fitzgerald vector, also in this case we have that neither the potential vector nor the scalar potential function are univocally determined, and thus also in this case we can exploit this indeterminateness in order to impose the gauge of Lorentz, which means to choose a priori the couple of potential functions that satisfies the condition

$$\nabla \cdot \mathbf{A} + j\omega\varepsilon_0\mu_0\phi(\mathbf{r}) = 0 \tag{11.24}$$

which reduces Eq. (11.23) to a Helmholtz equation as follows:

$$\nabla^2 \mathbf{A}(\mathbf{r}) + k_0^2 \mathbf{A}(\mathbf{r}) = -\mu_0 \mathbf{J}(\mathbf{r}) \tag{11.25}$$

Moreover, considering the divergence of Eq. (11.21) and substituting in the third Maxwell's equation (11.18), we obtain

$$\nabla \cdot \mathbf{E} = -j\omega\nabla \cdot \mathbf{A}(\mathbf{r}) - \nabla \cdot \nabla\phi(\mathbf{r}) = \frac{\rho}{\varepsilon_0} \Leftrightarrow \nabla^2\phi(\mathbf{r}) + j\omega\nabla \cdot \mathbf{A}(\mathbf{r}) = -\frac{\rho}{\varepsilon_0} \tag{11.26}$$

In Eq. (11.26) we have accounted for the fact that the scalar Laplacian operator is defined as the divergence of the gradient. Substituting Eq. (11.24) in Eq. (11.26), we obtain

$$\nabla^2\phi(\mathbf{r}) + k_0^2\phi(\mathbf{r}) = -\frac{\rho(\mathbf{r})}{\varepsilon_0} \tag{11.27}$$

So, under the gauge of Lorentz, also the scalar potential is the solution of a Helmholtz equation. The calculation steps for solving Eqs. (11.25) and (11.27) are reported in Appendix C. The solution is given by

$$\mathbf{A}(\mathbf{r}) = \mu_0 \iiint\limits_D \frac{\exp(-jk_0\|\mathbf{r}-\mathbf{r}'\|)}{4\pi\|\mathbf{r}-\mathbf{r}'\|} \mathbf{J}(\mathbf{r}') \, d\mathbf{r}' \qquad (11.28)$$

$$\phi(\mathbf{r}) = \frac{1}{\varepsilon_0} \iiint\limits_D \frac{\exp(-jk_0\|\mathbf{r}-\mathbf{r}'\|)}{4\pi\|\mathbf{r}-\mathbf{r}'\|} \rho(\mathbf{r}') \, d\mathbf{r}' \qquad (11.29)$$

Substituting (11.28) and (11.29) into (11.21), the electric field is given by

$$\mathbf{E}(\mathbf{r}) = -j\omega\mu_0 \iiint\limits_D \frac{\exp(-jk_0\|\mathbf{r}-\mathbf{r}'\|)}{4\pi\|\mathbf{r}-\mathbf{r}'\|} \mathbf{J}(\mathbf{r}') \, d\mathbf{r}' - \frac{1}{\varepsilon_0} \nabla \left(\iiint\limits_D \frac{\exp(-jk_0\|\mathbf{r}-\mathbf{r}'\|)}{4\pi\|\mathbf{r}-\mathbf{r}'\|} \rho(\mathbf{r}') \, d\mathbf{r}' \right)$$

$$(11.30)$$

Making use of the continuity equation

$$\nabla\cdot\mathbf{J} + j\omega\rho = 0 \Leftrightarrow \rho(\mathbf{r}) = -\frac{\nabla\cdot\mathbf{J}(\mathbf{r})}{j\omega} \qquad (11.31)$$

we have

$$\mathbf{E}(\mathbf{r}) = -j\omega\mu_0 \iiint\limits_D \frac{\exp(-jk_0\|\mathbf{r}-\mathbf{r}'\|)}{4\pi\|\mathbf{r}-\mathbf{r}'\|} \mathbf{J}(\mathbf{r}') \, d\mathbf{r}' + \frac{1}{j\omega\varepsilon_0} \nabla \left(\iiint\limits_D \frac{\exp(-jk_0\|\mathbf{r}-\mathbf{r}'\|)}{4\pi\|\mathbf{r}-\mathbf{r}'\|} \nabla'\cdot\mathbf{J}(\mathbf{r}') \, d\mathbf{r}' \right)$$

$$(11.32)$$

The gradient symbol in the second integral in Eq. (11.32) is primed because the derivations refer to the variable $\mathbf{r}'$. Performing the integration of the second term by parts, we achieve

$$\mathbf{E}(\mathbf{r}) = -j\omega\mu_0 \iiint\limits_D \frac{\exp(-jk_0\|\mathbf{r}-\mathbf{r}'\|)}{4\pi\|\mathbf{r}-\mathbf{r}'\|} \mathbf{J}(\mathbf{r}') \, d\mathbf{r}' + \frac{1}{j\omega\varepsilon_0} \nabla \left(\iiint\limits_D \nabla' \left(\frac{\exp(-jk_0\|\mathbf{r}-\mathbf{r}'\|)}{4\pi\|\mathbf{r}-\mathbf{r}'\|} \right) \cdot \mathbf{J}(\mathbf{r}') \, d\mathbf{r}' \right)$$

$$(11.33)$$

Now, it is easy to see that

$$\nabla' \frac{\exp(-jk_0\|\mathbf{r}-\mathbf{r}'\|)}{4\pi\|\mathbf{r}-\mathbf{r}'\|} = -\nabla \frac{\exp(-jk_0\|\mathbf{r}-\mathbf{r}'\|)}{4\pi\|\mathbf{r}-\mathbf{r}'\|} \qquad (11.34)$$

so that we have

$$\mathbf{E}(\mathbf{r}) = -j\omega\mu_0 \iiint\limits_D \frac{\exp(-jk_0\|\mathbf{r}-\mathbf{r}'\|)}{4\pi\|\mathbf{r}-\mathbf{r}'\|} \mathbf{J}(\mathbf{r}') \, d\mathbf{r}' + \frac{1}{j\omega\varepsilon_0} \nabla\nabla \cdot \left(\iiint\limits_D \frac{\exp(-jk_0\|\mathbf{r}-\mathbf{r}'\|)}{4\pi\|\mathbf{r}-\mathbf{r}'\|} \mathbf{J}(\mathbf{r}') \, d\mathbf{r}' \right)$$

$$(11.35)$$

Equation (11.35) allows us to express the dyadic Green's function. In fact, it can be compacted in the form

$$\mathbf{E}(\mathbf{r}) = -j\omega\mu_0 \iiint_D \mathbf{G}_a(\mathbf{r},\mathbf{r}')\mathbf{J}(\mathbf{r}')\,d\mathbf{r}' \tag{11.36}$$

on condition that

$$\mathbf{G}_a(\mathbf{r},\mathbf{r}') = \left(\mathbf{I} + \frac{1}{k_0^2}\nabla\nabla\right)\left(\frac{\exp(-jk_0\|\mathbf{r}-\mathbf{r}'\|)}{4\pi\|\mathbf{r}-\mathbf{r}'\|}\right) \tag{11.37}$$

where I is the identity matrix 3×3. Developing the spatial derivations in Eq. (11.37), the explicit expression of the terms of the free space dyadic Green's function, according to Eq. (11.3), are given by

$$G_{axx} = \frac{\exp(-jk_0\|\mathbf{r}-\mathbf{r}'\|)}{4\pi k_0^2}\left[\frac{3(x-x')^2}{\|\mathbf{r}-\mathbf{r}'\|^5} + j\frac{3(x-x')^2 k_0}{\|\mathbf{r}-\mathbf{r}'\|^4} - \frac{(x-x')^2 k_0^2 + 1}{\|\mathbf{r}-\mathbf{r}'\|^3} - j\frac{k_0}{\|\mathbf{r}-\mathbf{r}'\|^2} + j\frac{k_0^2}{\|\mathbf{r}-\mathbf{r}'\|}\right]$$
$$\tag{11.38}$$

$$G_{axy} = \frac{\exp(-jk_0\|\mathbf{r}-\mathbf{r}'\|)}{4\pi k_0^2}\left[\frac{3(x-x')(y-y')}{\|\mathbf{r}-\mathbf{r}'\|^5} + j\frac{3(x-x')(y-y')k_0}{\|\mathbf{r}-\mathbf{r}'\|^4} - \frac{(x-x')(y-y')k_0^2}{\|\mathbf{r}-\mathbf{r}'\|^3}\right]$$
$$\tag{11.39}$$

$$G_{axz} = \frac{\exp(-jk_0\|\mathbf{r}-\mathbf{r}'\|)}{4\pi k_0^2}\left[\frac{3(x-x')(z-z')}{\|\mathbf{r}-\mathbf{r}'\|^5} + j\frac{3(x-x')(z-z')k_0}{\|\mathbf{r}-\mathbf{r}'\|^4} - \frac{(x-x')(z-z')k_0^2}{\|\mathbf{r}-\mathbf{r}'\|^3}\right]$$
$$\tag{11.40}$$

$$G_{ayx} = G_{axy} \tag{11.41}$$

$$G_{ayy} = \frac{\exp(-jk_0\|\mathbf{r}-\mathbf{r}'\|)}{4\pi k_0^2}\left[\frac{3(y-y')^2}{\|\mathbf{r}-\mathbf{r}'\|^5} + j\frac{3(y-y')^2 k_0}{\|\mathbf{r}-\mathbf{r}'\|^4} - \frac{(y-y')^2 k_0^2 + 1}{\|\mathbf{r}-\mathbf{r}'\|^3} - j\frac{k_0}{\|\mathbf{r}-\mathbf{r}'\|^2} + j\frac{k_0^2}{\|\mathbf{r}-\mathbf{r}'\|}\right]$$
$$\tag{11.42}$$

$$G_{ayz} = \frac{\exp(-jk_0\|\mathbf{r}-\mathbf{r}'\|)}{4\pi k_0^2}\left[\frac{3(y-y')(z-z')}{\|\mathbf{r}-\mathbf{r}'\|^5} + j\frac{3(y-y')(z-z')k_0}{\|\mathbf{r}-\mathbf{r}'\|^4} - \frac{(y-y')(z-z')k_0^2}{\|\mathbf{r}-\mathbf{r}'\|^3}\right]$$
$$\tag{11.43}$$

$$G_{azx} = G_{axz} \tag{11.44}$$

$$G_{azy} = G_{ayz} \tag{11.45}$$

$$G_{azz} = \frac{\exp(-jk_0\|\mathbf{r}-\mathbf{r}'\|)}{4\pi k_0^2}\left[\frac{3(z-z')^2}{\|\mathbf{r}-\mathbf{r}'\|^5} + j\frac{3(z-z')^2 k_0}{\|\mathbf{r}-\mathbf{r}'\|^4} - \frac{(z-z')^2 k_0^2+1}{\|\mathbf{r}-\mathbf{r}'\|^3} - j\frac{k_0}{\|\mathbf{r}-\mathbf{r}'\|^2} + j\frac{k_0^2}{\|\mathbf{r}-\mathbf{r}'\|}\right]$$

$$(11.46)$$

11.6 THE PLANE WAVE SPECTRUM OF A 3D HOMOGENEOUS GREEN'S FUCNTION

We have seen that the three columns of Green's function represent three electric fields, and more precisely they are, apart from an unessential factor, the fields (calculated in the point $\mathbf{r} = (x,y,z)$) generated by an impulsive source (located in the point $\mathbf{r}' = (x',y',z')$) oriented along the x-, y-, and z-axis, respectively. Therefore, each of the columns can be expressed by means of its plane wave spectrum according to the calculations shown in Chapter 4 (see Section 4.4), and consequently also the overall dyadic Green's function can be expanded as an integral sum of plane waves. In particular, choosing the plane $z = z'$ as reference, analogously to Eq. (4.60), we can define the plane wave spectrum of Green's function as[2]

$$\hat{\mathbf{G}}_a(u,v,\mathbf{r}') = \int_{-\infty}^{+\infty}\int_{-\infty}^{+\infty} \mathbf{G}_a(x,y,z\to z',x',y',z')\exp(-jux)\exp(-jvy)\, dxdy \qquad (11.47)$$

However, the integral (11.47) is not immediate, because of the singular behavior of Green's function for $\mathbf{r} = \mathbf{r}'$. So, specific passages are required as illustrated in the following. In particular, let us start from the three-dimensional spatial Fourier transform of Eq. (11.17), calculated in three independent variables u, v, w. So, we retrieve the equality

$$\int_{-\infty}^{+\infty}\int_{-\infty}^{+\infty}\int_{-\infty}^{+\infty} \left(\nabla\times\nabla\times\mathbf{G}_a(\mathbf{r},\mathbf{r}') - k_0^2\mathbf{G}_a(\mathbf{r},\mathbf{r}')\right)\exp(-j(ux+vy+wz))\, dxdydz$$

$$(11.48)$$

$$= \int_{-\infty}^{+\infty}\int_{-\infty}^{+\infty}\int_{-\infty}^{+\infty} (\mathbf{I}\delta(\mathbf{r}-\mathbf{r}'))\exp(-j(ux+vy+wz))\, dxdydz$$

Let us label as $\mathbf{k}$ the symbolic vector $u i_x + v i_y + w i_z$, and let us label as $\hat{\hat{\mathbf{G}}}_a(\mathbf{k},\mathbf{r}')$ the three-dimensional transform of $\mathbf{G}_a(\mathbf{r},\mathbf{r}')$. It is straightforward to recognize that the nabla vector translates into $j\mathbf{k}$ in the transformed domain and that the formal vector product (which indicates the curl vector in the spatial domain) translates into an "authentic" vector product in the transformed domain. Consequently, we have

$$-\mathbf{k}\times\mathbf{k}\times\hat{\hat{\mathbf{G}}}_a(\mathbf{k},\mathbf{r}') - k_0^2\hat{\hat{\mathbf{G}}}_a(\mathbf{k},\mathbf{r}') = \mathbf{I}\exp(-j\mathbf{k}\cdot\mathbf{r}') \qquad (11.49)$$

---

[2] The reason why, unlike the 2D case, have considered z tending to z' instead of $z = z'$ will be clear in the following.

where the double vector product is applied on the columns of the dyadic Green's function and where the dot stands for scalar product.[3] Due to the general vector identity

$$v_1 \times v_2 \times v_3 = v_1(v_2 \cdot v_3) - v_3(v_1 \cdot v_2) \tag{11.50}$$

we can rewrite Eq. (11.49) as

$$-\mathbf{k}\left(\mathbf{k} \cdot \hat{\hat{\mathbf{G}}}_a(\mathbf{k}, \mathbf{r}')\right) + \|\mathbf{k}\|^2 \hat{\hat{\mathbf{G}}}_a(\mathbf{k}, \mathbf{r}') - k_0^2 \hat{\hat{\mathbf{G}}}_a(\mathbf{k}, \mathbf{r}')$$

$$= -\mathbf{k}\left(\mathbf{k} \cdot \hat{\hat{\mathbf{G}}}_a(\mathbf{k}, \mathbf{r}')\right) + \left(\|\mathbf{k}\|^2 - k_0^2\right) \hat{\hat{\mathbf{G}}}_a(\mathbf{k}, \mathbf{r}') = \mathbf{I} \exp(-j\mathbf{k} \cdot \mathbf{r}') \tag{11.51}$$

where the scalar product $\mathbf{k} \cdot \hat{\hat{\mathbf{G}}}_a(\mathbf{k}, \mathbf{r}')$ is meant with respect to the columns of the dyadic Green's function (so that the result is a vector) and the dyadic product $\mathbf{v_1 v_2}$ [that is meant applied to the first term in Eq. (11.51)] is meant in terms of components as

$$\mathbf{v_1 v_2} = \begin{pmatrix} v_{1x} \\ v_{1y} \\ v_{1z} \end{pmatrix} \left(v_{2x} v_{2y} v_{2z}\right) = \begin{pmatrix} v_{1x}v_{2x} & v_{1x}v_{2y} & v_{1x}v_{2z} \\ v_{1y}v_{2x} & v_{1y}v_{2y} & v_{1y}v_{2z} \\ v_{1z}v_{2x} & v_{1z}v_{2y} & v_{1z}v_{2z} \end{pmatrix} \tag{11.52}$$

Let us now calculate the scalar product of Eq. (11.51) times $\mathbf{k}$:

$$-\|\mathbf{k}\|^2\left(\mathbf{k} \cdot \hat{\hat{\mathbf{G}}}_a(\mathbf{k}, \mathbf{r}')\right) + \|\mathbf{k}\|^2\left(\mathbf{k} \cdot \hat{\hat{\mathbf{G}}}_a(\mathbf{k}, \mathbf{r}')\right) - k_0^2\left(\mathbf{k} \cdot \hat{\hat{\mathbf{G}}}_a(\mathbf{k}, \mathbf{r}')\right) = -k_0^2\left(\mathbf{k} \cdot \hat{\hat{\mathbf{G}}}_a(\mathbf{k}, \mathbf{r}')\right)$$

$$= (\mathbf{k} \cdot \mathbf{I}) \exp(-j\mathbf{k} \cdot \mathbf{r}') \Rightarrow \mathbf{k} \cdot \hat{\hat{\mathbf{G}}}_a(\mathbf{k}, \mathbf{r}') = -\frac{(\mathbf{k} \cdot \mathbf{I}) \exp(-j\mathbf{k} \cdot \mathbf{r}')}{k_0^2} \tag{11.53}$$

Substituting in Eq. (11.53) into Eq. (11.51), we obtain

$$\frac{\mathbf{k}(\mathbf{k} \cdot \mathbf{I}) \exp(-j\mathbf{k} \cdot \mathbf{r}')}{k_0^2} + \|\mathbf{k}\|^2 \hat{\hat{\mathbf{G}}}_a(\mathbf{k}, \mathbf{r}') - k_0^2 \hat{\hat{\mathbf{G}}}_a(\mathbf{k}, \mathbf{r}')$$

$$= \frac{\mathbf{kk} \exp(-j\mathbf{k} \cdot \mathbf{r}')}{k_0^2} + \left(\|\mathbf{k}\|^2 - k_0^2\right) \hat{\hat{\mathbf{G}}}_a(\mathbf{k}, \mathbf{r}') = \mathbf{I} \exp(-j\mathbf{k} \cdot \mathbf{r}') \tag{11.54}$$

$$\Rightarrow \hat{\hat{\mathbf{G}}}_a(\mathbf{k}, \mathbf{r}') = \frac{\left(\mathbf{I}k_0^2 - \mathbf{kk}\right) \exp(-j\mathbf{k} \cdot \mathbf{r}')}{k_0^2\left(\|\mathbf{k}\|^2 - k_0^2\right)}$$

---

[3] It is meant as a non-Hermitian scalar product: If $v_1 = v_{1x}i_x + v_{1y}i_y + v_{1z}i_z$ and $v_2 = v_{2x}i_x + v_{2y}i_y + v_{2z}i_z$ are two vectors, either real or complex, the non-Hermitian scalar product is defined as $v_1 \cdot v_2 = v_{1x}v_{2x} + v_{1y}v_{2y} + v_{1z}v_{2z}$, whereas the Hermitian scalar product is defined as $v_1 \circ v_2 = v_{1x}v_{2x}^* + v_{1y}v_{2y}^* + v_{1z}v_{2z}^*$, where "*" stands for conjugation. Clearly, the two scalar products are the same thing for real vectors, but not for complex vectors.

Consequently, inverting the Fourier transform, we can write

$$
\mathbf{G}_a(\mathbf{r},\mathbf{r}') = \frac{1}{(2\pi)^3} \int\limits_{-\infty}^{+\infty}\int\limits_{-\infty}^{+\infty}\int\limits_{-\infty}^{+\infty} \frac{\mathbf{I}k_0^2 - \mathbf{kk}}{k_0^2\left(\|\mathbf{k}\|^2 - k_0^2\right)} e^{j\mathbf{k}\cdot(\mathbf{r}-\mathbf{r}')} \, du\,dv\,dw
$$

$$
= \frac{1}{4\pi^2} \int\limits_{-\infty}^{+\infty} du \int\limits_{-\infty}^{+\infty} \frac{dv}{k_0^2} \left(\frac{1}{2\pi} \int\limits_{-\infty}^{+\infty} \frac{\mathbf{I}k_0^2 - \mathbf{kk}}{\left(u^2 + v^2 + w^2 - k_0^2\right)} e^{j\mathbf{k}\cdot(\mathbf{r}-\mathbf{r}')} \, dw \right)
$$

$$
= \frac{1}{4\pi^2} \int\limits_{-\infty}^{+\infty} du \int\limits_{-\infty}^{+\infty} \frac{dv}{k_0^2} \frac{1}{2\pi} \int\limits_{-\infty}^{+\infty} \frac{\begin{pmatrix} k_0^2 - u^2 & uv & uw \\ uv & k_0^2 - v^2 & vw \\ uw & vw & k_0^2 - w^2 \end{pmatrix}}{\left(u^2 + v^2 + w^2 - k_0^2\right)} e^{j\mathbf{k}\cdot(\mathbf{r}-\mathbf{r}')} \, dw \qquad (11.55)
$$

Equation (11.55) shows that the argument of the integral in dw does not tend to a null matrix for high values of w, because we have

$$
\lim_{w \to \pm\infty} \left(\frac{1}{\left(u^2 + v^2 + w^2 - k_0^2\right)} \begin{pmatrix} k_0^2 - u^2 & uv & uw \\ uv & k_0^2 - v^2 & vw \\ uw & vw & k_0^2 - w^2 \end{pmatrix} \right) = \begin{pmatrix} 0 & 0 & 0 \\ 0 & 0 & 0 \\ 0 & 0 & -1 \end{pmatrix} = -i_z i_z
$$

$$(11.56)$$

Subtracting and adding again the limit value (11.56) to the integrand in Eq. (11.55), we obtain

$$
\mathbf{G}_a(\mathbf{r},\mathbf{r}') = \frac{1}{(2\pi)^3} \int\limits_{-\infty}^{+\infty}\int\limits_{-\infty}^{+\infty}\int\limits_{-\infty}^{+\infty} \frac{\mathbf{I}k_0^2 - \mathbf{kk}}{k_0^2\left(\|\mathbf{k}\|^2 - k_0^2\right)} e^{j\mathbf{k}\cdot(\mathbf{r}-\mathbf{r}')} \, du\,dv\,dw
$$

$$
= \frac{1}{4\pi^2} \int\limits_{-\infty}^{+\infty} du \int\limits_{-\infty}^{+\infty} \frac{dv}{k_0^2} \frac{1}{2\pi} \int\limits_{-\infty}^{+\infty} \left[\left(\frac{\mathbf{I}k_0^2 - \mathbf{kk}}{\left(\|\mathbf{k}\|^2 - k_0^2\right)} + i_z i_z \right) e^{j\mathbf{k}\cdot(\mathbf{r}-\mathbf{r}')} \, dw \right]
$$

$$
- \frac{1}{4\pi^2} \int\limits_{-\infty}^{+\infty} du \int\limits_{-\infty}^{+\infty} \frac{dv}{k_0^2} \frac{1}{2\pi} \int\limits_{-\infty}^{+\infty} \left[i_z i_z e^{-j\mathbf{k}\cdot(\mathbf{r}-\mathbf{r}')} \, dw \right]
$$

$$
= \frac{1}{4\pi^2} \int\limits_{-\infty}^{+\infty} du \int\limits_{-\infty}^{+\infty} \frac{dv}{k_0^2} \frac{1}{2\pi} \int\limits_{-\infty}^{+\infty} \left[\left(\frac{\mathbf{I}k_0^2 - \mathbf{kk} + i_z i_z \left(\|\mathbf{k}\|^2 - k_0^2\right)}{\left(\|\mathbf{k}\|^2 - k_0^2\right)} \right) e^{j\mathbf{k}\cdot(\mathbf{r}-\mathbf{r}')} \, dw \right] - \frac{1}{k_0^2} i_z i_z \delta(\mathbf{r}-\mathbf{r}')
$$

$$(11.57)$$

This passage has allowed us to integrate separately the piece not integrable in classical terms (in fact it has been integrated in distributional terms). So, now the integrand in

Eq. (11.57) vanishes for $w \to \pm\infty$, and so it is licit to solve it by resorting to the theory of the residuals, in the same way followed for the (scalar) spectral integral in Appendix A. So, in the case at hand we obtain

$$\mathbf{G}_a(\mathbf{r},\mathbf{r}') = \frac{j}{8\pi^2 k_0^2}$$

$$\times \int_{-\infty}^{+\infty}\int_{-\infty}^{+\infty} \frac{\exp(ju(x-x'))\exp(jv(y-y'))\exp(jk_{z1}|z-z'|)}{k_{z10}} \left(\mathbf{I}k_0^2 - \mathbf{k}_0\mathbf{k}_0\right) du\,dv$$

$$-\frac{1}{k_0^2}\mathbf{i}_z\mathbf{i}_z\delta(\mathbf{r}-\mathbf{r}') \tag{11.58}$$

where, analogously to the notation followed in Chapter 4 [see Eq. (4.57)], we have

$$k_{z10}(u,v) = \sqrt{k_0^2 - u^2 - v^2} \tag{11.59}$$

where the imaginary part is meant as non-negative and where

$$\mathbf{k}_0 = \begin{cases} u\mathbf{i}_x + v\mathbf{i}_y + k_{z10}(u,v)\mathbf{i}_z, & z-z' > 0, \\ u\mathbf{i}_x + v\mathbf{i}_y - k_{z10}(u,v)\mathbf{i}_z, & z-z' < 0 \end{cases} \tag{11.60}$$

From a mathematical point of view, the different sign of the z-component of the wavevector descends from the fact that $|z-z'|$ is equal either to $z-z'$ or to $z'-z$, and so the integral path for the application of the theory of the residuals has to enclose either the pole at $w = \sqrt{k_0^2 - u^2 - v^2}$ or the pole at $w = -\sqrt{k_0^2 - u^2 - v^2}$ (where the square root is meant with a nonnegative imaginary part). From a physical point of view, the different sign of the z component of the wavevector is due to the fact that the field propagates in any case away from the sources.

In the integral in Eq. (11.58) we easily identify an inverse double Fourier transform, so that we have

$$\frac{j\exp(jk_{z1}|z-z'|)}{2k_0^2 k_{z1}}\left(\mathbf{I}k_0^2 - \mathbf{k}_0\mathbf{k}_0\right)$$

$$-\frac{1}{4\pi^2}\mathbf{i}_z\mathbf{i}_z \int_{-\infty}^{+\infty}\int_{-\infty}^{+\infty} \delta(\mathbf{r}-\mathbf{r}')\exp(ju(x-x'))\exp(jv(y-y'))\,dx\,dy$$

$$= \frac{j\exp(jk_{z1}|z-z'|)}{2k_0^2 k_{z1}}\left(\mathbf{I}k_0^2 - \mathbf{k}_0\mathbf{k}_0\right) - \frac{\mathbf{i}_z\mathbf{i}_z\delta(z-z')}{4\pi^2}$$

$$= \int_{-\infty}^{+\infty}\int_{-\infty}^{+\infty} \mathbf{G}_a(\mathbf{r},\mathbf{r}')\exp(-ju(x-x'))\exp(-jv(y-y'))\,dx\,dy \tag{11.61}$$

Comparing Eq. (11.61) with Eq. (11.47), we retrieve the expression of plane wave spectrum of Green's function in air, which is given by

$$\hat{\mathbf{G}}_a\left(u,v,\mathbf{r}'|_{z\to z'}\right) = \frac{j}{2k_0^2 k_{z10}}\left(\mathbf{l}k_0^2 - \mathbf{k}_0\mathbf{k}_0\right)\exp\left(-j(ux'+vy')\right)$$

$$= \begin{cases} \dfrac{j\exp(-jux')\exp(-jvy')}{2k_0^2 k_{1z0}}\begin{pmatrix} k_0^2-u^2 & -uv & -uk_{1z0} \\ -uv & k_0^2-v^2 & -vk_{1z0} \\ -uk_{1z0} & -vk_{1z0} & u^2+v^2 \end{pmatrix}, & z-z'>0, \\[4mm] \dfrac{j\exp(-jux')\exp(-jvy')}{2k_0^2 k_{1z0}}\begin{pmatrix} k_0^2-u^2 & -uv & uk_{1z0} \\ -uv & k_0^2-v^2 & vk_{1z0} \\ uk_{1z0} & vk_{1z0} & u^2+v^2 \end{pmatrix}, & z-z'<0 \end{cases} \tag{11.62}$$

Due to the impulse present in Eq. (11.61), the plane wave spectrum of Eq. (11.62) is bi-univocally related to Green's function for any $z \neq z'$ but not for $z = z'$.

The following calculation steps are of course the same for any homogeneous medium, on the condition that we make use of the wavenumber of that medium. Therefore, we can express the plane wave spectrum of a homogeneous Green's function $\mathbf{G}_s(\mathbf{r},\mathbf{r}')$ relative to an ideal world "made up of soil" as follows:

$$\hat{\mathbf{G}}_s\left(u,v,\mathbf{r}'|_{z\to z'}\right) = \frac{j}{2k_s^2 k_{z1s}}\left(\mathbf{l}k_s^2 - \mathbf{k}_s\mathbf{k}_s\right)\exp\left(-j(ux'+vy')\right)$$

$$= \begin{cases} \dfrac{j\exp(-jux')\exp(-jvy')}{2k_s^2 k_{1zs}}\begin{pmatrix} k_s^2-u^2 & -uv & -uk_{1zs} \\ -uv & k_s^2-v^2 & -vk_{1zs} \\ -uk_{1zs} & -vk_{1zs} & u^2+v^2 \end{pmatrix}, & z-z'>0, \\[4mm] \dfrac{j\exp(-jux')\exp(-jvy')}{2k_s^2 k_{1zs}}\begin{pmatrix} k_s^2-u^2 & -uv & uk_{1zs} \\ -uv & k_s^2-v^2 & vk_{1zs} \\ uk_{1zs} & vk_{1zs} & u^2+v^2 \end{pmatrix}, & z-z'<0 \end{cases} \tag{11.63}$$

where

$$k_{z1s}(u,v) = \sqrt{k_s^2 - u^2 - v^2} \tag{11.64}$$

with nonnegative imaginary part.

11.7 HALF-SPACE GREEN'S FUNCTIONS

The calculation of the homogeneous Green's functions is a preliminary step for the calculation of the half-space Green's functions. Indeed, since the homogeneous Green's function can be (and has been) expressed by mean of an integral superposition of plane waves, the half-space Green's function is achieved by considering the homogeneous Green's function as an incident "triple" of fields, wherein each plane wave component is reflected and refracted at the air–soil interface. The integral sum of the transmitted waves will provide the external Green's function, whereas the integral sum of the reflected waves, plus the incident (homogeneous) Green's function itself, will provide the internal Green's function.

However, the Fresnel coefficients, i.e. the reflection and transmission coefficient of the plane waves at the interface depend on the polarization of the incident field too (Franceschetti, 1997). Consequently, we will distinguish the TE component and the TM component in the homogeneous Green's function, and each component will be reflected and transmitted according to its own Fresnel coefficients.

So, the first step is to calculate the matrix decomposition of the plane wave spectrum of the homogeneous Green's functions such as

$$\hat{\mathbf{G}}_a(u,v) = \hat{\mathbf{G}}_{aTE}(u,v) + \hat{\mathbf{G}}_{aTM}(u,v),$$

$$\hat{\mathbf{G}}_s(u,v) = \hat{\mathbf{G}}_{sTE}(u,v) + \hat{\mathbf{G}}_{sTM}(u,v) \tag{11.65}$$

The decomposition of a plane wave impinging at the air soil interface along its TE and TM components is derived in Appendix D. Applying this decomposition to the columns of the homogeneous Green's function in air we achieve

$$\hat{\mathbf{G}}_{aTE}(u,v,z,\mathbf{r}') = j\frac{\exp(-jux')\exp(-jvy')}{2k_0^2 k_{z10}}\begin{pmatrix} \dfrac{k_0^2 v^2}{u^2+v^2} & \dfrac{-k_0^2 uv}{u^2+v^2} & 0 \\ \dfrac{-k_0^2 uv}{u^2+v^2} & \dfrac{k_0^2 u^2}{u^2+v^2} & 0 \\ 0 & 0 & 0 \end{pmatrix} \quad \forall z \tag{11.66}$$

$$\hat{\mathbf{G}}_{aTM}(u,v,z \rightarrow z',\mathbf{r}') = j\frac{\exp(-jux')\exp(-jvy')}{2k_0^2 k_{z10}} \times \begin{cases} \begin{pmatrix} \dfrac{u^2 k_{z10}^2}{u^2+v^2} & \dfrac{uvk_{z10}^2}{u^2+v^2} & -uk_{z10} \\ \dfrac{uvk_{z10}^2}{u^2+v^2} & \dfrac{v^2 k_{z10}^2}{u^2+v^2} & -vk_{z10} \\ -uk_{z10} & -vk_{z10} & u^2+v^2 \end{pmatrix}, & z-z' > 0, \\[2em] \begin{pmatrix} \dfrac{u^2 k_{z10}^2}{u^2+v^2} & \dfrac{uvk_{z10}^2}{u^2+v^2} & uk_{z10} \\ \dfrac{uvk_{z10}^2}{u^2+v^2} & \dfrac{v^2 k_{z10}^2}{u^2+v^2} & vk_{z10} \\ uk_{z10} & vk_{z10} & u^2+v^2 \end{pmatrix}, & z-z' < 0 \end{cases}$$

$$\tag{11.67}$$

The reflection and transmission TE and TM coefficients are retrieved in Appendix E. Applying in particular Eqs. (E.18) and (E.20) to the columns of the TE components of Green's function and applying Eqs. (E.32) and (E.34) to the columns of the TM component of Green's function, we have

$$
\hat{\mathbf{G}}_{as} = j\frac{\exp(-j(ux' + vy'))}{2k_0^2 k_{z10}} \left[T_{TE} \begin{pmatrix} \dfrac{k_0^2 v^2}{u^2 + v^2} & \dfrac{-k_0^2 uv}{u^2 + v^2} & 0 \\[3mm] \dfrac{-k_0^2 uv}{u^2 + v^2} & \dfrac{k_0^2 u^2}{u^2 + v^2} & 0 \\[3mm] 0 & 0 & 0 \end{pmatrix} \right.
$$

$$
\left. + T_{TM} \begin{pmatrix} \dfrac{u^2 k_{z10}^2}{u^2 + v^2} & \dfrac{uv k_{z10}^2}{u^2 + v^2} & uk_{z10} \\[3mm] \dfrac{uv k_{z10}^2}{u^2 + v^2} & \dfrac{v^2 k_{z10}^2}{u^2 + v^2} & vk_{z10} \\[3mm] 0 & 0 & 0 \end{pmatrix} + \frac{k_{z10}}{k_{z1s}} T_{TM} \begin{pmatrix} 0 & 0 & 0 \\ 0 & 0 & 0 \\ uk_{z10} & vk_{z10} & u^2 + v^2 \end{pmatrix} \right]
$$

$$
= j\frac{\exp(-j(ux' + vy'))}{2k_0^2 k_{z10}} \left[T_{TE} \begin{pmatrix} \dfrac{k_0^2 v^2}{u^2 + v^2} & \dfrac{-k_0^2 uv}{u^2 + v^2} & 0 \\[3mm] \dfrac{-k_0^2 uv}{u^2 + v^2} & \dfrac{k_0^2 u^2}{u^2 + v^2} & 0 \\[3mm] 0 & 0 & 0 \end{pmatrix} \right.
$$

$$
\left. + T_{TM} \begin{pmatrix} \dfrac{u^2 k_{z10}^2}{u^2 + v^2} & \dfrac{uv k_{z10}^2}{u^2 + v^2} & uk_{z10} \\[3mm] \dfrac{uv k_{z10}^2}{u^2 + v^2} & \dfrac{v^2 k_{z10}^2}{u^2 + v^2} & vk_{z10} \\[3mm] \dfrac{uk_{z10}^2}{k_{z1s}} & \dfrac{vk_{z10}^2}{k_{z1s}} & \dfrac{k_{z10}(u^2 + v^2)}{k_{z1a}} \end{pmatrix} \right]
$$

$$
= j\frac{\exp(-j(ux' + vy'))}{2k_0^2 k_{z10}} \left[\frac{2k_{z10}}{k_{z10} + k_{z1s}} \begin{pmatrix} \dfrac{k_0^2 v^2}{u^2 + v^2} & \dfrac{-k_0^2 uv}{u^2 + v^2} & 0 \\[3mm] \dfrac{-k_0^2 uv}{u^2 + v^2} & \dfrac{k_0^2 u^2}{u^2 + v^2} & 0 \\[3mm] 0 & 0 & 0 \end{pmatrix} \right.
$$

$$
\left. + \frac{2k_{z1s}\varepsilon_0}{k_{z1s} + k_{z10}\varepsilon_s} \begin{pmatrix} \dfrac{u^2 k_{z10}^2}{u^2 + v^2} & \dfrac{uv k_{z10}^2}{u^2 + v^2} & uk_{z10} \\[3mm] \dfrac{uv k_{z10}^2}{u^2 + v^2} & \dfrac{v^2 k_{z10}^2}{u^2 + v^2} & vk_{z10} \\[3mm] \dfrac{uk_{z10}^2}{k_{z1s}} & \dfrac{vk_{z10}^2}{k_{z1s}} & \dfrac{k_{z10}(u^2 + v^2)}{k_{z1s}} \end{pmatrix} \right] \tag{11.68}
$$

After summing the matrixes, the final result is

$$\hat{\mathbf{G}}_{as} = j\frac{\exp(-j(ux'+vy'))}{k_0^2(u^2+v^2)}$$

$$\times \begin{pmatrix} \dfrac{k_0^2 v^2}{k_{z10}+k_{z1s}} + \dfrac{\varepsilon_0 u^2 k_{z1s}k_{z10}}{k_{z1s}\varepsilon_0+k_{z10}\varepsilon_s} & \dfrac{-k_0^2 uv}{k_{z10}+k_{z1s}} + \dfrac{\varepsilon_0 uv k_{z1s}k_{z10}}{k_{z1s}\varepsilon_0+k_{z10}\varepsilon_s} & \dfrac{\varepsilon_0 u k_{z1s}(u^2+v^2)}{(k_{z1s}\varepsilon_0+k_{z10}\varepsilon_s)} \\[3mm] \dfrac{-k_0^2 uv}{k_{z10}+k_{z1s}} + \dfrac{\varepsilon_0 uv k_{z1s}k_{z10}}{k_{z1s}\varepsilon_0+k_{z10}\varepsilon_s} & \dfrac{k_0^2 u^2}{k_{z10}+k_{z1s}} + \dfrac{\varepsilon_0 v^2 k_{z1s}k_{z10}}{k_{z1s}\varepsilon_0+k_{z10}\varepsilon_s} & \dfrac{\varepsilon_0 v k_{z1s}(u^2+v^2)}{(k_{z1s}\varepsilon_0+k_{z10}\varepsilon_s)} \\[3mm] \dfrac{\varepsilon_0 u k_{z10}(u^2+v^2)}{(k_{z1s}\varepsilon_0+k_{z10}\varepsilon_s)} & \dfrac{\varepsilon_0 v k_{z10}(u^2+v^2)}{(k_{z1s}\varepsilon_0+k_{z10}\varepsilon_s)} & \dfrac{\varepsilon_0(u^2+v^2)^2}{k_{z1s}\varepsilon_0+k_{z10}\varepsilon_s} \end{pmatrix}$$

$$(11.69)$$

Equation (11.69) accounts for the fact that the source point is in air and the observation point is buried, so that we have $z - z' < 0$. For brevity of notation, let us label as $\mathbf{M}_{as1}(u,v)$ the matrix in the large parentheses in Eq. (11.69).

The air–soil Green's function is given by the inverse Fourier transform of Eq. (11.69) as follows:

$$\mathbf{G}_{as}(\mathbf{r},\mathbf{r'}) = \frac{j}{4\pi^2 k_0^2} \int_{-\infty}^{+\infty}\int_{-\infty}^{+\infty} \frac{\exp(j(u(x-x')+v(y-y')-k_{z10}z'-k_{z1s}z))}{(u^2+v^2)}\mathbf{M}_{as1}(u,v)\,dudv$$

$$(11.70)$$

For conventional reasons, it is more comfortable to express the quantity (11.70) as

$$\mathbf{G}_{as}(\mathbf{r},\mathbf{r'}) = \frac{j}{4\pi^2 k_0^2} \int_{-\infty}^{+\infty}\int_{-\infty}^{+\infty} \frac{\exp(j(-u(x-x')-v(y-y')-k_{z10}z'-k_{z1s}z))}{(u^2+v^2)}\mathbf{M}_{as}(u,v)\,dudv$$

$$(11.71)$$

with

$$\mathbf{M}_{as} = \begin{pmatrix} \dfrac{k_0^2 v^2}{k_{z10}+k_{z1s}} + \dfrac{\varepsilon_0 u^2 k_{z1s}k_{z10}}{k_{z1s}\varepsilon_0+k_{z10}\varepsilon_s} & \dfrac{-k_0^2 uv}{k_{z10}+k_{z1s}} + \dfrac{\varepsilon_0 uv k_{z1s}k_{z10}}{k_{z1s}\varepsilon_0+k_{z10}\varepsilon_s} & \dfrac{-\varepsilon_0 u k_{z1s}(u^2+v^2)}{(k_{z1s}\varepsilon_0+k_{z10}\varepsilon_s)} \\[3mm] \dfrac{-k_0^2 uv}{k_{z10}+k_{z1s}} + \dfrac{\varepsilon_0 uv k_{z1s}k_{z10}}{k_{z1s}\varepsilon_0+k_{z10}\varepsilon_s} & \dfrac{k_0^2 u^2}{k_{z10}+k_{z1s}} + \dfrac{\varepsilon_0 v^2 k_{z1s}k_{z10}}{k_{z1s}\varepsilon_0+k_{z10}\varepsilon_s} & \dfrac{-\varepsilon_0 v k_{z1s}(u^2+v^2)}{(k_{z1s}\varepsilon_0+k_{z10}\varepsilon_s)} \\[3mm] \dfrac{-\varepsilon_0 u k_{z10}(u^2+v^2)}{(k_{z1s}\varepsilon_0+k_{z10}\varepsilon_s)} & \dfrac{-\varepsilon_0 v k_{z10}(u^2+v^2)}{(k_{z1s}\varepsilon_0+k_{z10}\varepsilon_s)} & \dfrac{\varepsilon_0(u^2+v^2)^2}{k_{z1s}\varepsilon_0+k_{z10}\varepsilon_s} \end{pmatrix}$$

$$(11.72)$$

It is easy to check that the expressions (11.70) and (11.71) are equivalent, so from now on we will refer to (11.71).

Substituting Eq. (11.71) into Eq. (11.11), we retrieve the incident field in the soil due to a dipole oriented according to some direction $\hat{\mathbf{a}} \equiv \begin{pmatrix} a_x \\ a_y \\ a_z \end{pmatrix}$, which is the first "piece" to be put in the scattering equation (11.1):

$$\mathbf{E}_{inc}(\mathbf{r}) = \frac{\mu_0 \omega I_0 l}{4\pi^2 k_0^2} \int_{-\infty}^{+\infty}\int_{-\infty}^{+\infty} \frac{\exp(j(-u(x-x')-v(y-y')-k_{z10}z'-k_{z1s}z))}{(u^2+v^2)} \mathbf{M}_{as}(u,v) \begin{pmatrix} a_x \\ a_y \\ a_z \end{pmatrix} du\,dv$$

$$= \frac{\mu_0 \omega I_0 l}{4\pi^2 k_0^2} \left(\int_{-\infty}^{+\infty}\int_{-\infty}^{+\infty} \frac{\exp(j(-u(x-x')-v(y-y')-k_{z10}z'-k_{z1s}z))}{(u^2+v^2)} \mathbf{M}_{as}(u,v)\,du\,dv \right) \begin{pmatrix} a_x \\ a_y \\ a_z \end{pmatrix}$$

$$(11.73)$$

The Green's function $\mathbf{G}_{sa}(\mathbf{r},\mathbf{r}')$, with buried source and observation point in air, is derived in the same way, accounting for the fact that now the field propagates toward the positive side of the z-axis ($z < z'$) and the Fresnel coefficients from the soil to the air have to be accounted for (see Appendix E). The result is

$$\hat{\mathbf{G}}_{sa} = j\frac{\exp(-j(ux'+vy'))}{k_s^2(u^2+v^2)}$$

$$\times \begin{pmatrix} \dfrac{k_s^2 v^2}{k_{z10}+k_{z1s}} + \dfrac{\varepsilon_s u^2 k_{z1s}k_{z10}}{k_{z1s}\varepsilon_0 + k_{z10}\varepsilon_s} & \dfrac{-k_s^2 uv}{k_{z10}+k_{z1s}} + \dfrac{\varepsilon_s uv k_{z1s}k_{z10}}{k_{z1s}\varepsilon_0 + k_{z10}\varepsilon_s} & -\dfrac{\varepsilon_s u k_{z10}(u^2+v^2)}{(k_{z1s}\varepsilon_0 + k_{z10}\varepsilon_s)} \\[3mm] \dfrac{-k_s^2 uv}{k_{z10}+k_{z1s}} + \dfrac{\varepsilon_s uv k_{z1s}k_{z10}}{k_{z1s}\varepsilon_0 + k_{z10}\varepsilon_s} & \dfrac{k_s^2 u^2}{k_{z10}+k_{z1s}} + \dfrac{\varepsilon_s v^2 k_{z1s}k_{z10}}{k_{z1s}\varepsilon_0 + k_{z10}\varepsilon_s} & -\dfrac{\varepsilon_s v k_{z10}(u^2+v^2)}{(k_{z1s}\varepsilon_0 + k_{z10}\varepsilon_s)} \\[3mm] -\dfrac{\varepsilon_s u k_{z1s}(u^2+v^2)}{(k_{z1s}\varepsilon_0 + k_{z10}\varepsilon_s)} & -\dfrac{\varepsilon_s v k_{z1s}(u^2+v^2)}{(k_{z1s}\varepsilon_0 + k_{z10}\varepsilon_s)} & \dfrac{\varepsilon_s (u^2+v^2)^2}{k_{z1s}\varepsilon_0 + k_{z10}\varepsilon_s} \end{pmatrix}$$

$$(11.74)$$

Equation (11.74) expresses the plane wave spectrum of the external Green's function. Similarly to the position introduced with regard to Eq. (11.72), from now on we will label as $\mathbf{M}_{sa}(u,v)$ the matrix in large parentheses in Eq. (11.74). This allows us to express the external 3D Green's function in a compact way as follows:

$$\mathbf{G}_e(\mathbf{r},\mathbf{r}') = \mathbf{G}_{sa}(\mathbf{r},\mathbf{r}') = \frac{j}{4\pi^2 k_s^2} \int_{-\infty}^{+\infty}\int_{-\infty}^{+\infty} \frac{\exp(j(u(x-x')+v(y-y')-k_{z1s}z'-k_{z10}z))}{(u^2+v^2)} \mathbf{M}_{sa}(u,v)\,du\,dv$$

$$(11.75)$$

Finally, let us now retrieve the internal Green's function $\mathbf{G}_i(\mathbf{r},\mathbf{r}')$. From the discussion carried out in this chapter, it is by now clear that this function represents the set of the answers to an impulsive buried source, observed in any buried point. So, the internal Green's function is provided by the sum of the Green's function of a homogeneous soil (given by $\mathbf{G}_s(\mathbf{r},\mathbf{r}')$) plus the contribution back-reflected in the soil from the soil-air interface. So, let us start with the decomposition of $\mathbf{G}_s(\mathbf{r},\mathbf{r}')$ into its TE and TM components:

$$\hat{\mathbf{G}}_{sTE}(u,v,z,\mathbf{r}') = j\frac{\exp(-jux')\exp(-jvy')}{2k_s^2 k_{z1s}}\begin{pmatrix} \dfrac{k_s^2 v^2}{u^2+v^2} & \dfrac{-k_s^2 uv}{u^2+v^2} & 0 \\[2ex] \dfrac{-k_s^2 uv}{u^2+v^2} & \dfrac{k_s^2 u^2}{u^2+v^2} & 0 \\[2ex] 0 & 0 & 0 \end{pmatrix} \quad \forall z \qquad (11.76)$$

$$\hat{\mathbf{G}}_{sTM}(u,v,z\to z',\mathbf{r}') = j\frac{\exp(-jux')\exp(-jvy')}{2k_s^2 k_{z1s}} \times \begin{cases} \begin{pmatrix} \dfrac{u^2 k_{z1s}^2}{u^2+v^2} & \dfrac{uv k_{z1s}^2}{u^2+v^2} & -u k_{z1s} \\[2ex] \dfrac{uv k_{z1s}^2}{u^2+v^2} & \dfrac{v^2 k_{z1s}^2}{u^2+v^2} & -v k_{z1s} \\[2ex] -u k_{z1s} & -v k_{z1s} & u^2+v^2 \end{pmatrix}, & z-z'>0, \\[6ex] \begin{pmatrix} \dfrac{u^2 k_{z1s}^2}{u^2+v^2} & \dfrac{uv k_{z1s}^2}{u^2+v^2} & u k_{z1s} \\[2ex] \dfrac{uv k_{z1s}^2}{u^2+v^2} & \dfrac{v^2 k_{z1s}^2}{u^2+v^2} & v k_{z1s} \\[2ex] u k_{z1s} & v k_{z1s} & u^2+v^2 \end{pmatrix}, & z-z'<0 \end{cases}$$

$$(11.77)$$

which can be more compactly written as

$$\hat{\mathbf{G}}_{sTM}(u,v,z\to z',\mathbf{r}') = j\frac{\exp(-jux')\exp(-jvy')}{2k_s^2 k_{z1s}}$$

$$\times \begin{pmatrix} \dfrac{u^2 k_{z1s}^2}{u^2+v^2} & \dfrac{uv k_{z1s}^2}{u^2+v^2} & -\operatorname{sgn}(z-z')u k_{z1s} \\[2ex] \dfrac{uv k_{z1s}^2}{u^2+v^2} & \dfrac{v^2 k_{z1s}^2}{u^2+v^2} & -\operatorname{sgn}(z-z')v k_{z1s} \\[2ex] -u\operatorname{sgn}(z-z')k_{z1s} & -v\operatorname{sgn}(z-z')k_{z1s} & u^2+v^2 \end{pmatrix} \qquad (11.78)$$

where sgn stands for the sign function, equal to 1 if its argument is positive and equal to -1 if its argument is negative. At this point, based on the calculations presented in Appendix E [see, in particular, Eqs. (E.18), (E.19), (E.32), and (E.33)], and reminding ourselves that the soil–air reflection coefficients are the opposite of the air–soil ones (see again Appendix E), we obtain the plane wave spectrum of the "reflected Green's function", given by

$$
\hat{\mathbf{G}}_{rss} = j\frac{\exp(-j(ux' + vy'))}{2k_s^2 k_{z1s}} \left[-R_{TE} \begin{pmatrix} \dfrac{k_s^2 v^2}{u^2 + v^2} & \dfrac{-k_s^2 uv}{u^2 + v^2} & 0 \\[2mm] \dfrac{-k_s^2 uv}{u^2 + v^2} & \dfrac{k_s^2 u^2}{u^2 + v^2} & 0 \\[2mm] 0 & 0 & 0 \end{pmatrix} \right.
$$

$$
-R_{TM} \begin{pmatrix} \dfrac{u^2 k_{z1s}^2}{u^2 + v^2} & \dfrac{uv k_{z1s}^2}{u^2 + v^2} & -u\,\mathrm{sgn}(z - z')k_{z1s} \\[2mm] \dfrac{uv k_{z1s}^2}{u^2 + v^2} & \dfrac{v^2 k_{z1s}^2}{u^2 + v^2} & -v\,\mathrm{sgn}(z - z')k_{z1s} \\[2mm] 0 & 0 & 0 \end{pmatrix}
$$

$$
\left. +R_{TM} \begin{pmatrix} 0 & 0 & 0 \\ 0 & 0 & 0 \\ -u\,\mathrm{sgn}(z-z')k_{z1s} & -v\,\mathrm{sgn}(z-z')k_{z1s} & -(u^2 + v^2) \end{pmatrix} \right]
$$

$$
= j\frac{\exp(-j(ux' + vy'))}{2k_s^2 k_{z1s}^2} \left[-R_{TE} \begin{pmatrix} \dfrac{k_s^2 v^2}{u^2 + v^2} & \dfrac{-k_s^2 uv}{u^2 + v^2} & 0 \\[2mm] \dfrac{-k_s^2 uv}{u^2 + v^2} & \dfrac{k_s^2 u^2}{u^2 + v^2} & 0 \\[2mm] 0 & 0 & 0 \end{pmatrix} \right.
$$

$$
\left. -R_{TM} \begin{pmatrix} \dfrac{u^2 k_{z1s}^2}{u^2 + v^2} & \dfrac{uv k_{z1s}^2}{u^2 + v^2} & -u\,\mathrm{sgn}(z-z')k_{z1s} \\[2mm] \dfrac{uv k_{z1s}^2}{u^2 + v^2} & \dfrac{v^2 k_{z1s}^2}{u^2 + v^2} & -v\,\mathrm{sgn}(z-z')k_{z1s} \\[2mm] u\,\mathrm{sgn}(z-z')k_{z1s} & v\,\mathrm{sgn}(z-z')k_{z1s} & (u^2 + v^2) \end{pmatrix} \right] \tag{11.79}
$$

Substituting the reflection coefficient as retrieved in Appendix E, we obtain

$$\hat{\mathbf{G}}_{rss} = -j\frac{\exp(-j(ux'+vy'))}{2k_s^2 k_{z1s}}$$

$$\times \begin{pmatrix} \frac{k_s^2 v^2 (k_{z10}-k_{z1s})}{(u^2+v^2)(k_{z10}+k_{z1s})} + \frac{u^2 k_{z1s}^2 (k_{z1s}\varepsilon_0 - k_{z10}\varepsilon_s)}{(u^2+v^2)(k_{z1s}\varepsilon_0 + k_{z10}\varepsilon_s)} & \frac{-k_s^2 uv(k_{z10}-k_{z1s})}{(u^2+v^2)(k_{z10}+k_{z1s})} + \frac{uvk_{z1s}^2 (k_{z1s}\varepsilon_0 - k_{z10}\varepsilon_s)}{(u^2+v^2)(k_{z1s}\varepsilon_0 + k_{z10}\varepsilon_s)} & \frac{-u\,\mathrm{sgn}(z-z')k_{z1s}(k_{z1s}\varepsilon_0 - k_{z10}\varepsilon_s)}{(k_{z1s}\varepsilon_0 + k_{z10}\varepsilon_s)} \\[2mm] \frac{-k_s^2 uv(k_{z10}-k_{z1s})}{(u^2+v^2)(k_{z10}+k_{z1s})} + \frac{uvk_{z1s}^2 (k_{z1s}\varepsilon_0 - k_{z10}\varepsilon_s)}{(u^2+v^2)(k_{z1s}\varepsilon_0 + k_{z10}\varepsilon_s)} & \frac{k_s^2 u^2 (k_{z10}-k_{z1s})}{(u^2+v^2)(k_{z10}+k_{z1s})} + \frac{v^2 k_{z1s}^2 (k_{z1s}\varepsilon_0 - k_{z10}\varepsilon_s)}{(u^2+v^2)(k_{z1s}\varepsilon_0 + k_{z10}\varepsilon_s)} & \frac{-v\,\mathrm{sgn}(z-z')k_{z1s}(k_{z1s}\varepsilon_0 - k_{z10}\varepsilon_s)}{(k_{z1s}\varepsilon_0 + k_{z10}\varepsilon_s)} \\[2mm] \frac{u\,\mathrm{sgn}(z-z')k_{z1s}(k_{z1s}\varepsilon_0 - k_{z10}\varepsilon_s)}{(k_{z1s}\varepsilon_0 + k_{z10}\varepsilon_s)} & \frac{v\,\mathrm{sgn}(z-z')k_{z1s}(k_{z1s}\varepsilon_0 - k_{z10}\varepsilon_s)}{(k_{z1s}\varepsilon_0 + k_{z10}\varepsilon_s)} & \frac{(u^2+v^2)(k_{z1s}\varepsilon_0 - k_{z10}\varepsilon_s)}{(k_{z1s}\varepsilon_0 + k_{z10}\varepsilon_s)} \end{pmatrix}$$

$$(11.80)$$

Let us now label as $\mathbf{M}_{rss}$ the matrix in large parentheses. In this way, the contribution to the half-space Green's function coming from the reflection at the soil-air interface can be compactly written as

$$\mathbf{G}_{rss}(\mathbf{r},\mathbf{r}') = \frac{j}{4\pi^2 k_s^2} \int_{-\infty}^{+\infty}\int_{-\infty}^{+\infty} \frac{\exp(j(u(x-x')+v(y-y')-k_{z1s}(z+z')))}{k_{z1s}}\mathbf{M}_{rss}(u,v)\,du\,dv$$

$$(11.81)$$

Let us remind, at this point, after Eq. (11.37), that we can express compactly the Green's function of a homogeneous soil[4] as

$$\mathbf{G}_s(\mathbf{r},\mathbf{r}') = \left(\mathbf{I} + \frac{1}{k_s^2}\nabla\nabla\right)\left(\frac{\exp(-jk_s\|\mathbf{r}-\mathbf{r}'\|)}{4\pi\|\mathbf{r}-\mathbf{r}'\|}\right) \tag{11.82}$$

Equation (11.82) provides the direct contribution to the Green's function in a half space, with both a buried source and observation point. Eventually the internal Green's function of the problem is given by

$$\mathbf{G}_i(\mathbf{r},\mathbf{r}') = \mathbf{G}_{ss}(\mathbf{r},\mathbf{r}') = \mathbf{G}_{rss}(\mathbf{r},\mathbf{r}') + \mathbf{G}_s(\mathbf{r},\mathbf{r}')$$

$$\frac{j}{4\pi^2 k_s^2} \int_{-\infty}^{+\infty}\int_{-\infty}^{+\infty} \frac{\exp(j(u(x-x')+v(y-y')-k_{z1s}(z+z')))}{k_{z1s}}\mathbf{M}_{rss}(u,v)\,du\,dv$$

$$+ \left(\mathbf{I} + \frac{1}{k_s^2}\nabla\nabla\right)\left(\frac{\exp(-jk_s\|\mathbf{r}-\mathbf{r}'\|)}{4\pi\|\mathbf{r}-\mathbf{r}'\|}\right) \tag{11.83}$$

---

[4] Actually Eq. (11.37) refers to the free space, but the extension to the case of any homogeneous space is straightforward.

At this point, the scattering equations have been fully specified. In fact, they are given by Eqs. (11.1) and (11.2), with the incident field specified in Eq. (11.73), the external Green's function specified in Eq. (11.75), and the internal Green's function specified in Eq. (11.83).

The same procedure can be followed in order to calculate also the Green's function $\mathbf{G}_{aa}(\mathbf{r},\mathbf{r}')$, relative to the case of a half-space with both the source and observation points in air. However, this quantity does not enter the scattering equations, and we leave it as a possible exercise. The quantity $\mathbf{G}_{aa}(\mathbf{r},\mathbf{r}')$ is theoretically relevant if one wants to retrieve the scattered field data by subtraction of the incident field data from the total field data [see Eq. (11.13)]. However, as already emphasized in Chapter 7, this is not a robust procedure with experimental GPR data.

QUESTIONS

1. Is the 3D inverse scattering problem ill-posed? How could we prove this?
2. In the 2D case dealt with in Chapters 4 and 5, we have that the electric field is parallel to the direction of the source. Is it the same in 3D?
3. What is the main difference between the gauge of Lorentz applied in the calculations in Chapter 5 and that applied in Chapter 10?
4. In the 2D case we did not distinguish a reflection coefficient for TE polarization and another one for TM polarization. Why?
5. Is the GPR datum gathered under a 3D model a vector quantity for each observation point?

THREE-DIMENSIONAL DIFFRACTION TOMOGRAPHY

12.1 BORN APPROXIMATION AND DT IN 3D

The (first-order) Born approximation (BA) in 3D can be introduced in the same way as done in 2D, namely approximating the internal field with the incident one. The underlying physical rationale is the same as in 2D, and therefore it will be not repeated.

With the symbols introduced in Chapter 11, and with reference to Figure 11.1, this means that, under BA, the scattering equations reduce to

$$\mathbf{E}(\mathbf{r},\mathbf{r}_s) = \mathbf{E}_{inc}(\mathbf{r},\mathbf{r}_s), \quad \mathbf{r} \in D, \ \mathbf{r}_s \in \Sigma \tag{12.1}$$

$$\mathbf{E}_s(\mathbf{r}_0,\mathbf{r}_s) = k_s^2 \iiint_D \chi_e(\mathbf{r}')\mathbf{G}_e(\mathbf{r}_0,\mathbf{r}')\mathbf{E}_{inc}(\mathbf{r}',\mathbf{r}_s) \, d\mathbf{r}', \quad \mathbf{r}_0, \ \mathbf{r}_s \in \Sigma \tag{12.2}$$

Introduction to Ground Penetrating Radar: Inverse Scattering and Data Processing,
First Edition. Raffaele Persico.
© 2014 The Institute of Electrical and Electronics Engineers, Inc. Published 2014 by John Wiley & Sons, Inc.

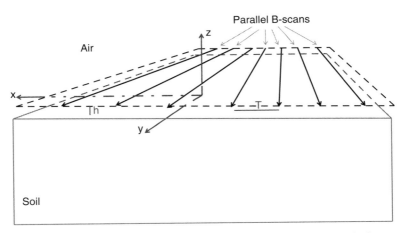

Figure 12.1. Geometrical scheme of the common offset C-scan. For color detail, please see color plate section.

The linear scattering equation (12.2) is a vector relationship. However, the GPR datum is in any case a scalar quantity, because it is essentially the tension value gathered by the receiving antenna. The correct relationship between the field and retrieved tension should account for the effective length of the receiving antenna, as illustrated in Chapter 4. However, as said in Chapter 11, we will approximate the received datum as the projection of the field in the observation point along the direction of the receiving dipole. Moreover, we will refer to the customary GPR prospecting performed in common offset (Daniels, 2004), so that we will assume that the transmitting and receiving antennas are two dipoles parallel to each other and also parallel to the air–soil interface. Therefore, with reference to Figure 12.1, it is comfortable to assume that a reference system such as the transmitting and receiving dipoles are directed either along the x-axis either along the y-axis. Let us focus on the case when the two dipoles are directed along the x-axis, are separated by a fixed offset Δ, and move sequentially along several lines parallel to the y-axis separated from each other by a fixed interline distance (transect) T and at a fixed height h (possibly and hopefully equal to zero, due to the considerations exposed in Chapter 9). This configuration is commonly called a C-scan (Daniels, 2004). Thus, resorting to Eqs. (11.4) and (11.11), the scalar scattered field datum under BA can be written as

$$E_{sx}(\mathbf{r}_0,\mathbf{r}_s) = -jk_s^2\mu_0\omega I_0 l \iint_D \chi_e(\mathbf{r}')i_x \cdot \mathbf{G}_{sa}(\mathbf{r}_0,\mathbf{r}')\mathbf{G}_{as}(\mathbf{r}',\mathbf{r}_s)\cdot i_x \, d\mathbf{r}', \quad \mathbf{r}_0, \mathbf{r}_s \in \Sigma \quad (12.3)$$

Substituting the expressions of $\mathbf{G}_{as}(\mathbf{r},\mathbf{r}_s)$ and $\mathbf{G}_{sa}(\mathbf{r}_0,\mathbf{r}')$ from Eqs. (11.71) and (11.75), we obtain

$$E_{sx}(\mathbf{r},\mathbf{r}_s) = \frac{-j\mu_0\omega I_0 l}{16\pi^4 k_0^2} \int\limits_{-\infty}^{+\infty}\int\limits_{-\infty}^{+\infty}\int\limits_{-\infty}^{+\infty} dx'\,dy'\,dz'\,\chi_e(x',y',z')$$

$$\times \int\limits_{-\infty}^{+\infty}\int\limits_{-\infty}^{+\infty} du_1\,dv_1 \int\limits_{-\infty}^{+\infty}\int\limits_{-\infty}^{+\infty} du_2\,dv_2 \frac{(i_x\cdot\mathbf{M}_{sa}(u_1,v_1)\mathbf{M}_{as}(u_2,v_2)\cdot i_x)}{(u_1^2+v_1^2)(u_2^2+v_2^2)}$$

$$\times \exp(j(u_1(x-x') + v_1(y-y') - u_2(x'-x_s) - v_2(y'-y_s) - (zk_{z10}(u_1,v_1)$$

$$+ z_s k_{z10}(u_2,v_2)) - (k_{z1s}(u_1,v_1) + k_{z1s}(u_2,v_2))z')), \quad \mathbf{r},\,\mathbf{r}_s \in \Sigma \qquad (12.4)$$

Due to the considered configuration, we have

$$y_s = y + \Delta,$$

$$x_s = x, \qquad (12.5)$$

$$z = z_s = h$$

Consequently the scalar scattered field datum is given by

$$E_{sx}(x,y;\omega,h,\Delta) = \frac{-j\mu_0\omega I_0 l}{16\pi^4 k_0^2} \int\limits_{-\infty}^{+\infty}\int\limits_{-\infty}^{+\infty}\int\limits_{-\infty}^{+\infty} dx'\,dy'\,dz'\,\chi_e(x',y',z')$$

$$\times \int\limits_{-\infty}^{+\infty}\int\limits_{-\infty}^{+\infty} du_1\,dv_1 \int\limits_{-\infty}^{+\infty}\int\limits_{-\infty}^{+\infty} du_2\,dv_2 \frac{(i_x\cdot\mathbf{M}_{sa}(u_1,v_1)\mathbf{M}_{as}(u_2,v_2)\cdot i_x)}{(u_1^2+v_1^2)(u_2^2+v_2^2)}$$

$$\times \exp(-j((u_1+u_2)x' + (v_1+v_2)y'))\exp(j((u_1+u_2)x$$

$$+ (v_1+v_2)y))\exp(-jz'((k_{z1s}(u_1,v_1) + k_{z1s}(u_2,v_2))))$$

$$\times \exp(jv_2\Delta)\exp(-jh(k_{z10}(u_1,v_1) + k_{z10}(u_2,v_2))), \quad \mathbf{r},\,\mathbf{r}_s \in \Sigma \quad (12.6)$$

In the 3D case, we will focus only on the case with data gathered at the air–soil interface ($h = 0$). That said, let us pose

$$u_1 + u_2 = p$$

$$v_1 + v_2 = q \qquad (12.7)$$

Thus, Eq. (12.6) is rewritten as

$$E_{sx}(x,y;\omega,0,\Delta) = \frac{-j\mu_0\omega I_0 l}{16\pi^4 k_0^2} \int_{-\infty}^{+\infty}\int_{-\infty}^{+\infty}\int_{-\infty}^{+\infty} dx'\,dy'\,dz'\chi_e(x',y',z')$$

$$\times \int_{-\infty}^{+\infty}\int_{-\infty}^{+\infty} dp\,dq \int_{-\infty}^{+\infty}\int_{-\infty}^{+\infty} du_2\,dv_2 \frac{(i_x\cdot\mathbf{M}_{sa}(p-u_2,q-v_2)\mathbf{M}_{as}(u_2,v_2)\cdot i_x)}{\left((p-u_2)^2+(q-v_2)^2\right)\left(u_2^2+v_2^2\right)}$$

$$\times \exp(-j(px'+qy'))\exp(j(px+qy))\exp(-jz'((k_{z1s}(p-u_2,q-v_2)$$

$$+k_{z1s}(u_2,v_2))))\exp(jv_2\Delta), \quad \mathbf{r},\,\mathbf{r}_s \in \Sigma \qquad (12.8)$$

In Eq. (12.8) we can recognize a double direct Fourier transform with respect to the contrast and a double inverse Fourier transform with respect to the data. Thus, from Eq. (12.8) we obtain

$$\hat{E}_{sx}(p,q;\omega,0,\Delta) = \frac{-j\mu_0\omega I_0 l}{4\pi^2 k_0^2} \int_{-\infty}^{+\infty} dz'\hat{\chi}_e(p,q,z')$$

$$\times \int_{-\infty}^{+\infty}\int_{-\infty}^{+\infty} \frac{(i_x\cdot\mathbf{M}_{sa}(p-u_2,q-v_2)\mathbf{M}_{as}(u_2,v_2)\cdot i_x)}{\left((p-u_2)^2+(q-v_2)^2\right)\left(u_2^2+v_2^2\right)}$$

$$\times \exp(-jz'((k_{z1s}(p-u_2,q-v_2)$$

$$+k_{z1s}(u_2,v_2))))\exp(jv_2\Delta)\,du_2\,dv_2, \quad \mathbf{r},\,\mathbf{r}_s \in \Sigma \qquad (12.9)$$

The double integral in $du_2\,dv_2$ is solved under the stationary phase approximation for high values of z', which amounts to assume targets that are not too shallow, analogously to the 2D case. There is a unique first-order stationary point at

$$u_2 = \frac{p}{2}, \quad v_2 = \frac{q}{2} \qquad (12.10)$$

Thus, under the above approximations we have

$$\hat{E}_{sx}(p,q;\omega,0,\Delta) = \int_{-\infty}^{+\infty} W(p,q;\omega,0,\Delta)\frac{\hat{\chi}_e(p,q,z')}{z'}\exp\left(-j2z'k_{z1s}\left(\frac{p}{2},\frac{q}{2}\right)\right)dz' \qquad (12.11)$$

where the spectral weight $W(p,q;\omega,0,\Delta)$ is given by

$$W(p,q;\omega,0,\Delta)$$

$$= \frac{\mu_0 \omega I_0 l}{\pi k_0^2} \frac{\left(4k_s^2 - p^2 - q^2\right)^{3/2} \left(i_x \cdot \mathbf{M}_{sa}\left(\frac{p}{2},\frac{q}{2}\right) \mathbf{M}_{as}\left(\frac{p}{2},\frac{q}{2}\right) \cdot i_x\right) \exp\left(-j\frac{p}{2}\Delta\right)}{\left(p^2 + q^2\right)^2 \sqrt{\left(4k_s^2 - p^2\right)\left(4k_s^2 - q^2\right)}}$$

$$= \frac{\mu_0 \omega I_0 l \exp\left(-j\frac{p}{2}\Delta\right) \left(4k_s^2 - p^2 - q^2\right)^{3/2}}{\pi k_0^2 (p^2 + q^2)^2 \sqrt{\left(4k_s^2 - p^2\right)\left(4k_s^2 - q^2\right)}} \sum_{n=1}^{3} \mathbf{M}_{sa1n}\left(\frac{p}{2},\frac{q}{2}\right) \mathbf{M}_{asn1}\left(\frac{p}{2},\frac{q}{2}\right)$$

$$= \frac{8\mu_0 \varepsilon_{sr} \omega I_0 l \exp\left(-j\frac{p}{2}\Delta\right) k_{z1s}^3}{\pi k_0^2 (p^2 + q^2)^2 \sqrt{\left(4k_s^2 - p^2\right)\left(4k_s^2 - q^2\right)}} \left[\left(\frac{k_0^2 q^2}{k_{z10} + k_{z1s}} + \frac{\varepsilon_0 p^2 k_{z1s} k_{z10}}{k_{z1s}\varepsilon_0 + k_{z10}\varepsilon_s}\right)^2\right.$$

$$\left. + \left(\frac{k_0^2 pq}{k_{z10} + k_{z1s}} - \frac{\varepsilon_0 pq k_{z1s} k_{z10}}{k_{z1s}\varepsilon_0 + k_{z10}\varepsilon_s}\right)^2 + \left(\frac{\varepsilon_0 p k_{z10}(p^2 + q^2)}{2(k_{z1s}\varepsilon_0 + k_{z10}\varepsilon_s)}\right)^2\right] \qquad (12.12)$$

The second member of Eq. (12.11) represents the Fourier transform of the auxiliary quantity

$$\chi_{e1}(x',y',z') = \frac{\chi_e(x',y',z')}{z'} \qquad (12.13)$$

calculated in the point

$$\eta(p,q) = p,$$

$$\xi(p,q) = q,$$

$$\varsigma(p,q) = 2\sqrt{k_s^2 - \frac{p^2}{4} - \frac{q^2}{4}} = \sqrt{4k_s^2 - p^2 - q^2} = \sqrt{4\frac{\omega^2}{c^2} - p^2 - q^2} \qquad (12.14)$$

In the end, the DT relationship in 3D is given by

$$\hat{E}_{sx}(p,q;\omega,0,\Delta) = W(p,q;\omega,0,\Delta)\hat{\tilde{\chi}}_{e1}(\eta(p,q),\xi(p,q),\varsigma(p,q)) \qquad (12.15)$$

with the spectral weight given by Eq. (12.12) and the coordinative transformation given by Eq. (12.14).

As can be seen, Eq. (12.15) is a conceptually straightforward extension of the DT relationship in 2D, expressed by Eq. (9.11).

12.2   IDEAL AND LIMITED-VIEW-ANGLE 3D RETRIEVABLE SPECTRAL SETS

Based on the coordinative transformation given in Eq. (12.14), relationship (12.15) provides a Fourier relationship between the spectrum of the data and that of the contrast normalized to the depth if and only if the soil is lossless and the couple (p,q) verifies the inequality $p^2 + q^2 \le 4k_s^2$. This means that the 2D ideal (i.e., relative to an infinite observation line) visible interval in 3D becomes the ideal visible circle, centered in the origin and with ray equal to $2k_s$ in the plane (p,q). It is characterized by the inequality

$$C_v = (p,q) : p^2 + q^2 \le 4k_s^2 \tag{12.16}$$

This ideal circle corresponds to the case of an infinite observation plane with effective view angle ranging from $-\pi/2$ to $\pi/2$ along any horizontal direction.

At any fixed frequency, from Eq. (12.14) we recognize that the visible circle is transformed in a curved surface in the space (η,ξ,ς), whose equation is given by

$$\eta^2 + \xi^2 + \varsigma^2 = 4k_s^2 \tag{12.17}$$

Equation (12.17) is the equation of a sphere centered in the origin and with ray $2k_s$ in the space (η,ξ,ς). It is easy to recognize, in particular, that the visible circle C_v corresponds to the half-sphere of Eq. (12.17) enclosed in the half-space $\varsigma \ge 0$.

At variance of the frequency, the spectral retrievable set in 3D is given by the points (η,ξ,ς) enclosed between the two half spheres centered in the origin with minimum ray $2k_{s\,min}$ and maximum ray $2k_{s\,max}$, respectively. This means that the retrievable spectral set is limited. In particular, we have that the 3D retrievable spectral set is given by the solid of revolution of the homologous 2D set around the ς-axis.

It is important to outline that the spectral weight tends to zero all over the bound of the visible circle, so that the "actual" retrievable spectral set is never equal to the ideal one, in the sense that it would not be equal to the ideal set even if the measurement plane surface were unlimited, analogously to what happened in 2D (see Section 9.2).

With regard to the case of a limited view angle, the equivalent in 3D of Section 9.3 should be developed. The procedure is conceptually straightforward but of course quite long. In particular, the maximum view angle (meant at the moment in a merely geometrical sense) obviously depends on the shape of the bound of the observation surface and in general is not the same along any horizontal direction.

In the (common) case that the area of interest is large with respect to the square of the central wavelength in the soil, and on condition that the shape of this area is not extremely elongated, it makes a sense to introduce the concept of an effective maximum view angle, which can be averagely estimated from the data in the same way shown in the 2D case. Rigorously, the maximum view angle is in general a quantity depending on the particular horizontal direction, and in particular it is in general not the same along the x-axis or along the y-axis. Physically, this is because the radiation patterns of the antennas are in general not symmetrical around the z-axis. In particular, in the considered case (a Hertzian dipole as source and a mere "projector" as receiver) this can be recognized from the fact that in Eq. (12.12) the spectral weight is not a function of $p^2 + q^2$, nor do we have $W(p,q) = W(q,p)$.

If a particularly refined analysis is required or desired, we might estimate from the data the maximum effective view $\theta_{e\,\max y}$ along the y-axis (i.e., the axis along which the B-scans develops) and the effective maximum view angle $\theta_{e\,\max x}$ along the x-axis (theoretically, this can be retrieved joining the traces of the different B-scans at the same value of y). In this case, the visible set can be approximated by the canonical ellipsis with axes $4k_s \sin(\theta_{e\,\max y})$ and $4k_s \sin(\theta_{e\,\max x})$ in the plane (p,q). The fact that the visible effective set in general is not circular drives to the consideration that (rigorously) the retrievable spectral set is not rotationally symmetric around the axis ς. This means, on one side, that the optimal spatial step along y and along x, namely the spatial step between the traces along any B-scan and the transect between any two adjacent B-scans, are not rigorously equal to each other, and this also means that the resolution achievable along x is rigorously not the same as that expected along y or along any other horizontal direction. However, $\theta_{e\,\max x}$ and $\theta_{e\,\max y}$ are expected of the same order, and since we look for an order of magnitude with regard to both the spatial step to adopt and the resolution that we can hope to achieve, we will refer just to a $\theta_{e\,\max}$ that will be assumed equal along any horizontal direction. Operatively, it is more simple to evaluate $\theta_{e\,\max}$ from the B-scans, gathered along the y-axis in the considered case. Under this assumption, the effective visible set accounting for the finite view angle is circular, and its equation is

$$C_{ve} = (p,q) : p^2 + q^2 \leq 4k_s^2 \sin^2(\theta_{emax}) \qquad (12.18)$$

Inequality (12.18) describes the circle centered in the origin and with ray equal to $2k_s \sin(\theta_{e\,\max})$.

If the frequency is between $f_{\min}$ and $f_{\max}$, we will achieve the spectral set as the intersection (in the half-space $\varsigma > 0$) of volume enclosed between the two spheres centered in the origin with ray $2k_{s\,\min}$ and $2k_{s\,\max}$ and the cone of angular aperture $\theta_{e\,\max}$. Mathematically, the spectral set S is described as follows:

$$S = S_1 \cap S_2 \cap S_3 \cap S_4 \qquad (12.19)$$

where

$$S_1 = (\eta,\xi,\varsigma) : \varsigma \geq 0,$$
$$S_2 = (\eta,\xi,\varsigma) : \eta^2 + \xi^2 + \varsigma^2 \geq 4k_{s\min}^2,$$
$$S_3 = (\eta,\xi,\varsigma) : \eta^2 + \xi^2 + \varsigma^2 \leq 4k_{s\max}^2, \qquad (12.20)$$
$$S_4 = (\eta,\xi,\varsigma) : \eta^2 + \xi^2 \leq \varsigma^2 \frac{\sin^2(\theta_{emax})}{\cos^2(\theta_{emax})}$$

A quantitative representation of the retrievable spectral set is given in Figure 12.2.

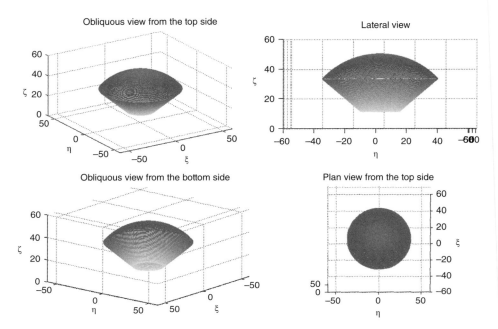

<u>Figure 12.2.</u> Quantitative representation of the effective 3D retrievable spectral set. Involved parameters: $f_{min} = 200$ MHz, $f_{max} = 600$ MHz, $\varepsilon_{sr} = 4$, and $\theta_{e\,max} = 0.8481$ radians ($\sin(\theta_{e\,max}) = 0.75$).

12.3 SPATIAL STEP AND TRANSECT

The spatial step and the transect or, in other terms, the spatial needed steps along the direction of the movement of the antennas and along the horizontal direction orthogonal to this are driven by the Nyquist criterion. Therefore, the spatial step both along the x- and y-axis has to be not larger than 2π times the inverse of the diameter of the visible circle, both along the x-axis and along the y-axis. This step is the maximum one allowed in order to guarantee an aliasing-free reconstruction of the effective visible circle. Of course, we are mainly interested in mutlifrequency prospecting and so, conservatively, the bound will be referred to the maximum involved frequency. Consequently, in the end we have

$$\text{spatial step} = s = \frac{2\pi}{2(2k_{smax}\sin(\theta_{emax}))} = \frac{2\pi}{4\sin(\theta_{emax})\dfrac{2\pi}{\lambda_{smin}}} = \frac{\lambda_{smin}}{4\sin(\theta_{emax})},$$

$$\text{transect} = T = \frac{2\pi}{2(2k_{smax}\sin(\theta_{emax}))} = \frac{2\pi}{4\sin(\theta_{emax})\dfrac{2\pi}{\lambda_{smin}}} = \frac{\lambda_{smin}}{4\sin(\theta_{emax})}$$

$$(12.21)$$

Analogously to the 2D case, the needed spatial step (and also the needed transect) can be relaxed if the targets of interest are not very shallow.

With regard to the spatial step, the 3D result confirms the 2D result [see Eq. (9.49)]. Instead, the evaluation of the needed transect makes sense only within a 3D model. From a practical point of view, in most cases it is not a problem to guarantee the Nyquist rate with regard to the spatial step along the B-scan, because the step guaranteed by the odometer is in most cases automatically narrower than the Nyquist rate. Instead, to guarantee the Nyquist rate with regard to the transect is in most cases unpractical, at least with a classical (single antenna) GPR system. To mitigate this problem, in many cases a grid of measurements along two orthogonal C-scans, along both the x-axis and the y-axis (Conyers, 2004), is taken. In this way, the combination (i.e., the multiplication, pixel by pixel, after some suitable interpolation) of the two images achieved from the two C-scans mitigates the "necessarily too large" transect.

On the other hand, it is also to be outlined that, in most cases, even if the transect is larger than that theoretically required, the user can achieve meaningful and above all useful results (the published case histories are endless because indeed a transect larger than the Nyquist rate is the praxis). This happens for two reasons:

1. Based on the DT equation (12.15), the spectrum of the field within the visible circle is proportional to the spectrum of the object function. The effective visible circle is approximately the maximum possible extension of the spatial spectrum of the field at a given frequency. However, the actual extension of the spectrum of the field might be even narrower, which relaxes the anti-aliasing requirements. This can happen if the targets of interest are larger than the Nyquist rate. In other words, we really need the Nyquist rate only if we have to achieve the maximum available resolution, which fortunately does not occur in all the case histories.

2. The effective visible spectrum depends also on the depth of the targets of interest, because of the sine of the maximum effective view angle, which customarily decreases versus the depth, as said. Thus, the Nyquist rate is "fully" needed only if the targets of interest are small and shallow. In particular, for deeper targets the achievable performances get unavoidably degraded but, on the other hand, also the sampling requirements get consequently relaxed.

12.4 HORIZONTAL RESOLUTION

Raffaele Persico and Raffaele Solimene

In order to estimate the horizontal (and then the vertical) resolution, we will follow the same steps already implemented in the 2D case (see Section 9.5). In particular, neglecting the effect of the spectral weighting function, we will approximate the reconstruction of a point-like target—that is, the point spread function, with the inverse Fourier transform of the spectrum of the object function restricted to the retrievable spectral set.

Still, in order to achieve a closed-form solution, we will approximate the retrievable spectral set as a cylinder. The basis of the cylinder is the visible circle at the central frequency—that is, the circle with equation

$$\eta^2 + \xi^2 = 4k_{sc}^2 \sin^2(\theta_{e\max}) \tag{12.22}$$

The height of the cylinder ranges between $2k_{s\min}$ and $2k_{s\max}$ along the ς-axis. In Figure 12.3, the approximated cylindrical spectral set is quantitatively represented, in comparison with the actual spectral set. The involved parameters are the same as in Figure 12.2. Let us refer to this volume as Cyl. This cylinder is the solid of revolution around the ς-axis of the dashed rectangle that approximates the spectral set in 2D (see Figure 9.11).

At this point, neglecting also the unessential normalization of the contrast to the depth [see Eq. (12.13)], let us consider a point-like target described as

$$\chi_e(x',y',z') = \chi_0 \delta(x' - x_0)\delta(y' - y_0)\delta(z' - z_0) \tag{12.23}$$

Its point spread function is given by

$$\chi_{er}(x',y',z') = \frac{\chi_0}{8\pi^3} \iint_{Cyl} \exp(j(x' - x_0)\eta) \exp(j(y' - y_0)\xi) \exp(j(z' - z_0)\varsigma)\, d\eta d\xi d\varsigma \tag{12.24}$$

It is convenient to solve in polar coordinates the double integral, so we write

$$\begin{aligned}
\eta &= \rho\cos(\varphi), \\
\xi &= \rho\sin(\varphi), \\
x' - x_0 &= \rho_1 \cos(\varphi_1), \\
y' - y_0 &= \rho_1 \sin(\varphi_1)
\end{aligned} \tag{12.25}$$

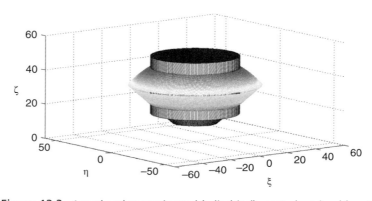

Figure 12.3. Actual and approximated (cylindrical) spectral retrievable set.

where

$$\rho = \sqrt{\eta^2 + \xi^2},$$

$$\rho_1 = \sqrt{(x'-x_0)^2 + (y'-y_0)^2},$$

$$\varphi = \begin{cases} \arcsin\left(\dfrac{\xi}{\sqrt{\eta^2+\xi^2}}\right) & \text{if } \eta \geq 0, \\[4mm] \pi - \arcsin\left(\dfrac{\xi}{\sqrt{\eta^2+\xi^2}}\right) & \text{if } \eta < 0 \end{cases} \tag{12.26}$$

$$\varphi_1 = \begin{cases} \arcsin\left(\dfrac{y'-y_0}{\sqrt{(x'-x_0)^2+(y'-y_0)^2}}\right) & \text{if } x'-x_0 \geq 0, \\[4mm] \pi - \arcsin\left(\dfrac{y'-y_0}{\sqrt{(x'-x_0)^2+(y'-y_0)^2}}\right) & \text{if } x'-x_0 < 0 \end{cases}$$

Thus, Eq. (12.24) can be rewritten as

$$\chi_{er}(x',y',z') = \frac{\chi_0}{8\pi^3} \int_{2k_{s\,\min}}^{2k_{s\,\max}} \exp(j(z'-z_0)\varsigma)\,d\varsigma \int_0^{2k_{sc}\sin(\theta_{e\max})} \rho\,d\rho \int_0^{2\pi} \exp(j\rho\rho_1 \cos(\varphi-\varphi_1))\,d\varphi =$$

$$= \frac{\chi_0}{8\pi^3} \int_{2k_{s\,\min}}^{2k_{s\,\max}} \exp(j(z'-z_0)\varsigma)\,d\varsigma \int_0^{2k_{sc}\sin(\theta_{e\max})} \rho\,d\rho \int_0^{2\pi} \exp(j\rho\rho_1 \cos(\varphi))\,d\varphi =$$

$$= \frac{\chi_0}{8\pi^3}(k_{s\,\max}-k_{s\,\min})\exp(j2k_{sc}(z'-z_0))\operatorname{sinc}((k_{s\,\max}-k_{s\,\min})(z'-z_0)) \times$$

$$\times \int_0^{2k_{sc}\sin(\theta_{e\max})} \rho\,d\rho \int_0^{2\pi} \exp(j\rho\rho_1 \cos(\varphi))\,d\varphi \tag{12.27}$$

In particular, in Eq. (12.27) we have exploited the periodicity of the argument of the third integral, which makes the result independent of φ_1.

Based on the calculation steps developed in Appendix F, we can solve the integral (12.27) in terms of the Bessel's function of first kind and order one (Abramowitz and Stegun, 1972) as follows:

$$\chi_{er}(x',y',z')$$
$$= \frac{4\chi_0(k_{s\,max}-k_{s\,min})\exp(j2k_{sc}(z'-z_0))k_{sc}^2\sin^2(\theta_{e\,max})}{\pi^2}$$

$$\times \operatorname{sin} c((k_{s\,max}-k_{s\,min})(z'-z_0))\frac{J_1\left(2k_{sc}\sin(\theta_{e\,max})\sqrt{(x'-x_0)^2+(y'-y_0)^2}\right)}{2k_{sc}\sin(\theta_{e\,max})\sqrt{(x'-x_0)^2+(y'-y_0)^2}}$$

$$(12.28)$$

In Figure 12.4, the function $J_1(x)/x$ is represented versus x: It is an even function somehow similar to a sinc function. In particular, the first zeroes of the function are at $x = 3.84$. Thus, the extension of the main lobe of the function is provided by the following equation:

$$2k_{sc}\sin(\theta_{emax})\sqrt{(x'-x_0)^2+(y'-y_0)^2}$$
$$=2\times3.84 \Rightarrow \sqrt{(x'-x_0)^2+(y'-y_0)^2}=\frac{3.84}{k_{sc}\sin(\theta_{e\,max})}=\frac{3.84}{\pi}\frac{\lambda_{sc}}{2\sin(\theta_{e\,max})} \quad (12.29)$$

Similarly to the 2D case, this distance also provides the horizontal resolution, which therefore is given by

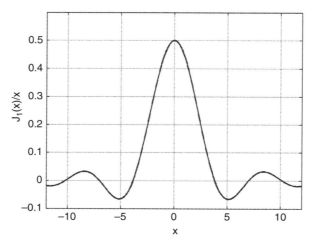

Figure 12.4. Graph of $J_1(x)/x$.

$$HR = \frac{3.84}{\pi} \frac{\lambda_{sc}}{2\sin(\theta_{e\,max})} \approx 1.22 \frac{\lambda_{sc}}{2\sin(\theta_{e\,max})} \qquad (12.30)$$

Since what we can retrieve is (only) an order of magnitude, Eq. (12.30) essentially confirms the 2D result of Eq. (9.44), namely the fact that best retrievable horizontal resolution is expected of the order of one-half of the internal wavelength for shallow targets and is expected to degrade progressively versus the depth.

12.5 VERTICAL RESOLUTION, FREQUENCY AND TIME STEPS

The calculation of the vertical resolution is fully analogous to the 2D case. This stems from the fact that, in the calculation of the point spread function, the integral in $d\varsigma$ is decoupled from the integrals along the other spectral variables and is exactly the same in 2D and 3D [see Eqs. (9.42) and (12.28)]. Consequently, the vertical achievable resolution can be quantified by means of Eqs. (9.45) and (9.46), here repeated for a more comfortable reading:

Lossless case

$$VR = \frac{c}{B} \approx \frac{c}{f_c} = \lambda_{sc} \qquad (12.31)$$

Lossy case (perturbative approach)

$$VR = \frac{c}{B_e} \approx \frac{c}{f_{ce}} = \lambda_{sce} \qquad (12.32)$$

With regard to the needed frequency step, the calculations are straightforwardly derived from the 2D dealing of Section 9.8, with the only difference that condition (9.53) is replaced by its equivalent in 3D

$$\Delta\varsigma(\eta,\xi) = \sqrt{4(k_s + \Delta k_s)^2 - \eta^2 - \xi^2} - \sqrt{4k_s^2 - \eta^2 - \xi^2} \qquad (12.33)$$

whose minimum versus η and ξ is achieved at $(\eta,\xi) = (0,0)$.

Thus, also in this case we will not repeat the passages, referring the reader to Section 9.8. Here, we will just repeat the result, which is [see Eq. (9.52)]

$$\Delta f = \frac{c_0}{2b\sqrt{\varepsilon_{sr}}} \qquad (12.34)$$

where b is the vertical extent of the investigation domain and ε_{sr} is the relative permittivity of the soil (we are not considering the case of magnetic soils in this chapter).

Let us also remind that this is the frequency step for processing the data, which is possibly different from the frequency step needed to gather the data in the frequency domain, for which the reader is referred to Chapter 3.

Finally, with regard to the time step, the dealing remains the same as in Section 9.9, and also in this case there is no difference with respect to the 2D case. In particular, the time step needed for gathering the data is given in any case by Eq. (3.39), and the time step needed to process the data in the time domain after filtering them for any reason (in particular, for some possible narrow band interferences) is given by Eq. (9.54).

QUESTIONS

1. Does the ideal (with infinite observation line) 2D visible interval at fixed frequency translate into a square range in the analogous 3D case?
2. Is the 3D retrievable spectral set the rotation volume of the 2D spectral set around its symmetry axis?
3. Is the horizontal resolution in 3D better than that available in 2D, because of the more accurate model?
4. Is the Nyquist criterion an indispensable requirement with regard to the spatial step of the data along both the x and y horizontal directions?

13

THREE-DIMENSIONAL MIGRATION ALGORITHMS

13.1 3D MIGRATION FORMULAS IN THE FREQUENCY DOMAIN

In this section we will provide 3D migration formulas for the case developed in the previous chapter, namely data gathered at the air–soil interface in common offset on a nonmagnetic soil and without magnetic targets. The source is assimilated to a Hertzian dipole, and the received signal is approximated as the projection of the field along the direction of the receiving dipole in the observation point.

The migration in 3D is based on the inversion of Eq. (12.15), which (omitting the dependence of the spectral weight on the height of the measurement line and on the offset) is formally written as

$$\hat{\tilde{\chi}}_{e1}(\eta,\xi,\varsigma) = \frac{1}{W(p(\eta,\xi,\varsigma),q(\eta,\xi,\varsigma);\omega(\eta,\xi,\varsigma))}\hat{\tilde{E}}_{SX}(p(\eta,\xi,\varsigma),q(\eta,\xi,\varsigma);\omega(\eta,\xi,\varsigma)) \quad (13.1)$$

Where the spatial spectral weight W is provided by Eq. (12.12) and where the inversion is defined only for (η,ξ,ς) belonging to the spectral retrievable set, which automatically provides also a regularization of the result. The formal expression of the coordinative

---

Introduction to Ground Penetrating Radar: Inverse Scattering and Data Processing,
First Edition. Raffaele Persico.
© 2014 The Institute of Electrical and Electronics Engineers, Inc. Published 2014 by John Wiley & Sons, Inc.

transformation from a volume in the space (η,ξ,ς) to a volume in the space (p,q,ω) is achieved by inverting Eq. (12.14), which provides

$$p = \eta,$$
$$q = \xi,$$
$$\omega = \frac{c}{2}\sqrt{\eta^2 + \xi^2 + \varsigma^2}$$

(13.2)

where c is the propagation velocity of the electromagnetic waves in the soil. Thus Eq. (13.1) can be more simply rewritten as

$$\hat{\hat{\chi}}_{e1}(\eta,\xi,\varsigma) = \frac{1}{W\left(\eta,\xi;\frac{c}{2}\sqrt{\eta^2 + \xi^2 + \varsigma^2}\right)}\hat{E}_{sx}\left(\eta,\xi;\frac{c}{2}\sqrt{\eta^2 + \xi^2 + \varsigma^2}\right)$$

(13.3)

The object function is provided by the 3D inverse Fourier transform of Eq. (13.1):

$$\chi_{e1}(x',y',z') = \frac{1}{8\pi^3}\iiint_{Sp\_set} \frac{1}{W\left(\eta,\xi,\frac{c}{2}\sqrt{\eta^2 + \xi^2 + \varsigma^2}\right)}\hat{E}_{sx}\left(\eta,\xi;\frac{c}{2}\sqrt{\eta^2 + \xi^2 + \varsigma^2}\right)$$

$$\times \exp(j(\eta x' + \xi y' + \varsigma z'))\, d\eta d\xi d\varsigma$$

(13.4)

where the integration is performed over the retrievable spectral set.

Equation (13.4) requires an interpolation of the spectrum of the data, but it is a three-dimensional inverse Fourier transform and thus can be implemented by means of fast IFFT algorithms.

Alternatively, the integration can be performed in the variables η, ξ, and ω, by means of the substitution

$$\frac{c}{2}\sqrt{\eta^2 + \xi^2 + \varsigma^2} = \omega \Rightarrow \varsigma = \sqrt{\frac{4\omega^2}{c^2} - \eta^2 - \xi^2} \Rightarrow d\varsigma = \frac{4\omega d\omega}{\sqrt{\frac{4\omega^2}{c^2} - \eta^2 - \xi^2}}$$

(13.5)

In this way, we can rewrite Eq. (13.4) as

$$\chi_{e1}(x',y',z') = \frac{1}{4\pi^3}\iiint_{Sp\_set_1} \frac{\omega \hat{E}_{sx}(\eta,\xi;\omega)}{W(\eta,\xi;\omega)\sqrt{\frac{4\omega^2}{c^2} - \eta^2 - \xi^2}}$$

$$\times \exp\left(j\left(\eta x' + \xi y' + \sqrt{\frac{4\omega^2}{c^2} - \eta^2 - \xi^2}z'\right)\right)d\eta d\xi d\omega$$

(13.6)

Equation (13.6) does not require any interpolation of the spectrum of the data, but it is not an inverse Fourier transform, and so the integral cannot be calculated by means of fast IFFT algorithms. In Eq. (13.6) the integration domain has been labeled as the Sp_set_1 because it is the spectral retrievable set in the space (η,ξ,ω) straightforwardly specified from the spectral retrievable set Sp_set in the space (η,ξ,ς), which coincides with the set S characterised by Eqs. (12.19) and (12.20).

We can achieve an integral expression where the extremes of the integration are explicated, by passing to polar coordinatives. In particular, we can substitute in Eq. (13.4)

$$\eta = \rho\cos(\varphi)\sin(\theta_1),$$
$$\xi = \rho\sin(\varphi)\sin(\theta_1), \qquad (13.7)$$
$$\varsigma = \rho\cos(\theta_1)$$

where the symbol θ_1 has been adopted in order not to confuse the integration variable with the maximum effective view angle. Substituting into Eq. (13.4) and putting in the integral also the Jacobian of the transformation $\rho^2\sin(\theta_1)$, we obtain

$$\chi_{el}(x',y',z') = \frac{1}{8\pi^3}\int_0^{2\pi}d\varphi\int_{\frac{\pi}{2}-\theta_{e\max}}^{\frac{\pi}{2}}\sin(\theta_1)d\theta_1\int_{2k_{s\min}}^{2k_{s\max}}\rho^2 d\rho$$

$$\times\frac{\hat{E}_{sx}\left(\rho\cos(\varphi)\sin(\theta_1),\rho\sin(\varphi)\sin(\theta_1),\frac{c}{2}\rho\right)}{W\left(\rho\cos(\varphi)\sin(\theta_1),\rho\sin(\varphi)\sin(\theta_1),\frac{c}{2}\rho\right)}$$

$$\times\exp(j(\rho\cos(\varphi)\sin(\theta_1)x'+\rho\sin(\varphi)\sin(\theta_1)y'+\rho\cos(\theta_1)z')) \quad (13.8)$$

Equation (13.8) requires some interpolation of the data and is not an inverse Fourier transform, so it is computationally less convenient with respect to the previous expressions. However, it allows us to write in a closed form the extremes of the integrals and allows us to appreciate more immediately the effect (and the relevance) of the maximum view angle.

Equations (13.4), (13.6), and (13.8), are three alternative expressions for the f–k migration in 3D. Similarly to the 2D case, a simplified but robust expression of the migration can be achieved by neglecting the effect of the spectral weight W and extending the spectral integral to the entire space (η,ξ,ς), or alternatively the entire space (η,ξ,ω), relying on the fact that the spectrum of the data automatically vanishes outside the visible circle. In particular, assimilating to a unitary (but not dimensionless) factor F the quantity $\dfrac{2\omega}{W(\eta,\xi;\omega)\sqrt{\dfrac{4\omega^2}{c^2}-\eta^2-\xi^2}}$ in Eq. (13.6), we can write:

$$\chi_{el}(x',y',z') \approx \frac{F}{8\pi^3} \int_{-\infty}^{+\infty} \int_{-\infty}^{+\infty} \int_{-\infty}^{+\infty} \hat{E}_{sx}(\eta,\xi;\omega) \exp\left(j\left(\eta x' + \xi y' + \sqrt{\frac{4\omega^2}{c^2} - \eta^2 - \xi^2}\,z'\right)\right) d\eta d\xi d\omega$$

(13.9)

or alternatively in the space (η,ξ,ς)

$$\chi_{el}(x',y',z') \approx \frac{cF}{16\pi^2} \int_{-\infty}^{+\infty} \int_{-\infty}^{+\infty} \int_{-\infty}^{+\infty} \frac{\varsigma \hat{E}_{sx}\left(\eta,\xi;\frac{c}{2}\sqrt{\eta^2 + \xi^2 + \varsigma^2}\right)}{\sqrt{\eta^2 + \xi^2 + \varsigma^2}} \exp(j(\eta x' + \xi y' + \varsigma z')) d\eta d\xi d\varsigma$$

(13.10)

The expressions (13.9) and (13.10) correspond to classical expressions of the f–k migration in 3D [see, e.g., Eqs. (67a) and (67b) in Stolt (1978), under the same clarifications provided in Chapter 10 with regard to the 2D case].

13.2   3D MIGRATION FORMULAS IN THE TIME DOMAIN

In order to express the migration formulas versus the data in space and time domain, we can start from Eq. (13.6). In particular, it is comfortable to define

$$g(\eta,\xi,\omega) = \begin{cases} \dfrac{\omega}{4\pi^3 W(\eta,\xi;\omega)\sqrt{\dfrac{4\omega^2}{c^2} - \eta^2 - \xi^2}}, & (\eta,\xi,\omega) \in Sp\_set_1 \\ 0, & (\eta,\xi,\omega) \notin Sp\_set_1 \end{cases}$$

(13.11)

So that Eq. (13.6) can be more compactly rewritten as

$$\chi_{el}(x',y',z') = \int_{-\infty}^{+\infty} \int_{-\infty}^{+\infty} \int_{-\infty}^{+\infty} g(\eta,\xi,\omega)\hat{E}_{sx}(\eta,\xi;\omega) \times$$

$$\times \exp\left(j\left(\eta x' + \xi y' + \sqrt{\frac{4\omega^2}{c^2} - \eta^2 - \xi^2}\,z'\right)\right) d\eta d\xi d\omega \quad (13.12)$$

Now, the spectrum $\hat{E}_{sx}(\eta,\xi;\omega)$ is expressed versus the data in space and time domain as

$$\hat{E}_{sx}(\eta,\xi;\omega) = \int_{-\infty}^{+\infty} \int_{-\infty}^{+\infty} \int_{-\infty}^{+\infty} E_{sx}(x,y,t) \exp(-j(\eta x + \xi y)) \exp(-j\omega t)\, dxdydt \quad (13.13)$$

Substituting Eq. (13.13) into Eq. (13.12), we obtain

$$\chi_{e1}(x',y',z') = \int_{-\infty}^{+\infty}\int_{-\infty}^{+\infty}\int_{-\infty}^{+\infty} g(\eta,\xi,\omega)\exp\left(j\left(\eta x' + \xi y' + \sqrt{\frac{4\omega^2}{c^2} - \eta^2 - \xi^2}z'\right)\right) \times$$

$$\times \int_{-\infty}^{+\infty}\int_{-\infty}^{+\infty}\int_{-\infty}^{+\infty} E_{sx}(x,y,t)\exp(-j(\eta x + \xi y))\exp(-j\omega t)\,dxdydtd\eta d\xi d\omega =$$

$$= \int_{-\infty}^{+\infty}\int_{-\infty}^{+\infty}\int_{-\infty}^{+\infty} g_1(x-x',y-y',z',t)E_{sx}(x,y,t)\,dxdydt \qquad (13.14)$$

where the function $g_1(x - x', y - y', z', t)$ is defined as

$$g_1(x-x',y-y',z',t) = \int_{-\infty}^{+\infty}\int_{-\infty}^{+\infty}\int_{-\infty}^{+\infty} g(\eta,\xi,\omega)\left[-j\left(\omega t + \eta(x-x') + \xi(y-y') + \right.\right.$$

$$\left.\left. -z'\sqrt{\frac{4\omega^2}{c^2} - \eta^2 - \xi^2}\right)\right]d\eta d\xi d\omega \qquad (13.15)$$

In general, the integral (13.15) cannot be solved in a closed form; thus, in general, integral (13.14) cannot be solved in closed form either. However, if we assume that the function $g(\eta,\xi,\omega)$ is unitary, then we can close the integrals. In fact, in this case we can approximate

$$\chi_{e1}(x',y',z') = \int_{-\infty}^{+\infty}\int_{-\infty}^{+\infty}\int_{-\infty}^{+\infty} g_2(x-x',y-y',z',t)E_{sx}(x,y,t)\,dxdydt \qquad (13.16)$$

with

$$g_2(x-x',y-y',z',t) = F\int_{-\infty}^{+\infty} \exp(-j\omega t)\,d\omega$$

$$\times \int_{-\infty}^{+\infty}\int_{-\infty}^{+\infty} \exp\left[-j\left(\eta(x-x') + \xi(y-y') - \sqrt{\frac{4\omega^2}{c^2} - \eta^2 - \xi^2}z'\right)\right]d\eta d\xi$$

$$\qquad (13.17)$$

where F is a unitary but not dimensionless factor. Equation (13.17) can also be put in the form

$$g_2(x-x',y-y',z',t) = F\frac{\partial}{\partial z'}\int_{-\infty}^{+\infty}\exp(-j\omega t)\,d\omega\,\times$$

$$\times\int_{-\infty}^{+\infty}\int_{-\infty}^{+\infty}\frac{\exp\left[-j\left(\eta(x-x')+\xi(y-y')-\sqrt{\dfrac{4\omega^2}{c^2}-\eta^2-\xi^2}\,z'\right)\right]}{-j\sqrt{\dfrac{4\omega^2}{c^2}-\eta^2-\xi^2}}\,d\eta\,d\xi$$

$$(13.18)$$

At this point, let us remind that in the 3D case the axis z' is directed upward—this is, in the air half-space—we have that z' in Eq. (13.18) is a negative quantity, and so the function can again be rewritten as

$$g_2(x-x',y-y',z',t) = jF\frac{\partial}{\partial z'}\int_{-\infty}^{+\infty}\exp(-j\omega t)\,d\omega\,\times$$

$$\times\int_{-\infty}^{+\infty}\int_{-\infty}^{+\infty}\frac{\exp\left[-j\eta(x-x')-j\xi(y-y')-j\sqrt{\dfrac{4\omega^2}{c^2}-\eta^2-\xi^2}\,|z'|\right]}{\sqrt{\dfrac{4\omega^2}{c^2}-\eta^2-\xi^2}}\,d\eta\,d\xi$$

$$(13.19)$$

At this point, the integral in Eq. (13.19) can be solved in $d\eta\,d\xi$ by using the calculations shown in Appendix A. In particular, based on Eq. (A.19) we have

$$\int_{-\infty}^{+\infty}\int_{-\infty}^{+\infty}\frac{\exp\left[-j\eta(x-x')-j\xi(y-y')-j\sqrt{\dfrac{4\omega^2}{c^2}-\eta^2-\xi^2}\,|z'|\right]}{\sqrt{\dfrac{4\omega^2}{c^2}-\eta^2-\xi^2}}\,d\eta\,d\xi$$

$$=\frac{2j\pi\exp\left(j\dfrac{2\omega}{c}\sqrt{(x-x')^2+(y-y')^2+z'^2}\right)}{\sqrt{(x-x')^2+(y-y')^2+z'^2}}$$

$$(13.20)$$

Substituting Eq. (13.20) into Eq. (13.19), we obtain

$$g_2(x-x',y-y',z',t) = -2\pi F \frac{\partial}{\partial z'} \int_{-\infty}^{+\infty} \frac{\exp\left(-j\omega\left(t-\frac{2}{c}\sqrt{(x-x')^2+(y-y')^2+z'^2}\right)\right)}{\sqrt{(x-x')^2+(y-y')^2+z'^2}}\, d\omega$$

$$= -4\pi^2 F \frac{\partial}{\partial z'} \frac{\delta\left(t-\frac{2}{c}\sqrt{(x-x')^2+(y-y')^2+z'^2}\right)}{\sqrt{(x-x')^2+(y-y')^2+z'^2}} \qquad (13.21)$$

Substituting Eq. (13.21) into Eq. (13.16), eventually we obtain

$$\chi_{e1}(x',y',z') = -4\pi^2 F \frac{\partial}{\partial z'} \int_{-\infty}^{+\infty}\int_{-\infty}^{+\infty}\int_{-\infty}^{+\infty} \frac{\delta\left(t-\frac{2}{c}\sqrt{(x-x')^2+(y-y')^2+z'^2}\right)}{\sqrt{(x-x')^2+(y-y')^2+z'^2}} E_{sx}(x,y,t)\, dxdydt$$

$$= -4\pi^2 F \frac{\partial}{\partial z'} \int_{-\infty}^{+\infty}\int_{-\infty}^{+\infty} \frac{E_{sx}\left(x,y,t=\frac{2}{c}\sqrt{(x-x')^2+(y-y')^2+z'^2}\right)}{\sqrt{(x-x')^2+(y-y')^2+z'^2}}\, dxdy \qquad (13.22)$$

Apart from an unessential factor, Eq. (13.22) is the classical 3D migration formula in the time domain, as reported (for example) in Schneider (1978, page 53).

Performing the derivative under the sign of integral, we can still write

$$\chi_{e1}(x',y',z') \approx -4\pi^2 F \frac{\partial}{\partial z'} \int_{-\infty}^{+\infty}\int_{-\infty}^{+\infty} \frac{E_{sx}\left(x,y,t=\frac{2r}{c}\right)}{r}\, dxdy$$

$$= -4\pi^2 F \int_{-\infty}^{+\infty}\int_{-\infty}^{+\infty} \left(\frac{2z'}{cr^2}\frac{\partial}{\partial t}E_{sx}(x,y,t)\bigg|_{t=\frac{2r}{c}} - \frac{z'E_{sx}\left(x,y,t=\frac{2r}{c}\right)}{r^3}\right) dxdy \qquad (13.23)$$

where $r = \sqrt{(x-x')^2+(y-y')^2+z'^2}$ is the distance between the source point $(x,y,0)$ and the investigation point (x',y',z'). The quantity z'/r is easily recognized to be just the view angle for the current source and investigation points, so that Eq. (13.23) can be also rewritten as

$$\chi_{e1}(x',y',z') = 4\pi^2 F \int\limits_{-\infty}^{+\infty} \int\limits_{-\infty}^{+\infty} \frac{2\cos(\theta)}{cr} \left. \frac{\partial E_{sx}(x,y,t)}{\partial t} \right|_{t=\frac{2r}{r}} dxdy$$

$$-4\pi^2 F \int\limits_{-\infty}^{+\infty} \int\limits_{-\infty}^{+\infty} \frac{\cos(\theta) E_{sx}\left(x,y,t=\dfrac{2r}{c}\right)}{r^2} dxdy \qquad (13.24)$$

The signa are linked to the fact that $z' < 0$ whereas we mean θ as the (acute) view angle, so that $\cos(\theta) = -z'/r$. Some commercial code, such as e.g., the GPRSLICE, allow to perform 3D migrations. The reader will find an example in Section 15.9.

13.3 3D VERSUS 2D MIGRATION FORMULAS IN THE TIME DOMAIN

We think it might be of interest, at this point, to expose the physical reason why in the formulas of the 2D migration in the time domain [Eqs. (10.35) and (10.36)] the datum is integrated versus the time, whereas in the homologous 3D formulas [Eqs. (13.22) and (13.23)] the integration along the time disappears.[1]

The reason is illustrated by means of Figure 13.1. In particular, in a 2D geometry (panel A), we have depicted a 2D source constituted by a filamentary current at the air–soil interface and a point-like 2D target, parallel to each other and both indefinitely long along one of the axes. For the sake of clarity, the air–soil interface has been made fully transparent in Figure 13.1. Moreover, the observation point along the axis orthogonal to the axis of invariance is evidenced too.

Let us now assume that the source is a filamentary current and radiates an electromagnetic pulse in a nondispersive soil. This means that the filament is crossed by a temporally impulsive electrical current. This situation can be also viewed as the contemporary radiation of an infinite series of (3D) electromagnetic pulses emanated by a series of adjacent elementary Hertzian dipoles: The single pieces of the blue dashed line in Figure 13.1 can be interpreted as representative of this series. Let us now concentrate on the cross section of the target at $x = 0$ according to the Cartesian system in Figure 13.1. The incident field in this point will be not temporally impulsive, because the waves radiated by all the equivalent Hertzian dipoles that compose the 2D source propagate at the same finite velocity. So, the elementary Hertzian dipole d_1 (see Figure 13.1, panel A) at $x = 0$ will be the first one to illuminate the cross section at $x = 0$ of the target at hand, at the instant time r_{st}/c (where r_{st} is the minimum distance between the source and the cross section of the target at $x = 0$ and where c is the propagation velocity in the soil). Immediately afterward, at the time $r_{st}/c + dt$, the cross section will be illuminated by the two "adjacent" equivalent Hertzian dipoles d_{2a} and d_{2b}, placed at $x = dx$ and $x = -dx$ along the filamentary source, respectively. Then, at the time

---

[1] In particular, this also means that the 2D formula is not merely given by the 3D formula with one less spatial integration.

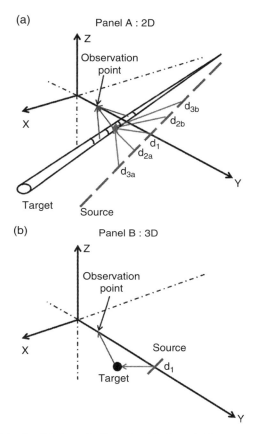

Figure 13.1. Pictorial of a 2D (panel A) and a 3D (panel B) geometry with an impulsive point-like source and a point-like target. For color detail, please see color plate section.

$r_{st}/c + 2dt$ the cross section at hand is illuminated by the two dipoles d_{3a} and d_{3b}, placed at $x = 2dx$ and $x = -2dx$ along the filamentary source, and so on. The subsequent contributions are progressively weaker because of the geometrical spreading and because of the possible losses in the propagation medium, so the field incident in the cross section of the target at $x = 0$ vanishes versus the time. However, theoretically its duration time ranges from the time instant r_{st}/c up to $+\infty$ The same reasoning also holds with regard to the field scattered from the 2D target and so, in the end, the scattered field received in the observation point ranges in time from the instant $(r_{st} + r_{to})/c$ [where r_{to} is the distance between the cross section of the target at $x = 0$ and the observation point (that also is placed at $x = 0$)] up to the time $+\infty$. If we neglect the offset between the source and observation points along the y-axis (as implicitly done when putting equal to 1 the spectral weight), then we have that the scattered field is a vanishing quantity theoretically observed from the time instant $2r/c$ (where $r = r_{st} = r_{to}$) up to the time instant $+\infty$. In the end, we can say that a trace of the actual three-dimensionality of the "world" inexorably remains also within a 2D model.

Instead, within a 3D model (Figure 13.1, panel B), if we assume a point-like impulsive source (i.e., a Hertzian dipole crossed by a temporally impulsive current) and a point-like target, then we will measure an impulsive scattered field in the observation point.

QUESTIONS

1. Are the 2D migration formulas obtained from the 3D ones with the suppression of one of the integrals?
2. Could 3D migration formulas be obtained also if the transmitting and receiving antennas were orthogonal to each other?
3. Suppose we have at our disposal an ideal array of dipole antennas. The dipoles are at a $\lambda_{smin}/4$ distance from each other and are orthogonal to the direction of the B-scans. Does this make the gathering of orthogonal B-scans useless, in principle?
4. Is there some way to gather in a unique going-through all the information gathered with two orthogonal sets of B-scans?

14

THE SINGULAR VALUE DECOMPOSITION

14.1 THE METHOD OF MOMENTS

Let us start from the 2D scattering equation under the Born approximation without considering magnetic contrasts. The data are gathered in common offset configuration and the height of the data is possibly greater than zero. So, the starting point is Eq. (9.55), repeated here for readability.

$$E_s(x;\omega) = k_s^2 \iint_D G_e(x,x',h,z';\omega) E_{inc}(x+\Delta,x',h,z';\omega)\chi_e(x',z')\,dx'dz', \quad x \in \Sigma \quad (14.1)$$

where the involved symbols have been already explained in Chapter 9. Equation (14.1) can be discretized by means of the method of moments (MoM). MoM is a well-known subject, so its description will be very brief. Let us just state that the kind of MoM used here is based on point matching in both spatial and frequency domains. Thus, we will expand the contrast in a finite set of basis functions $\varphi_k(x',z')$ $k = 1, \ldots, K$, assuming that K is high enough to allow a good representation of the targets of interest. That said, the contrast is expressed as

Introduction to Ground Penetrating Radar: Inverse Scattering and Data Processing,
First Edition. Raffaele Persico.

$$\chi_e(x',z') \cong \sum_{k=1}^{K} \alpha_k \varphi_k(x',z') \tag{14.2}$$

Under the assumption (14.2), to retrieve the contrast means to retrieve its expansion coefficients $\alpha_1, \ldots \alpha_K$. Substituting Eq. (14.2) in Eq. (14.1) and particularizing it to a finite set of points $x_1, \ldots, x_M$ and $\omega_1, \ldots, \omega_N$, we have a finite number of linear equations as follows:

$$E_s(x_1;\omega_1) = \sum_{k=1}^{K} \alpha_k k_{s1}^2 \iint_D G_e(x_1,x',h,z';\omega_1) E_{inc}(x_1 + \Delta, x', h, z';\omega_1) \varphi_k(x',z') \, dx' dz',$$

$$E_s(x_2;\omega_1) = \sum_{k=1}^{K} \alpha_k k_{s1}^2 \iint_D G_e(x_2,x',h,z';\omega_1) E_{inc}(x_2 + \Delta, x', h, z';\omega_1) \varphi_k(x',z') \, dx' dz',$$

$\ldots$

$$E_s(x_M;\omega_1) = \sum_{k=1}^{K} \alpha_k k_{s1}^2 \iint_D G_e(x_M,x',h,z';\omega_1) E_{inc}(x_M + \Delta, x', h, z';\omega_1) \varphi_k(x',z') \, dx' dz',$$

$$E_s(x_1;\omega_2) = \sum_{k=1}^{K} \alpha_k k_{s2}^2 \iint_D G_e(x_1,x',h,z';\omega_2) E_{inc}(x_1 + \Delta, x', h, z';\omega_2) \varphi_k(x',z') \, dx' dz',$$

$$E_s(x_2;\omega_2) = \sum_{k=1}^{K} \alpha_k k_{s2}^2 \iint_D G_e(x_2,x',h,z';\omega_2) E_{inc}(x_2 + \Delta, x', h, z';\omega_2) \varphi_k(x',z') \, dx' dz',$$

$\ldots$

$$E_s(x_M;\omega_N) = \sum_{k=1}^{K} \alpha_k k_{sN}^2 \iint_D G_e(x_M,x',h,z';\omega_N) E_{inc}(x_M + \Delta, x', h, z';\omega_M) \varphi_k(x',z') \, dx' dz'$$

$$\tag{14.3}$$

Equation (14.3) can be resumed in a general way as

$$\boldsymbol{A}\boldsymbol{\alpha} = \boldsymbol{d} \tag{14.4}$$

where $\boldsymbol{d}$ is the column vector of the data $d_1, \ldots, d_{Nd} = E_s(x_1,\omega_1), \ldots, E_s(x_M,\omega_N), N_d = NM$ is the number of data, and $\boldsymbol{\alpha} = \alpha_1, \ldots, \alpha_K$ is the column vector of the unknown expansion coefficients of the contrast.

The size of the matrix $\boldsymbol{A}$ is $(N_d, K) = (NM, K)$; that is, it has as many rows as the data and as many columns as the unknowns. The matrix $\boldsymbol{A}$ is given by the piling of the submatrixes relative to all the exploited frequencies. For each frequency ω_n, the relative submatrix $\boldsymbol{S}$ is characterized as

$$S(i,j) = k_{sn}^2 \iint_D G_e(x_i, x', h, z'; \omega_n) E_{inc}(x_i + \Delta, x', h, z'; \omega_n) \varphi_j(x', z') \, dx' dz' \qquad (14.5)$$

Of course, the wavenumber has been indexed after the frequency. Three-dimensional problems can be discretized in a similar way: In this case, the data column is achieved by reordering after a unique index the values of the field in the points (x_i, y_j) and at the frequency ω_k, whereas the unknown is expanded in this case along a three-dimensional basis as follows:

$$\chi_e(x', y', z') \cong \sum_{k=1}^K \alpha_k \varphi_k(x', y', z') \qquad (14.6)$$

Formally, the number of unknowns is still given by K and the number of data is still equal $N_d = NM$, where N is the number of observation points and M is the number of frequencies. Of course, however, K and N are expected quite larger in a three-dimensional problem than in a two-dimensional one, which makes three-dimensional inversions computationally more demanding than two-dimensional ones.

14.2 REMINDERS ABOUT EIGENVALUES AND EIGENVECTORS

The SVD of a rectangular matrix will be introduced in this chapter as an extension of the basic theory of the eigenvalues and eigenvectors of a square matrix. So, preliminarily, some reminders about the eigenvalues and eigenvectors are provided in relationship to matrix inversions. In particular, the basic theory of the linear transformations between two complex vector spaces, equipped with the usual scalar Hermitian (see footnote 2 in Section 11.6) product for complex vectors, is taken for granted. Of course, also the basic concepts about eigenvalues and eigenvectors are taken for granted.

That said, let us consider a square [let's say sized (N,N)] maximum rank matrix $\boldsymbol{M}$, and let $\lambda_1, \ldots, \lambda_N$ be its eigenvalues ordered after a decreasing modulus. Possibly, some of the eigenvalues can be equal to each other; that is, the case of multiple eigenvalues is enclosed too. Let be $\boldsymbol{m}_1, \ldots, \boldsymbol{m}_N$ be the unitary norm eigenvectors corresponding to the eigenvalues. The set $\boldsymbol{m}_1, \ldots, \boldsymbol{m}_N$ constitutes (either "naturally" for different eigenvalues or by construction thanks to the Gram–Schmidt procedure) an orthonormal basis for C^N. Consequently, we have

$$\langle \boldsymbol{m}_h, \boldsymbol{m}_k \rangle = \delta_{h,k} \qquad (14.7)$$

where the symbol $\langle \boldsymbol{m}_h, \boldsymbol{m}_k \rangle$ stands for the Hermitian scalar product in $[C^K]$ and $\delta_{h,k}$ is the Kronecker's delta function, equal to 1 if $h = k$ and 0 if $h \neq k$.

Let us now write

$$\boldsymbol{U} = (\boldsymbol{m}_1, \ldots, \boldsymbol{m}_n) \qquad (14.8)$$

The matrix U is a unitary[1] matrix, sized (N,N), whose columns are the eigenvectors. Equation (14.7) implies that

$$U^+ U = I \qquad (14.9)$$

where U^+ is the adjoint (i.e., the transposed and conjugate) of U and I is the identity matrix, sized (N,N). This shows that a unitary matrix is invertible and its inverse coincides with its adjoint, which is in turn a unitary matrix.[2]

Now, from the definition of eigenvalues and eigenvectors, we have

$$M m_1 = \lambda_1 m_1$$
$$M m_2 = \lambda_2 m_2 \qquad (14.10)$$
$$\ldots$$
$$M m_N = \lambda_N m_N$$

which in matrix terms can be expressed as

$$M U = U \Lambda \Rightarrow M = U \Lambda U^{-1} = U \Lambda U^+ \qquad (14.11)$$

where Λ is a diagonal square matrix [also sized (N,N)] whose non-null elements are the eigenvalues.

At this point, let us consider a generic algebraic system with a square maximum rank matrix:

$$M x = y \qquad (14.12)$$

Since the matrix has by hypothesis maximum rank, we can formally express the unique exact solution of the system as

$$x = M^{-1} y \qquad (14.13)$$

However, this solution might be unstable if the condition number—that is, the modulus of the ratio between the first (largest level) and the last (smaller level) eigenvalues—is much larger than 1. We can understand this by exploiting the eigenvalue decomposition (14.11), as follows:

$$M x = y \Rightarrow U \Lambda U^+ x = y \Rightarrow x = U \Lambda^{-1} U^+ y \qquad (14.14)$$

---

[1] A matrix is said to be unitary when its columns are orthogonal to each other and have unitary norm.
[2] This also means that the property of being unitary can be indifferently defined with regard to the rows or to the columns of the matrix.

where Λ^{-1} is the inverse of Λ and is a diagonal matrix whose elements along the diagonal are the inverse of the eigenvalues.

In terms of eigenvalues and eigenvectors, the solution shown in Eq. (14.14) is written as

$$x = \sum_{n=1}^{N} \frac{1}{\lambda_n} \langle y, m_n \rangle m_n \qquad (14.15)$$

This makes clear that Eq. (14.14) expresses the solution after a change of basis, and the adopted basis is just constituted by the eigenvectors.

This also suggests a possible regularization scheme. In fact, the terms that can generate instability are those relative to the smallest eigenvalues. In particular, inserting in Eq. (14.15) the data affected by any kind of equivalent *noise*[3] Eq. (14.15) evolves into

$$x = \sum_{n=1}^{N} \frac{1}{\lambda_n} \langle y_0 + \text{noise}, m_n \rangle m_n = \sum_{n=1}^{N} \frac{1}{\lambda_n} \langle y_0, m_n \rangle m_n + \sum_{n=1}^{N} \frac{1}{\lambda_n} \langle \text{noise}, m_n \rangle m_n \qquad (14.16)$$

where y_0 is the vector of the ideal (noise-free) data. The stability can be guaranteed by arresting the sum to some, $N_1 \leq N$, thus eliminating the contribution related to the smallest eigenvalues:

$$x \approx \sum_{n=1}^{N_1} \frac{1}{\lambda_n} \langle y, m_n \rangle m_n \qquad (14.17)$$

Of course, the choice of N_1 is dictated by a trade-off between accuracy and robustness of the solution, and so it should account for the level of the involved uncertainties affecting the data. In a matrix form, the regularized solution can be expressed by

$$x = U \Lambda_R^{-1} U^+ y \qquad (14.18)$$

where Λ_R^{-1} is the regularized inverse of Λ; that is, it is a diagonal matrix where the elements on the diagonal are equal to

$$\Lambda_R^{-1}(i,i) = \begin{cases} \dfrac{1}{\lambda_i}, & i \leq N_1, \\ 0, & i > N_1 \end{cases} \qquad (14.19)$$

---

[3] This includes any source of uncertainty for the data, as for example the model error and the interferences, further than the "actual" thermal noise generated by the receiver.

At this point, we can extend the discussion also to the case of a square matrix that does not show the maximum rank. In fact, this can be seen in a sense as the limit case for a very ill-conditioned matrix. Indeed, while very low eigenvalues lead to a progressively more instable solution, if we loosely admit the concept of null eigenvalues,[4] they lead to an indeterminate (which somehow means completely instable) solution. In this case, in order to have any solution, the regularization is mandatory, and the inversion formulas are still provided by Eqs.(14.17) and (14.18), taking for granted that the choice of N_1 cuts out the "naturally null" eigenvalues plus possibly the smallest non-null ones. The eigenvectors relative to null eigenvalues are just the eigenvectors belonging to the null space of the linear transformations. Of course, the regularized solution is not an exact solution, in the case of both a maximum rank matrix and a nonmaximum rank matrix.

14.3   THE SINGULAR VALUE DECOMPOSITION

In GPR problems, after linearizing the scattering equations, the algebraic system is in general rectangular; that is, the number of equations in general is not equal to the number of unknowns. In particular, in most cases, GPR systems work with an odometer, as said, and this means that usually we don't choose apriori the number of gathered data. Of course, we might resample the data so to have as many data as unknowns, but in general this is not a convenient operation, because an increased number of data can help in amortizing the noise. Thus, in general we have to deal with the problem of solving rectangular linear algebraic systems in a regularized way, which requires an extension of the eigenvalue theory; this extension is the singular value decomposition (Bertero and Boccacci, 1998). In these conditions, we are not guaranteed that an exact solution even exists, and we will look for a least square solution.

So, with reference to Eq. (14.4), we are now looking for a vector α such as the quantity

$$F(\alpha) = \|A\alpha - d\|^2 \qquad (14.20)$$

is minimum. Such a quantity is achieved by identifying the vector α such as the residual $A\alpha - d$ is orthogonal to all the elements of the range of the linear transformation identified by the matrix A. In fact, in this case, for any chosen α_1 belonging to the space of the unknowns C^K, we have

$$\langle A\alpha - d, A\alpha_1 \rangle = 0 \Rightarrow \alpha_1^+ A^+ A\alpha - \alpha_1^+ A^+ d = 0 \quad \forall \alpha_1 \in C^K \qquad (14.21)$$

Such an α makes minimum $F(\alpha)$. In fact, given any $\alpha_1 \in C^K$, we have

---

[4] We mean them just as the null elements of the main diagonal of the matrix Λ.

$$\begin{aligned}
F(\alpha_1) &= \|A\alpha_1 - d\|^2 = \|A\alpha_1 - A\alpha + A\alpha - d\|^2 = \|A(\alpha_1 - \alpha) + A\alpha - d\|^2 \\
&= \langle A(\alpha_1 - \alpha) + A\alpha - d, A(\alpha_1 - \alpha) + A\alpha - d \rangle \\
&= \|A(\alpha_1 - \alpha)\|^2 + \langle A\alpha - d, A(\alpha_1 - \alpha) \rangle + \langle A(\alpha_1 - \alpha), A\alpha - d \rangle + \|A\alpha - d\|^2 \\
&= \|A(\alpha_1 - \alpha)\|^2 + \|A\alpha - d\|^2 \geq \|A\alpha - d\|^2 = F(\alpha)
\end{aligned} \tag{14.22}$$

because $A(\alpha_1 - \alpha)$ is an element of the range of the linear transformation, and so it is by definition orthogonal to $A\alpha - d$.

Moreover, condition (14.21) can be verified only if $A^+ A\alpha - A^+ d$ is a null vector, because α_1 is an arbitrary vector. This means that α has to be an exact solution of the algebraic system

$$A^+ A\alpha = A^+ d \tag{14.23}$$

Let us note that $A^+ A$ is a square matrix sized (K,K), and this makes it possible to resort to the eigenvalue theory for the solution of the system (14.23). In a first moment, let now suppose that $A^+ A$ is invertible (it is easy to test that this happens if and only if A has maximum rank). In this case, there exists a unique least square solution of the problem, expressed by

$$\alpha = (A^+ A)^{-1} A^+ d \tag{14.24}$$

The matrix $(A^+ A)^{-1} A^+$ is customarily called the pseudo-inverse of A. However, since the underlying physical problem is ill-posed, the stability of the solution has to be guaranteed in the case where we look either for an exact solution or for a least square solution. In particular, the matrix $A^+ A$ is in general ill-conditioned, and its eigenvalues can show a meaningful dynamic range. Thus, a regularization is needed. In order to perform it, let us start noting that the eigenvalues of the matrix $A^+ A$ are real and nonnegative quantities. This can be shown by calculating the square norm of the Am_i, given by the scalar product of the vector times itself:

$$\langle Am_i, Am_i \rangle = m_i^+ A^+ Am_i = \lambda_i m_i^+ m_i = \lambda_i = \|Am_i\|^2 \geq 0 \tag{14.25}$$

Consequently, we can order the eigenvalues of $A^+ A$ in a decreasing series of nonnegative numbers, which we will call $\sigma_1^2, \sigma_2^2, \ldots, \sigma_K^2$. The nonnegative square roots of these eigenvalues $\sigma_1, \sigma_2, \ldots, \sigma_k$ will be taken as the singular values of the matrix A.[5]

At this point, let us consider the quantities

$$n_i = \frac{1}{\sigma_i} Am_i \tag{14.26}$$

---

[5] This choice is conventional: For example, the singular values might be chosen negative on condition to reverse the corresponding singular vector.

By definition, we have

$$\mathbf{A}m_i = \sigma_i n_i \qquad (14.27)$$

Moreover, we have

$$\mathbf{A}^+ (\sigma_i n_i) = \sigma_i \mathbf{A}^+ n_i = \mathbf{A}^+ \mathbf{A} m_i = \sigma_i^2 m_i \Rightarrow \mathbf{A}^+ n_i = \sigma_i m_i \qquad (14.28)$$

which is the dual of Eq. (14.27). Moreover, from Eqs. (14.27) and (14.28), we have

$$\mathbf{A}\mathbf{A}^+ n_i = \mathbf{A}(\sigma_i m_i) = \sigma_i \mathbf{A} m_i = \sigma_i^2 n_i \qquad (14.29)$$

This shows that the matrix $\mathbf{A}\mathbf{A}^+$ has the same eigenvalues as $\mathbf{A}^+ \mathbf{A}$, and the vectors n_i are the eigenvectors of $\mathbf{A}\mathbf{A}^+$. Let us specify that $\mathbf{A}\mathbf{A}^+$ is a square matrix sized (N_d, N_d). Thus, the size of this matrix is different from the size of $\mathbf{A}^+ \mathbf{A}$. However, having the same (non null) eigenvalues, the two matrixes have the same rank. Therefore, in order to be guaranteed that $\mathbf{A}^+ \mathbf{A}$ achieves its maximum rank, as supposed up to now, at the moment we will suppose that $N_d \geq K$; that is, we will suppose that the data are more than the unknowns. So, when saying that the two matrixes have the same eigenvalues, we mean more precisely that they have the same K non-null eigenvalues, whereas $\mathbf{A}\mathbf{A}^+$ will have $N_d - K$ extra null eigenvalues. It can be also seen that the K eigenvectors n_i result automatically orthonormal to each other. In fact,

$$\langle n_i, n_j \rangle = \frac{1}{\sigma_i \sigma_j} \langle \mathbf{A} m_i, \mathbf{A} m_j \rangle = \frac{1}{\sigma_i \sigma_j} m_j^+ \mathbf{A}^+ \mathbf{A} m_i = \frac{\sigma_i^2}{\sigma_i \sigma_j} \langle m_i, m_j \rangle = \frac{\sigma_i}{\sigma_j} \delta_{i,j} = \delta_{i,j} \qquad (14.30)$$

The interested reader can easily verify that also $\mathbf{A}\mathbf{A}^+$ is a Hermitian matrix. The vectors m_i are by definition the singular vectors in the space of the unknowns, whereas the vectors n_i are by definition the singular vectors in the space of the data. The singular value decomposition (SVD) of the matrix $\mathbf{A}$ is eventually the triple $\{m_i, \sigma_i, n_i\} i = 1, \ldots, K$. As shown, its evaluation can be recast as the calculation of the eigenvalues and eigenvectors of suitable auxiliary square matrixes. At this point, we can provide a matrix factorization based on SVD. In fact, from Eq. (14.28), we have

$$\mathbf{A}^+ (n_1, n_2, \ldots, n_K) = (m_1, m_2, \ldots, m_K) \begin{pmatrix} \sigma_1 & 0 & 0..0 \\ 0 & \sigma_2 & 0..0 \\ & \cdots & \\ 0 & 0 & ...\sigma_K \end{pmatrix} \Rightarrow \mathbf{A}^+ \mathbf{V} = \mathbf{U}\mathbf{\Sigma} \qquad (14.31)$$

where $\mathbf{V}$ is the matrix of the singular vectors in the space of the data, $\mathbf{U}$ is the matrix of singular vectors in the space of the unknowns, and $\mathbf{\Sigma}$ is the diagonal matrix of the singular values.

Consequently, being V a unitary matrix, we have

$$A^+ = U\Sigma V^+ \tag{14.32}$$

On the other hand, based on the classical eigenvalue theory resumed in the previous section, the square matrix A^+A can be factorized as

$$A^+A = U\Sigma^2 U^+ \tag{14.33}$$

where Σ^2 is a diagonal matrix where $\Sigma^2(i,i) = \sigma_i^2$. Therefore, the inverse of A^+A can be factorized [see Eq. (14.14)] as

$$(A^+A)^{-1} = U\Sigma^{-2} U^+ \tag{14.34}$$

where Σ^{-2} is a diagonal matrix where $\Sigma^{-2}(i,i) = 1/\sigma_i^2$.

Substituting Eqs. (14.32) and (14.34) in Eq. (14.24), we can express the least square solution as

$$\alpha = U\Sigma^{-2}U^+ U\Sigma V^+ d = U\Sigma^{-1}V^+ d \tag{14.35}$$

where Σ^{-1} is a diagonal matrix where $\Sigma^{-1}(i,i) = 1/\sigma_i$.

The least square solution (14.35) can be equivalently expressed as

$$\alpha = \sum_{i=1}^{K} \frac{1}{\sigma_i} \langle d, n_i \rangle m_i \tag{14.36}$$

Equation (14.36) makes it clear that the SVD is based on a double change of basis, both in the space of the unknowns and in the space of the data. The ill-conditioning is expressed by the high ratio between the first (maximum) and the last (minimum) singular value. As shown in the previous section with regard to the exact solution of a square algebraic system, also the least square solution can be regularized by thresholding the singular values,[6] so that, choosing $N_1 < K$, we have a regularized solution that can be expressed as

$$\alpha \approx \sum_{i=1}^{N_1} \frac{1}{\sigma_i} \langle d, n_i \rangle m_i \tag{14.37}$$

Similarly to the case of square matrixes, at this point we can remove the hypothesis of maximum rank matrix. In particular, we can account for this situation by choosing

---

[6] Also other kinds of regularizations are possible, where the contributions of the smallest singular values are progressively dumped instead of being abruptly thresholded, but we will not deal with them.

N_1 such that the null singular values are cut out, further than possibly the smallest non-null ones.

Finally, under the viewpoint of a regularized reconstruction, even the hypothesis that the system is overdetermined can be removed. In fact, when performing a regularization of the kind (14.37), the actual number of unknowns looked for is in the end N_1. However, please note that this does not mean that we can a priori impose the problem looking formally for N_1 unknowns: actually, the problem has to be discretized suitably, in order to guarantee that the singular vectors do not (meaningfully) depend on the discretization itself. In other words, the singular vectors have to approximate quantities independent of the discretization, which are the singular functions of the scattering operator (Bertero and Boccacci, 1998).

In other words, the question is not only how many unknowns we should look for, but also what kind of unknowns we should look for. The optimal choice for the unknowns is, as shown, the coefficients of the object function along the singular functions, because this allows us to separate the "invertible" part form the "noninvertible" part of the linear relationship at hand. However, the singular functions are not known in a closed form, and so the initial (trial) representation of the object function has to be refined enough in order to represent them adequately. In Section 15.4, the reader will find some exercises on the number of trial unknowns.

In conclusion, the SVD is a way to solve the problem in a regularized way and it is calculable from two correlated eigenvalue–eigenvector problems. This fact also provides the formal way for the calculation of the SVD. However, the formal "classical" calculation of the eigenvalues (performed by means of the roots of the characteristic polynomial of the matrix) and of the relative eigenvectors is practicable only for small matrixes. In inverse scattering problems applied to GPR data, the matrix might have thousands of rows and columns. This makes it necessary to make use of suitable numerical algorithms for the evaluation of the SVD. Numerical algorithms for SVD are the object of a specific research field (Golub and van Loan, 1996), underlying all the computational available SVD routines.

14.4 THE STUDY OF THE INVERSE SCATTERING RELATIONSHIP BY MEANS OF THE SVD

The singular value decomposition provides not only a method for the solution of the problem but also a possible method for the analysis the problem. In particular, even if numerically, the SVD can help us to understand the characteristics of the scattering operator. In particular, the class of retrievable object functions is characterized by the span of the singular functions relative to the the singular values smaller than or at most equal to the chosen threshold. This means that the regularization influences the class of retrievable targets and that there is a trade-off between the details that we can retrieve (related to the dimension of the space where the solution is looked for) and the robustness of the solution (provided by the ratio between the smallest and the maximum retained singular value). The two exigencies are contrasting to each other, and the choice of the threshold is a compromise between them. The optimal choice of the threshold (i.e., the level of the regularization) is in general a nontrivial problem.

Some solutions have been proposed (Bertero and Boccacci, 1998), but they refer to linear problems where the data are affected by white noise. Actually, in GPR prospecting we are not guaranteed that the data are affected (only) by white noise and in general we don't have at our disposal an objective measure of the level of the noise [i.e., we don't know the signal-to-noise ratio (SNR), because customarily we cannot switch on the receiver of a GPR system without switching on its transmitter too]. Of course, it is even more improbable to have at our disposal a measure of the statistics of the noise. Moreover, the regularization should account not only for the "real" noise (i.e., the noise produced in the electronics of the receiving system) but also for several further kinds of disturbances, provided (for example) by the clutter (i.e., the impossibility to insert in a deterministic inversion model the roughness of the surface), the model error (i.e., the fact that we are modeling a 3D vector nonlinear world by means of a linear, possibly even 2D and scalar, model), and possible interferences by other electromagnetic sources (in particular, radio and TV transmissions and mobile phones). Thus, the comprehensive "equivalent noise" is in general "colored" and has an unknown level, and it is rigorously even correlated to the signal itself (because of the model error). Finally, it is also worth emphasizing that the noise superposed to the signal depends on our exigencies, because it also depends on the depth range that we need to investigate: Of course the signal reflected by deepest targets is usually weaker and thus noisier than that reflected by the shallowest targets.

In these conditions, it is practically impossible to predict an optimal level for the regularization. On the other hand, the SVD allows us to easily retrieve the reconstruction at several levels of regularization, so that a heuristic choice of the best one is possible based on the experience of the human operator. Incidentally, this also means that the personal experience regarding the kind of problem at hand is essential: Physics–mathematics cannot automatically provide the best solution and, in particular, the best interpretation.

At any rate, whatever the adopted regularization method, the SVD can help in understanding the class of retrievable profiles and thus can provide an important insight about the possibilities of our reconstruction strategy.

For example, in order to know whether the spatial step or the frequency step is sufficiently narrow, we can compare the curves[7] of the singular values (some exercises are proposed in the next chapter); in particular, a comparative analysis between singular values and singular functions can back up, correct, or possibly deny the conclusions retrieved from DT. In particular, let us stress that the SVD analysis accounts for the linear scattering operator "as it is," without the further assumption of a lossless media and of electrically deep targets.

A quantity possibly useful for comprehension purposes is given by the spectral content (SPEC) of the singular functions' (in the space of the unknowns) upper threshold. In particular, if N_1 is the number of singular functions upper threshold, and $u_n(x,z)$

---

[7] The curve of the singular values is a common but not rigorous term (because the singular values are a sequence) often used in literature. It is generically meant as a smooth curve that interpolates the singular values.

$[u_n(x,y,z)$ if we adopt a 3D model] is the nth singular function in the space of the unknowns, the spectral content is defined as (Persico et al., 2005, Persico, 2006, Persico and Soldovieri, 2010)

$$SPEC = \sum_{N=1}^{N_1} \left|\hat{\hat{u}}_n(\eta,\zeta)\right| \quad \text{within a 2D approach}$$

$$SPEC = \sum_{N=1}^{N_1} \left|\hat{\hat{u}}_n(\eta,\xi,\zeta)\right| \quad \text{within a 3D approach}$$

(14.38)

that is, the spectral content is the sum of the moduli of the spectra of the singular functions in the space of the unknowns. The spectral set is a tool that can provide an insight about the spatial frequencies retrievable in realistic cases. In particular, we can hope to retrieve correctly only the spectral components of the object function where the spectral content is meaningful. Please note that the support of the spectral content does not provide the spatial frequencies surely retrievable, but its information is to be meant in negative, in the sense that it shows the spatial frequencies surely not retrievable. Notwithstanding, it can be an immediate (even if incomplete) way to visualize the filtering properties of the operator in a framework that overcomes the constraints imposed by DT.

Another quantity worth considering is the spatial content (SPAC) of the singular functions in the space of the unknowns, defined as

$$SPAC = \sum_{N=1}^{N_1} |u_n(x,z)| \quad \text{within a 2D approach,}$$

$$SPAC = \sum_{N=1}^{N_1} |u_n(x,y,z)| \quad \text{within a 3D approach}$$

(14.39)

where the symbol have the same meaning as in Eq. (14.38). To show the usefulness of the spatial content, let us start from the fact that it is intuitive (and can be easily shown numerically) that the singular functions relative to lower and lower singular values have their support (meant under some energetic criterion) centered on progressively deeper buried levels. This just means that, due to the progressive attenuation of the GPR signal (due both to the geometrical spreading and above all to the losses), the deeper layers of the soil provide a weaker "echoes" and are progressively more difficult to be retrieved. This fact is not accounted for by the spectral content. The spatial content can help in understanding whether the applied regularization is excessive. In particular, given a target at depth d, we cannot retrieve it if the support of the spatial content is not extended at least up to the depth d.

In Section 15.5 the reader will find some exercises on the spectral and the spatial content. Finally, he/she will find some 2D reconstructions based on numerical regularized SVD throughout Chapter 15, along with some examples of 3D reconstructions based on numerical regularized SVD in Section 15.10.

QUESTIONS

1. Does the SVD provide a model more refined than that provided by a migration?
2. Does a more refined model necessarily provide a better result?
3. Is the computational burden only a CPU time problem?
4. Does a nontruncated SVD ideally provide the exact solution of a linear problem?
5. Can we control the investigated depth range changing the available regularization parameters making use of a migration algorithm?
6. Can we control the investigated depth range changing the available regularization parameters making use of an SVD algorithm?

15

NUMERICAL AND EXPERIMENTAL EXAMPLES

15.1 EXAMPLES WITH REGARD TO THE MEASURE OF THE PROPAGATION VELOCITY

In this section, some examples of measure of propagation velocity based on the diffraction curves are shown. In order to introduce the examples, let us preliminarily emphasize that, of course, the real GPR pulses cannot have a zero duration, because the band of the system (in particular the band of the antennas) is never infinite. Therefore, the diffraction curves introduced in Chapter 2 are didactic abstractions. In the real word, the diffraction curves have some "thickness," and this constitutes an unavoidable source of uncertainty, both with regard to the propagation velocity of the waves and with regard to the depth of the buried targets.

15.1.1 Common Offset Interfacial Data with Null Offset on a Homogeneous Soil

The first example is about the method of the diffraction hyperbolae in common offset configuration and is implemented with simulated data. The simulation has been implemented with the GPRMAX code (Giannopulos, 2003), based on the method of

Introduction to Ground Penetrating Radar: Inverse Scattering and Data Processing,
First Edition. Raffaele Persico.
© 2014 The Institute of Electrical and Electronics Engineers, Inc. Published 2014 by John Wiley & Sons, Inc.

the finite differences in time domain FDTD (Kunz and Luebbers, 1993). The simulated "ground truth" is given in Figure 15.1.

The target is a metallic (perfect electric conductor) cylindrical pipe with ray equal to 1 cm. The depth of the center is 50 cm, whereas the abscissa of the center is 1.3 m after the starting position of the B-scan. The soil shows a relative dielectric permittivity equal to 5, a relative magnetic permeability equal to 1, and an electric conductivity equal to 0.01 S/m. A B-scan that is 2.5 m long is considered, with spatial step of 2.5 cm. The source is a Ricker pulse with nominal central frequency of 500 MHz. The offset between source and observation point is equal to zero. The data are represented in Figure 15.2, after zero timing and interface muting.

Let us now evaluate the propagation velocity from the data. Didactically, let us first do this in a "manual" fashion—that is, by means of a simple homemade code. The procedure has consisted in the following steps:

1. A vector of data has been identified by choosing N points $(x_1,t_1), \ldots, (x_n,t_n)$, where for every position x_n (corresponding to a radar trace) the relative return time t_n has been chosen as the time corresponding to the maximum modulus of the radar trace (after zero timing and interface muting). The couple corresponding to the minimum return time has been assumed as the couple (x_0,t_0) in relationship with Eq. (2.2).

2. A heuristic regularization has been applied to the data, retaining only those traces for which the value of the maximum along the trace was not smaller than 0.1 times the value of the global maximum level, which is achieved at the point (x_0,t_0).

3. A vector of trial values for the propagation velocity has been set. In particular, we have spanned the range $[c_1,c_2] = [0.33 \times 10^8$ m/s, 3×10^8 m/s], with a step of 267,000 m/s, corresponding to 1/1000 of the investigated range. The initial trial value corresponds to the propagation velocity of the electromagnetic waves

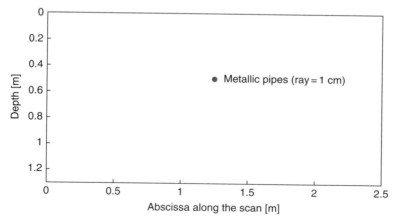

Figure 15.1. Geometry of the simulated scenario.

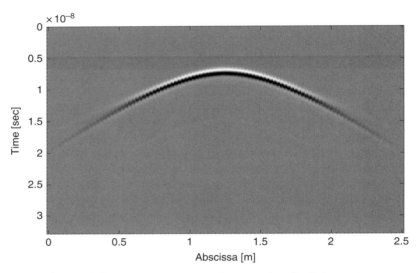

Figure 15.2. The data relative to the ground truth of Figure 15.1.

in fresh water, whereas the final value corresponds to the value of the propagation velocity in free space.

4. For each trial value c_{trial} of the propagation velocity, a vector of model data has been calculated according to Eq. (2.2) as

$$t_{\text{mod}n} = \frac{2}{c_{\text{trial}}} \sqrt{(x_n - x_0)^2 - \left(\frac{c_{\text{trial}} t_0}{2}\right)^2} \tag{15.1}$$

5. A cost function has identified as the least square difference between the retained data and corresponding model data. Thus, the cost function is given by

$$f(c_{\text{trial}}) = \sum_{n=1}^{N} \sqrt{(t_n - t_{\text{mod}n})^2} \tag{15.2}$$

6. The result has been achieved as the values of c_{trial} that makes minimum the cost function.

In Figure 15.3 the graph of the cost function is provided, with a zoom about its (unique) minimum. As it can be seen, the minimum is reached at the value $c_{\text{trial}} = 1.33 \times 10^8$ m/s. Let us remind that the data had been simulated imposing a relative dielectric permittivity equal to 5, a conductivity of 0.01 S/m and no magnetic property to the soil. This means that we can neglect the conductivity for the evaluation of the velocity, so that the actual propagation velocity is $c = c_0/\sqrt{5} = 1.34 \times 10^8$ m/s. Therefore, the propagation velocity has been estimated with an excellent precision.

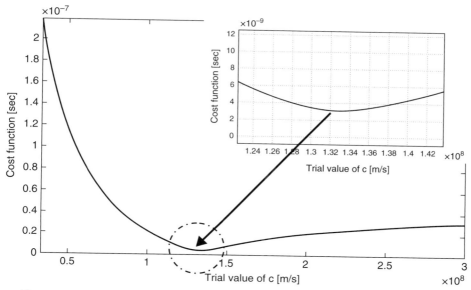

<u>Figure 15.3.</u> The cost function for a numerical evaluation of the propagation velocity.

As said, the proposed exercise is didactic: In the real practice, such a procedure would be time-consuming, because each of the data points t_n should be identified "almost manually" (in particular, the maximum value within the "hyperbola thickness" in general is not the same as the maximum value along the trace). However, there are commercial code can calculate the propagation velocity from the shape of the diffraction hyperbolas in a more automatic and faster fashion. This is theoretically less refined than the least square minimization just shown, but it is essentially the same in conceptual terms and in terms of available precision in the field. In particular, we have imported the data of the presented exercise in Reflexw (Sandmeier, 2003).

The evaluation in this case is graphical, achieved by superposing a hyperbola relative to a trial value of the propagation velocity to the apex of the diffraction curve. In Figure 15.4, a screen of this code is shown, where five trial hyperbolas are superposed to the diffraction hyperbola of the data. The trial values for the propagation velocity range from 1.16×10^8 m/s (corresponding to the narrower hyperbola) to 1.56×10^8 m/s. (corresponding to the larger hyperbola) with a step of 0.1×10^8 m/s. The central hyperbola clearly provides the best matching between model and data. It corresponds to the value 1.36×10^8 m/s, in good agreement with the actual propagation velocity.

15.1.2 Common Offset Interfacial Data on a Wall, Neglecting the Offset Between the Antennas

Let us now show an example with experimental data. The data are those gathered in the situation depicted in Figure 2.1, where a metallic road is placed and then removed from one side of the wall. Let us emphasize that even if this time the scenario is not constituted

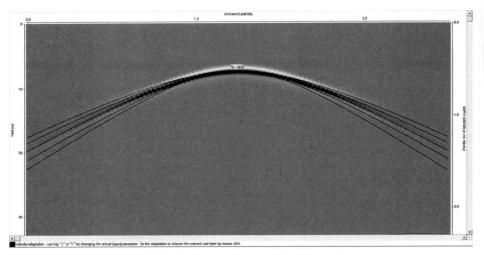

<u>Figure 15.4.</u> Evaluation of the propagation velocity performed graphically in Reflexw.

by a homogeneous half-space, the propagation from the antennas to the target and vice versa occurs in the wall, so that the diffraction hyperbola is substantially the same that one would achieve in a soil made up of the same material constituting the wall (that is concrete). The data were gathered with a Ris Hi-mode system, equipped with antennas at nominal central frequency of 2000 MHz. The scan was performed in continuous mode, downward along a vertical B-scan. The metallic road was placed in horizontal position behind the wall, in order to maximize the reflected signal from it (Conyers, 2004). The results are provided in Figure 15.5.

Of course, a real case presents a much more confused scenario. In particular, on purpose we didn't choose a controlled test site, and we just gathered the data on the wall of a dismissed hovel: The several anomalies visible in the wall are probably internal reinforcement metallic bars.

However, the "extra" diffraction hyperbola due to the marker (represented by the circle) is clearly identified, even if its left-hand half-branch is masked by some other internal target. In the case at hand the thickness of the wall was directly measurable (see Figure 2.1), and so we know that this thickness is 40 cm. Now, since the top of the hyperbola occurs at 5.3 ns, this provides a propagation velocity of 1.5×10^8 m/s, which is a value plausible for a concrete, even if slightly smaller than the average expected values, presented in tables (for example) in Conyers (2004), Daniels (2004), and Jol (2009). For comparison, also the shape of the diffraction hyperbola has been exploited. We have made use for this of the Reflexw code, and the result was 1.40×10^8 m/s. This case requires the watchfulness to match only the right-hand side of the hyperbola, because its right-hand side is not visible. Actually, there is a sort of branch of hyperbola also on the left side of the bar, but it is a feature visible also in absence of the bar, and so it is not due to the bar but to something else (even if we don't

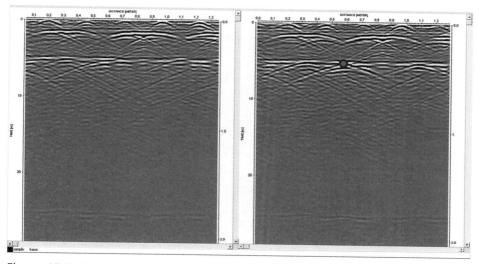

Figure 15.5. The experimental example relative to the photographs shown in Figure 2.1. **Left-hand image**: The signal achieved without the auxiliary target. **Right hand image**: The signal achieved with the auxiliary target.

know what). It is not simple to determine the better estimation, because the wall is not perfectly homogeneous. So, we might choose the average value 1.45×10^8 m/s. The discrepancy between the two retrieved values and the central value is 3.45%.

15.1.3 Interfacial Common Offset Data on a Homogeneous Soil: The Effect on the Offset Between the Antennas

Let us now present a simulation depicting the effect of the offset. In particular, we propose a simulation with realistic parameters, where the source is a Ricker pulse at central frequency equal to 2 GHz and the offset between source and observation point is 5 cm. We have considered a small metallic target at the depth of 5 cm. The soil shows a relative dielectric permittivity $\varepsilon_{sr} = 5$ and an electric conductivity $\sigma = 0.01$ S/m. The soil does not show magnetic properties. The spatial step is 1.25 cm. In Figure 15.6 several diffraction curves achieved with a null offset are represented, superposed to the data, whereas in Figure 15.7 several diffraction curves achieved considering the actual offset are superposed to the data. The data have been simulated by means of GPRMAX. In both cases, the solid lines range from the trial values $\varepsilon_{sr} = 1$ (top curve) to the trial value $\varepsilon_{sr} = 7$ (lowest line), and in both cases the best matching with the data is heuristically achieved for the diffraction curve relative to $\varepsilon_{sr} = 3.5$ (dotted line). This means that we retrieve the propagation velocity with an error of about 19.5%. The result is less good than that achieved in the previous simulations. This is due to some near-field effects and possibly also due to some numerical problem resulting from the closeness of the target to the

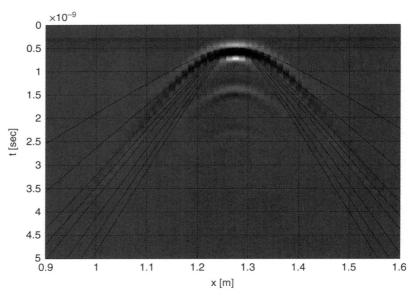

Figure 15.6. Diffraction curves neglecting the offset. **Solid lines from top to bottom:** Curves relative to trial relative permittivities ranging from 1 to 7. **Dotted line:** Heuristic best matching curve relative to the trial value 3.5.

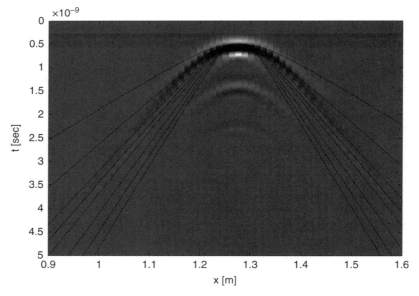

Figure 15.7. Diffraction curves considering the offset. **Solid lines from top to bottom:** Curves relative to trial relative permittivities ranging from 1 to 7. **Dotted line:** Heuristic best matching curve relative to the trial value 3.5.

interface. However, to consider a deeper target would have made obvious the result, unless we had simulated an improbably large offset. The relevant result is that the offset does not have too much effect on the result.

15.1.4 Noninterfacial Common Offset Data with a Null Offset Between the Antennas

Let us now show some results relative to the effect of the height of the observation line: In particular, we have considered a small metallic target at the depth of 0.5 m and an observation line at the height of 0.5 m above the air–soil interface. The soil has the same characteristics of the previous exercise.

The source is a Ricker pulse with central frequency 500 MHz and the spatial step of the data is 5 cm. In Figure 15.8 the effect of the refraction at the interface is neglected, whereas in Figure 15.9 it is accounted for. The set of diffraction curves are analogous to that considered in the previous exercise. Figures 15.8 and 15.9 allow some observations: First of all, we see that the height of the observation line has a consistent weight, and to neglect it can lead to meaningful errors. However, it is also important to stress that, at variance of the trial propagation velocity, the diffraction curves considering the height are "closer" to each other than in the case of data gathered at the air–soil interface. In other terms, if one calculated the cost function [i.e., the homologous of Eq. (15.2)]

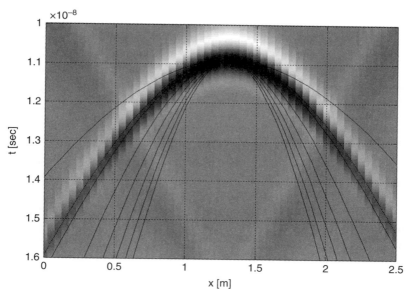

Figure 15.8. Diffraction curves neglecting the height of the observation line. **Solid lines from top to bottom**: Curves relative to trial relative permittivities ranging from 1 to 7. **Dotted line**: Heuristic best matching curve relative to the trial value 1.8.

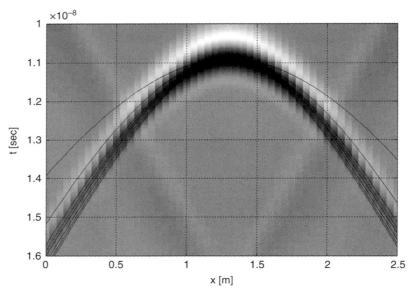

Figure 15.9. Diffraction curves considering the height of the observation line. **Solid lines from top to bottom:** Curves relative to trial relative permittivities ranging from 1 to 7. **Dotted line:** Heuristic best matching curve relative to the trial value 3.5.

in the case of data gathered at some distance from the air–soil interface, he/she would see a much flatter behavior regarding the minimum with respect to the case of interfacial data, which makes it more difficult to perform a reliable measurement. The physical reason has been exposed in Section 2.3.4.

15.1.5 Common Midpoint Data

Let us now show an example with common midpoint data. In particular, let us in a first moment erase the effect of the air–soil interface. The data are obtained with GPRMAX, and the reference scenario is the same as in the example of the common (null) offset relative to Figures 15.1 and 15.2. The data start over the target and extend along the same observation line of the previous example, with the same spatial step. We have kept also the same Cartesian reference system of the previous example, for comparison. The result is the half-hyperbola shown in Figure 15.10. This time, the evaluation of the propagation velocity is achieved from the slant of the tail of the diffraction hyperbola. In Figure 15.10 we have represented the same diffraction curve in both panels, for sake of graphical clarity. However, in the lower panel we have gridded the image and have superposed the tangent to the farthest (from the vertex) visible point of the hyperbola.

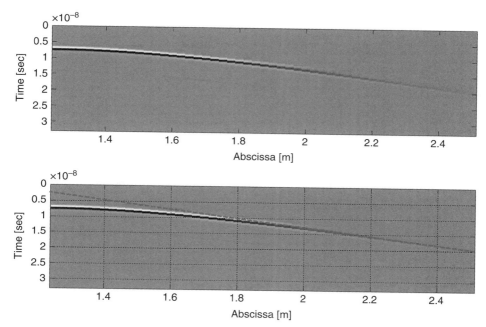

Figure 15.10. The diffraction curve retrieved from common midpoint data. The curve is the same in both the upper and the lower panels, where a grid and the tangent line in the "tail" point have been superposed.

The extreme points of the dashed line in Figure 15.10 are $(x, t) = (1.25$ m, 3 ns) on the left-hand side and $(x, t) = (2.5$ m, 21 ns) on the right-hand side. So, the slant of the line is given by

$$s = \frac{(21-3) \times 10^{-9}\,\mathrm{s}}{(2.5-1.25)\,\mathrm{m}} = 14.4 \times 10^{-9}\,\mathrm{s/m} \qquad (15.3)$$

On the basis of Eq. (2.3), the propagation velocity can be estimated as

$$c = \frac{2}{s} = 1.39 \times 10^{8}\,\mathrm{m/s} \qquad (15.4)$$

Let us remind ourselves that the actual value for this example was 1.34×10^{8} m/s, so the percentage error is 3.7%.

At this point, let us reconsider the complete data—that is, the data of the same CMP without erasing the interface contribution and the direct coupling. The amplitude of the signal received from all the points has been made at the same level, thereby mitigating the effect of the natural high dynamic range, due to the fact that the antennas get progressively farther from each other. This is equivalent to radiating some more power when the antennas gets farther apart. The result is shown in Figure 15.11: The difference

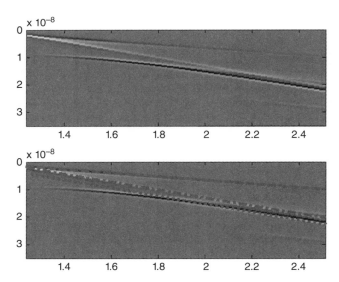

Figure 15.11. Measure of the propagation velocity from complete CMP data.

between the direct signal in air and that in the soil is evident, and it is also evident that direct signal received through the soil provides a line almost parallel to the tangent to the diffraction hyperbola (far from the vertex) relative to the buried target.

We have then zoomed apart (this is not shown here) the image in Figure 15.11 and have seen that the line relative to the path in air passes for the two points (1.3 m, 2 ns) and (1.6 m, 4 ns), whereas the second line passes for the two points (1.5 m, 6 ns) and (1.9 m,12 ns). This allows us to retrieve the slants of the two lines, from which the two (definitely satisfying) values $c_0 = 3 \times 10^8$ m/s and $c = 1.33 \times 10^8$ m/s are worked out.

15.2 EXERCISES ON SPATIAL STEP AND HORIZONTAL RESOLUTION

In this section, we propose some examples about the needed spatial step and the correlated achievable horizontal resolution. To construct a first example, let us consider a 2D investigation domain D sized 2×2 m^2, starting from the depth of 0.5 m. The relative permittivity of the soil is equal to $\varepsilon_{sr} = 5$, and the electric conductivity is equal to 10^{-3} S/m. The frequency band ranges from 200 to 710 MHz and the frequency step is 15 MHz. The investigation domain has been discretized by means of 51 complex Fourier harmonics along the horizontal direction and 45 step functions along the depth. The observation line Σ is at the air–soil interface and has the same length as the investigation domain. The offset between the source and the observation point is equal to zero.

In Figure 15.12 the reference geometry for this simulation is shown, whereas in Figure 15.13 the singular values of the discretized linear scattering operator are shown at variance of the spatial step, having fixed the frequency step at 15 MHz.

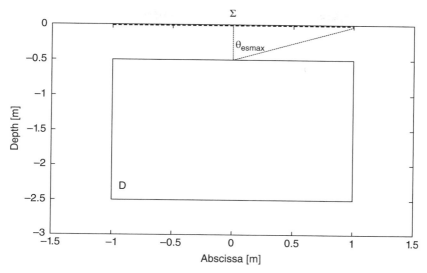

Figure 15.12. Maximum view angle on the top of the investigation domain. Due to the limited size of D, it is approximated with the maximum geometrical view angle.

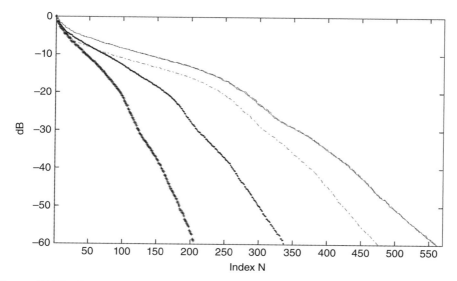

Figure 15.13. Singular values at variance of the spatial step: crosses (lower curve): $s = 40$ cm; dots: $s = 20$ cm; dots and dashes: $s = 10$ cm; solid line: $s = 5$ cm; dashes (almost indistinguishable from the solid line): $s = 2.5$ cm.

In particular, the spatial step is progressively halved starting from 40 cm to 20 cm, to 10 cm, to 5 cm, and to 2.5 cm. From Figure 15.13, it is clearly seen that the curve of the singular values grows meaningfully up to the spatial step of 5 cm. However, passing from 5 to 2.5 cm the curve keeps substantially unchanged. This means that the quantity of information is saturated; that is, adding more measurement points doesn't add a meaningful quantity of information. It is implicit that the validity of this reasoning is also related to the dynamic range of the singular values taken into consideration. In other words, one has to establish the level up to which the singular values are meaningful. As widely stressed in Chapter 14, from a practical point of view, it is difficult to establish how many singular values should be accounted for in a nonheuristic way. On the basis of the specific practical experiences of this author, however, the heuristically optimal choice in most cases lies within the range between -20 and -40 dB. Consequently, the interval considered in Figure 15.13 (which extends up to -60 dB) can be practically viewed as a conservative range. Let us now compare this result with those achieved under DT in Chapter 9. In particular, let us first note that, in the case at hand, the minimum involved wavelength in the soil (neglecting the losses) is equal to $\lambda_{smin} = c_0/f_{max}\sqrt{\varepsilon_{sr}} = 18.9$ cm. Moreover, in the case at hand we are considering a relatively small investigation domain, and therefore we can approximate the effective maximum view angle with the geometrical maximum view angle. The sine of the maximum view angle with respect to the shallowest part of the investigation domain (shown in Figure 15.12) is equal to $\sin(\theta_{esmax}) = 1/\sqrt{1^2 + 0.5^2} = 0.89$. Therefore, the optimal spatial step under the DT approximation in the case at hand [see Eq. (9.49)] is equal to $s = \lambda_{smin}/4\sin(\theta_{esmax}) = 5.3$ cm. This means that, according to DT, a spatial step meaningfully smaller than 5 cm is redundant, whereas a step meaningfully larger is inadequate and causes a loss of information. So, the behavior of the singular values is in noticeable agreement with the predictions achieved under a DT model.

Let us now show some reconstruction examples. The spatial step of the data is 5 cm and the model is built up as just previously described. The data have been achieved from FDTD simulations performed with GPRMAX, and the central frequency of the exploited Ricker pulse was heuristically chosen at 320 MHz, so that the -6 dB band of the data was extended from about 200 to 710 MHz. In the six reconstructions shown in Figure 15.14, we propose the case of two circular, electrically small (ray = 1 cm) metallic targets buried at the depth of 55 cm (meant as the depth of the centers), progressively shifted toward each other from 35 to 10 cm (meant as the distance of the two centers). The reconstruction has been achieved from an inversion of the linear scattering operator performed by means of a singular value decomposition regularized with a threshold at -20 dB on the singular values. As can be seen, the two targets are well distinguished from each other if their reciprocal distance is equal to or larger than 20 cm. Instead, if their distance is equal to 15 cm, then their reconstructed images touch each other and they are hardly distinguishable. Finally, if their distance is equal to or smaller than 10 cm, they are definitely fused into a unique target. In the case at hand, the DT resolution is of the order of 16.6 cm [see Eq. (9.43)], in noticeable agreement with the achieved results.

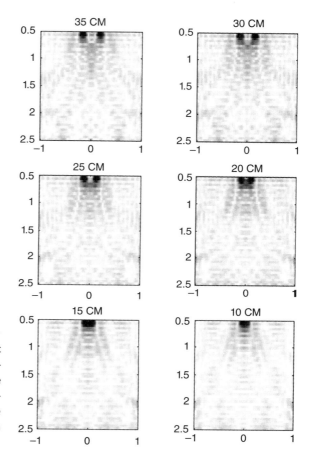

Figure 15.14. Reconstruction by inversion of two circular targets at the same depth (55 cm) progressively closer to each other. The title of each panel refers to the distance between the centers of the two targets. The axes are in meters.

In order to show the worsening of the horizontal resolution versus the depth, we have repeated the example shifting the two targets 1 m below, at the depth of 155 cm. The sine of the maximum view angle for these two targets is restricted to 0.51, which leads to a DT resolution of about 29 cm [see Eq. (9.43)]. Figure 15.15 is the analogous of Figure 15.14 for these deeper targets, and it shows results in good agreement with the DT previsions. In particular, the two spots are hardly distinguished at 30 cm and collapse into a unique spot for smaller values of the spacing between them.

The data exploited in Figures 15.14 and 15.15 are noiseless. In order to show the robustness versus the noise, in Figures 15.16 and 15.17 the same cases of Figures 15.15 and 15.16 are proposed, with the only difference being the addition of a white Gaussian noise to the data. The signal-to-noise ratio, referring to the total field data, is 60 dB. The comparison between Figure 15.14 and Figure 15.16 and between Figure 15.15 and Figure 15.17 shows that the inversion is quite robust against noise.

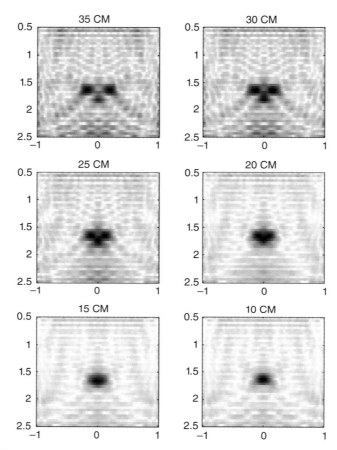

Figure 15.15. Reconstruction by inversion of two circular targets at the same depth (155 cm) progressively closer to each other. The title of each panel refers to the distance between the centers of the two targets. The axes are in meters.

At this point, let us investigate the dependence of the resolution on the degree of regularization. In particular, in Figures 15.18 and 15.19 the same noisy data of Figures 15.16 and 15.17 are inverted, thresholding the singular values at −40 dB instead of −20 dB. Figures 15.18 and 15.19 show clearly the trade-off between resolution and robustness. In particular, in the case of the shallower targets, we can say that the lower threshold allows a marginal improvement of the resolution. However, in the case of the deeper targets, the lower threshold makes the reconstruction much poorer. It is interesting to stress that this happens at a parity of signal-to-noise ratio, because the echoes from the deeper targets are weaker than those from analogous targets at a shallower depth. Indeed, the signal-to-noise ratio is an integral parameter referring to the entire recorded time range (either real or synthetic). During this time, however, the instantaneous signal power gets progressively lower whereas the instantaneous noise power remains of the same

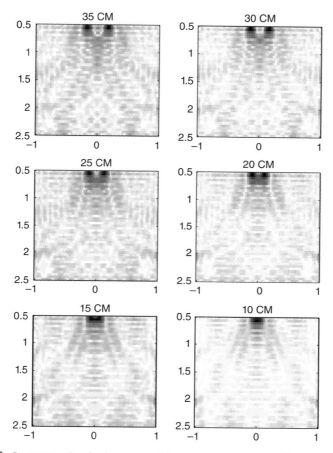

Figure 15.16. Reconstruction by inversion of two circular targets at the same depth (55 cm) progressively closer to each other. The title of each panel refers to the distance between the centers of the two targets. The axes are in meters. The data are noisy with SNR = 60 dB, evaluated on the total field.

order. This is sometimes loosely but explicatively expressed saying that the signal-to-noise ratio gets worse versus the depth of the targets.

This confirms the prevision (see Section 14.4) that in most cases the optimal threshold should be chosen heuristically on the basis of the results achieved with several threshold levels or more in general under several regularization degrees. At the same time, these examples also show that, in any case, the resolution improvements possibly achievable from a weaker regularization are expected to be marginal with respect to the DT predictions.

In Figures 15.20 and 15.21 the same noisy data of the previous examples have been migrated in the time domain. The migration has been performed with the commercial

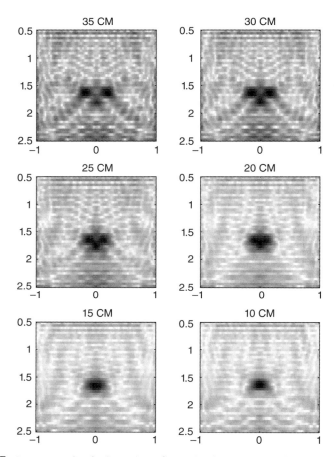

Figure 15.17. Reconstruction by inversion of two circular targets at the same depth (155 cm) progressively closer to each other. The title of each panel refers to the distance between the centers of the two targets. The axes are in meters. The data are noisy with SNR = 60 dB, evaluated on the total field.

code Reflexw (Sandmeier, 2003). More precisely, after importing the raw noisy data in Reflexw, a data-driven processing has been performed as follows:

1. The zero time has been moved of 2.75 ns.
2. A background removal up to 4 ns on all the traces has been performed in order to erase the residual air–soil interface.
3. The data have been filtered by means of a Butterworth filter in order to clean some noise. The chosen cutoff frequencies for the filters have been set at 200 and 710 MHz, respectively, in order to make the migration-based processing homologous to the inversion.

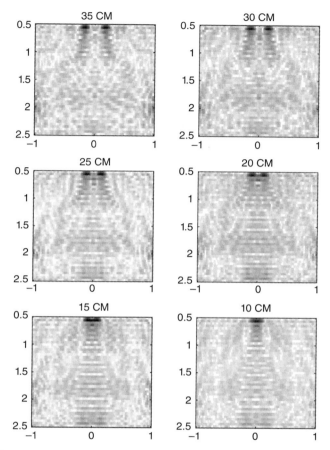

Figure 15.18. Reconstruction by inversion of two circular targets at the same depth (55 cm) progressively closer to each other. The title of each panel refers to the distance between the centers of the two targets. The axes are in meters. The data are noisy with SNR = 60 dB, evaluated on the total field. The SVD is thresholded at −40 dB.

4. A gain versus the depth has been applied in order to make more visible the targets. The value of the gain has been set at 5 dB/ns up to a maximum gain equal to 40 dB.
5. Kirchhoff migration on 25 traces has been applied, on the basis of the wideness of the diffraction hyperbolas.
6. After exporting the migrated data in MATLAB, time–depth conversion has been achieved from the propagation velocity retrieved from the diffraction hyperbolas, and the same spatial window shown for the inversion results has been adopted.

The results of Figures 15.20 and 15.21 basically drive to the same considerations of the inversion results. Namely, the resolution is of the same order as that foreseen within

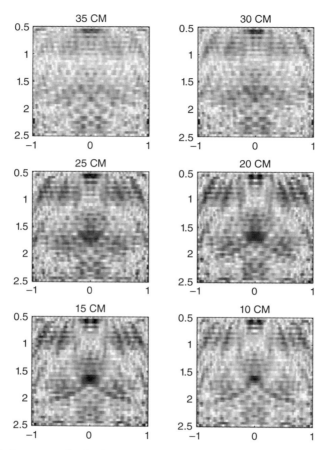

Figure 15.19. Reconstruction by inversion of two circular targets at the same depth (155 cm) progressively closer to each other. The title of each panel refers to the distance between the centers of the two targets. The axes are in meters. The data are noisy with SNR = 60 dB, evaluated on the total field. The SVD is thresholded at –40 dB.

DT, and it degrades versus the depth. The artifact due to the crossing point between the diffraction hyperbolas is evident both within the inversion-based and migration-based result.

Before ending this section, we found it interesting to show a case where the effective maximum view angle is different from the geometrical maximum view angle. This example has been achieved by just adopting a time bottom scale of 30 ns in the simulated data (in the previous examples the bottom scale was 40 ns in all cases). We have simulated two targets at the depth of 1.95 m, with the same measurement parameter of the other examples shown in this section. Now, with reference to panel A in Figure 15.22, since the observation line is 2 m long, if we calculate the maximum view angle geometrically with respect to the initial or the final point of the scan and

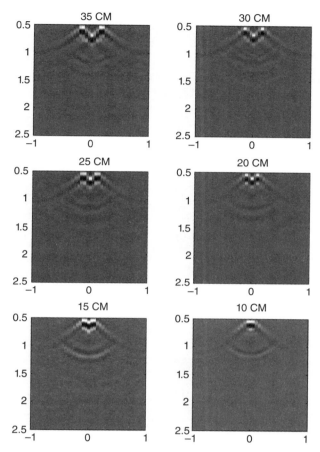

Figure 15.20. Reconstruction by Kirchhoff migration of two circular targets at the same depth (55 cm) progressively closer to each other. The title of each panel refers to the distance between the centers of the two targets. The axes are in meters. The data are noisy with SNR = 60 dB, evaluated on the total field.

with respect to the midpoint between the targets, it is easy to recognize that the sine of the geometrical maximum view angle is equal to $\sin(\theta_{emax}) = 1/\sqrt{1+1.95^2} = 0.456$ ($\theta_{emax} = 27.13$ degrees). In this case, the hypotenuse of the triangle ABC is long $\sqrt{1+1.95^2} = 2.191$ m (see Figure 15.22, panel A) and, according to Eq. (9.43), the expected horizontal resolution should be of the order of 32 cm. However, the longest time up to which we are now recording the signal is 30 ns, and therefore the maximum hypotenuse that we have to consider for the triangle ABC restricts to (see Figure 15.22, panel B) $AC = 0.5c_0 \times 30 \times 10^{-9}s./\sqrt{\varepsilon_{sr}} = 2.0125$ m. Consequently, the segment BC, that represents the maximum abscissa under which we still see some tail of the diffraction hyperbola restricts from 1 m to $AB = \sqrt{2.0125^2 - 1.95^2}$ m $= 0.498$ m, and coherently

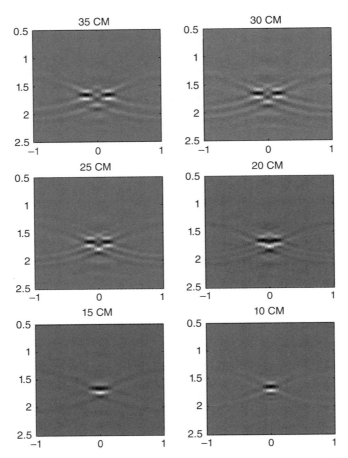

Figure 15.21. Reconstruction by Kirchhoff migration of two circular targets at the same depth (155 cm) progressively closer to each other. The title of each panel refers to the distance between the centers of the two targets. The axes are in meters. The data are noisy with SNR = 60 dB, evaluated on the total field.

the sine of the effective view angle restricts to $\sin(\theta_{emax}) = (BC/AC)(0.4977/2.0125) = 0.247$ ($\theta_{emax} = 14.32$ degrees). Accordingly, based on Eq. (9.43) the expected horizontal resolution degrades to about 58 cm.

In Figures 15.23 and 15.24 the reconstruction of two circular small (ray = 1 cm) metallic targets at the same depth (1.95 m) is shown. The distance between the two targets is labeled on each subfigure. We have considered distances larger than in the previous figures, because of the foreseen loss of resolution. The reconstructions in Figure 15.23 are achieved from a linear inversion performed by SVD and regularized by truncation of the singular values with threshold at −20 dB. The reconstructions of Figure 15.24 are achieved from the same migration-based processing adopted in Figures 15.20 and 15.21. The synthetic data are affected by a Gaussian white noise with SNR = 60 dB,

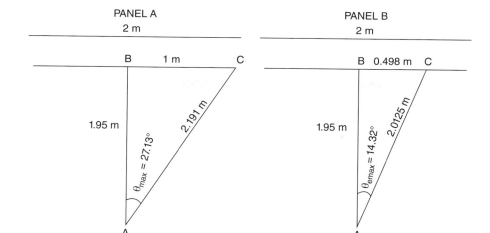

Figure 15.22. Quantitative pictorial of the restriction of the effective maximum view angle due to the reduction of the time bottom scale from 40 ns to 30 ns.

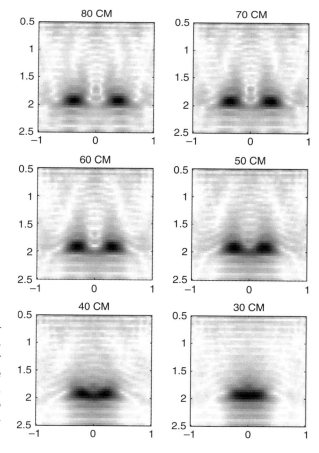

Figure 15.23. Reconstruction by inversion of two circular targets at the same depth (195 cm). The panels are labeled after the distance between the centers. The axes are in meters. SNR = 60 dB with respect to the total field. The SVD is thresholded at −20 dB.

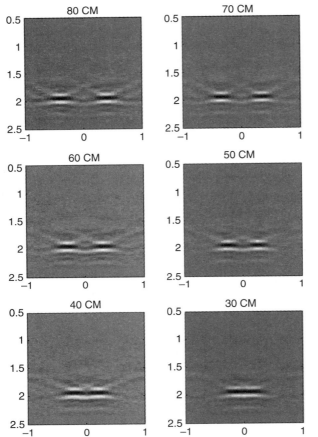

Figure 15.24. Reconstruction by Kirchhoff migration of two circular targets at the same depth (195 cm) progressively closer to each other. The title of each panel refers to the distance between the centers of the two targets. The axes are in meters. The data are noisy with SNR = 60 dB, evaluated on the total field.

evaluated with respect to the total field. From both Figures 15.23 and 15.24, we can appreciate the loss of resolution due to the restriction of the effective maximum view angle, which appears completely fused into each other if their distance is 30 cm and appears fully detached from each other only if their distance is 60 cm or more. As a corollary, this exercise also shows that the chosen time bottom scale should be longer than the expected (if any) time depth of the targets of interests, because we need the whole tails of the diffraction hyperbolas in order to focus at best the buried objects.

15.3 EXERCISES ON FREQUENCY STEP AND VERTICAL RESOLUTION

Some exercises on the frequency step and on the vertical resolution can be proposed on the basis of the same examples proposed in the previous section. In particular, let us remind ourselves that we have made use of a matrix calculated in the frequency

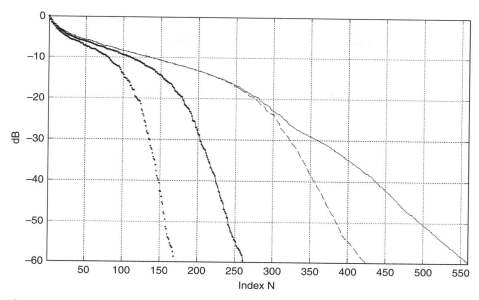

Figure 15.25. Behavior of the singular values at variance of the frequency step. Crosses (lower curve): $\Delta f = 127.5$ MHz. Dots: $\Delta f = 63.75$ MHz. Dashes: $\Delta f = 30$ MHz. Solid (upper curve): $\Delta f = 15$ MHz.

band 200–710 MHz, sampled with a frequency step equal to 15 MHz. Moreover, the investigation domain ranges along the depth from 0.5 to 2.5 m. This means that, according with the DT model [Eq. (9.52)], the frequency step should be not larger than $\Delta f = c_0/2b\sqrt{\varepsilon_{sr}\mu_{sr}} = 33.54$ MHz. Thus, the first question is whether or not the chosen frequency step of 15 MHz was redundant. In order to answer this question, in Figure 15.25 we show the behavior of the singular values while progressively enlarging the frequency step from 15 to 30, 63.75, and 127.5 MHz.[1] The other parameters are unchanged with respect to the simulations of the previous section, and in particular the spatial step is 5 cm.

From Figure 15.25, we see that the suitable frequency step is actually dependent on the level of the allowed regularization. In particular, if the singular values are to be thresholded at −20 dB, then a frequency step of 15 MHz is redundant and the DT bound of 30 MHz is adequate. However, if the data are particularly "good" and allows us to make use of a weaker regularization, then it can be useful to reduce the frequency step with respect to the DT previsions. The evaluation of this possibility is case-dependent but, as said, to the best of our experience it is quite hard that the GPR data allow to achieve a good reconstruction with a regularization threshold lower than −40 dB. In order to test the effect of these considerations versus the quality of the reconstruction and in particular versus the vertical resolution, we now propose some examples with two

---

[1] This means to consider 5, 9, 18, and 35 frequencies equally spaced in the band 200–710 MHz, respectively.

circular and electrically small (ray = 1 cm) metallic targets superposed to each other at the same abscissa but at different depths. The data are noisy with white Gaussian noise, and the signal-to-noise ratio with respect to the total field is equal to SNR = 60 dB. The soil parameters are the same as the previous examples, as well as the number and dislocation of data. The frequency step is 15 MHz. In Figure 15.26, some inversion results are shown: The upper target is buried at 55 cm and the distance between the upper and the lower target is labeled on each panel. The SVD is thresholded at −20 dB. In Figure 15.27, the analogous results for two deeper targets of the same kind are shown; this time the upper target is at the depth of 155 cm. Also for Figure 15.27 the SVD is thresholded at −20 dB. The comparison between Figures 15.26 and 15.27 shows that the evaluation

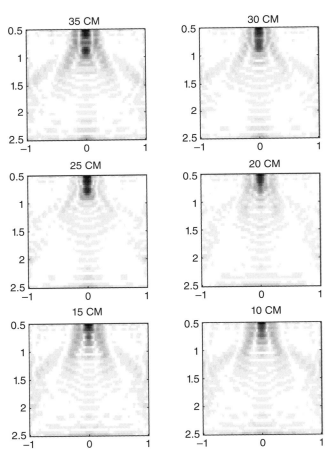

Figure 15.26. Reconstruction by inversion of two circular superposed targets (the upper one buried at 55 cm) progressively closer to each other. The title of each panel refers to the distance between the centers of the two targets. The axes are in meters. The data are noisy with SNR = 60 dB, evaluated on the total field. The SVD is thresholded at −20 dB.

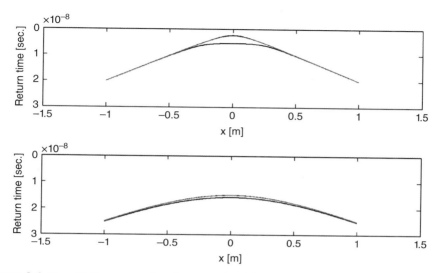

Figure 2.6 Quantitative comparison between the diffraction curve with offset equal to 0 cm (blue line), 10 cm (red line), and 50 cm (black line). The propagation velocity is 10^8 m/s. The depth of the target is 12.5 in the upper panel and 75 cm in the lower panel.

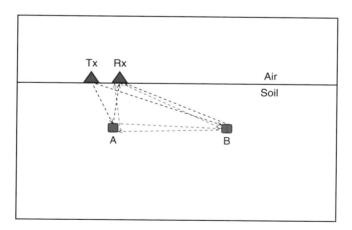

Figure 8.1 Schematic for the physical reason of the nonlinearity of the scattering. The black arrows represent the direct contributions of the two small buried targets separately considered. The red arrows represent the contribution of the mutual interaction between A and B to the scattered field.

Introduction to Ground Penetrating Radar: Inverse Scattering and Data Processing,
First Edition. Raffaele Persico.
© 2014 The Institute of Electrical and Electronics Engineers, Inc. Published 2014 by John Wiley & Sons, Inc.

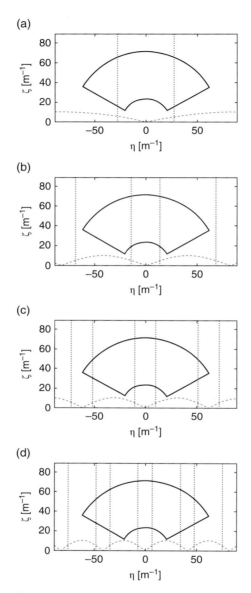

Figure 9.19 Red dashed lines: the modulus of the erasing function (emphasized by a factor 5 for graphic reasons). Blue dot lines: the erasing belts deriving from the erasing function. Solid black line: the spectral set without the differential effect. The parameters are: f_{min} = 200 MHz, f_{max} = 600 MHz, ε_{sr} = 4, μ_{sr} = 2. **Panel a:** Δ = 3.83 cm. **Panel b:** Δ = 7.65 cm. **Panel c:** Δ = 10.21 cm. **Panel d:** Δ = 15.31 cm. The erased belt is calculated according to Eq. (9.75).

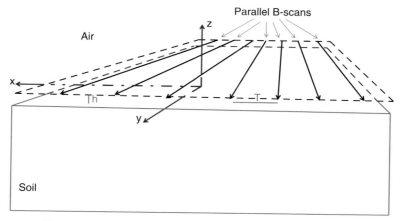

Figure 12.1 Geometrical scheme of the common offset C-scan.

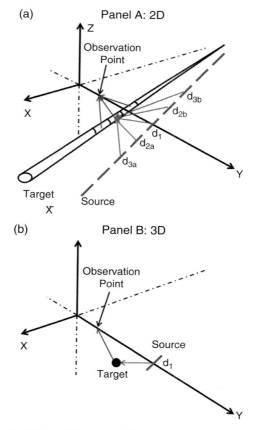

Figure 13.1 Pictorial of a 2D (panel A) and a 3D (panel B) geometry with an impulsive point-like source and a point-like target.

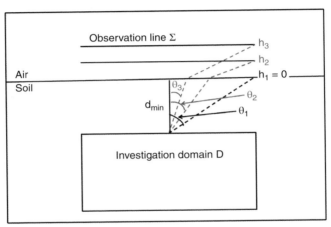

Figure 15.40 Pictorial of the reduction of the maximum view angle when increasing the height of the observation line.

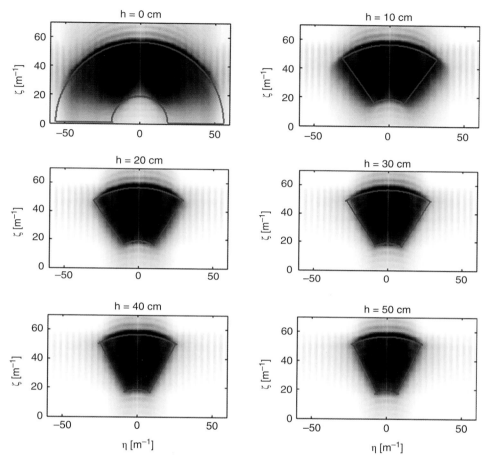

Figure 15.41 Spectral content of the singular functions (threshold of the singular value at −20 dB), together with the DT bound of the spectral retrievable set (red lines) at variance of the height *h* of the observation line.

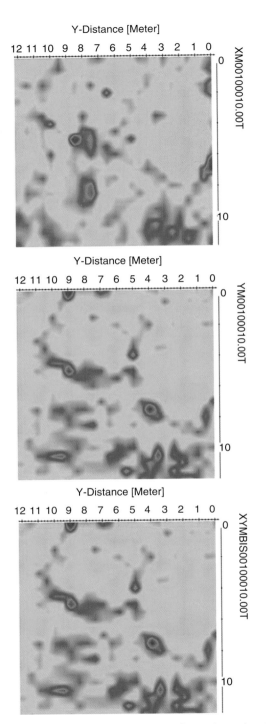

Figure 15.72 Depth slices at 47 cm obtained with Reflexw through a 2D migration-based processing. **Upper panel**: Result obtained from the B-scans parallel to only the *x*-axis. **Middle panel**: Result obtained from the B-scans parallel to only the *y*-axis. **Lower panel**: Result obtained from the B-scans parallel to both the *x*-axis and the *y*-axis. The axes are in meters.

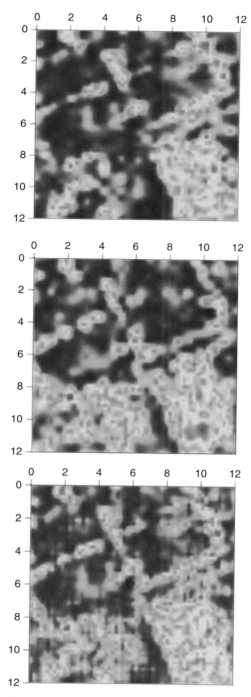

Figure 15.73 Depth slices at 47 cm obtained with GPRSLICE through 2D migration-based processing. **Upper panel**: Result obtained from the B-scans parallel to only the x-axis. **Middle panel**: Result obtained from the B-scans parallel to only the y-axis. **Lower panel**: Result obtained from the B-scans parallel to both the x-axis and the y-axis. The axes are in meters.

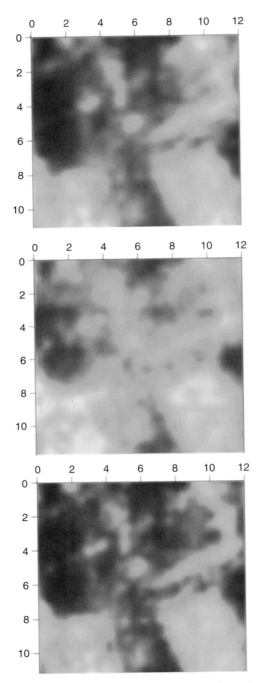

Figure 15.74 Depth slices at 47 cm obtained with GPRSLICE through a 3D migration-based processing. **Upper panel**: Result obtained from the B-scans parallel to only the x-axis. **Middle panel**: Result obtained from the B-scans parallel to only the y-axis. **Lower panel**: Result obtained from the B-scans parallel to both the x-axis and the y-axis. The axes are in meters.

of the vertical resolution is more "tricky" than that of the horizontal resolution. In particular, in the proposed examples it is clear that the upper target partially masks the lower one, whose image appears faded. Then, as said in Chapter 9 (see Section 9.6) there is an interference term between the two targets that depends on the position of the targets and of the source.

In particular, this can cause some anomalies such as, with reference to Figure 15.26, an artifact in the middle between the two targets when these are at the distance of 35 cm or, with reference to Figure 15.27, the fact that the two targets seem better distinguishable from each other when their distance is 30 cm instead of 35 cm.

The DT vertical resolution in the case at hand is of the order of 26 cm according to Eq. (9.45). The rougher equation (9.46) yields to 29 cm. In the case at hand, the dispersion

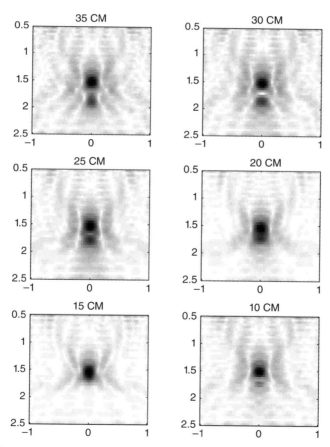

Figure 15.27. Reconstruction by inversion of two circular superposed targets (the upper one buried at 155 cm) progressively closer to each other. The title of each panel refers to the distance between the centers of the two targets. The axes are in meters. The data are noisy with SNR = 60 dB, evaluated on the total field. The SVD is thresholded at −20 dB.

of the soil is not meaningful and we do not appreciate a different band between the signal coming from a target at 55 cm and a target at 155 cm. As we can see, the achieved results are in good agreement with the DT previsions, and in particular we distinguish the two targets from each other if their distance is equal to 25 cm or larger. Moreover, in the case at hand, unlike the case of the horizontal resolution, we don't see any meaningful loss of resolution versus the depth, and actually the vertical resolution seems even better for the deeper targets of Figure 15.27 than for the shallower targets of Figure 15.26.

In order to show the dependence of the resolution on the regularization, in Figures 15.28 and 15.29 we show the results achieved thresholding the SVD at −40 dB instead of −20 dB, in the same cases of Figure 15.26 and 15.27, respectively.

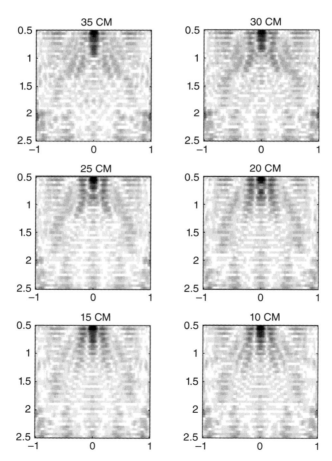

Figure 15.28. Reconstruction by inversion of two circular superposed targets (the upper one buried at 55 cm) progressively closer to each other. The title of each panel refers to the distance between the centers of the two targets. The axes are in meters. The data are noisy with SNR = 60 dB, evaluated on the total field. The SVD is thresholded at −40 dB.

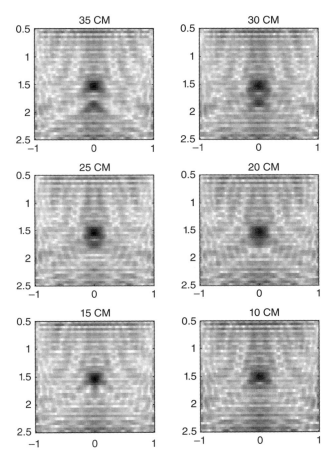

Figure 15.29. Reconstruction by inversion of two circular superposed targets (the upper one buried at 155 cm) progressively closer to each other. The title of each panel refers to the distance between the centers of the two targets. The axes are in meters. The data are noisy with SNR = 60 dB, evaluated on the total field. The SVD is thresholded at –40 dB.

Figures 15.28 and 15.29 make us appreciate that, in the case at hand, no meaningful improvement is achieved from a weaker regularization. This results is at least partially due to the masking of the deeper target of behalf of the shallower one. In particular, the results of Figures 15.28 and 15.29 indirectly show that a frequency step of 30 MHz would have been "sufficient" for the inversion at hand. In particular, when thresholding at –20 dB, the number of upper threshold singular values are substantially the same for $\Delta f = 30$ MHz and $\Delta f = 15$ MHz (see Figure 15.25). This shows in turn that the DT-based previsions about the needed frequency step [see Eq. (9.52)] provide a reasonable order of magnitude.

In Figures 15.30 and 15.31 the data of Figures 15.28 and 15.29 are focused by a migration-based algorithm. The details of the processing are the same of those described

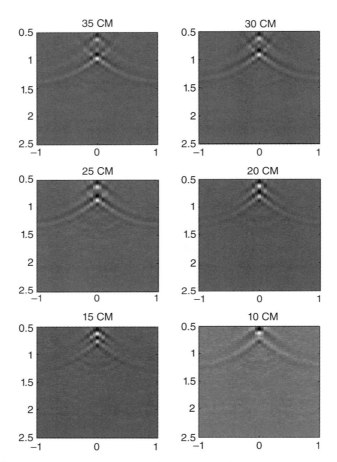

Figure 15.30. Reconstruction by Kirchhoff migration of two circular superposed targets (the upper one buried at 55 cm) progressively closer to each other. The title of each panel refers to the distance between the centers of the two targets. The axes are in meters. The data are noisy with SNR = 60 dB, evaluated on the total field.

for Figures 15.20 and 15.21. From Figures 15.30 and 15.31 we see that the migration-based algorithm provides some more "tails" starting from the focused targets but, on the other hand, the capability of distinguishing the two targets seems even better than those provided by the inversion algorithm.

To sum up, the vertical resolution is intrinsically a concept definable in a way less rigorous than the horizontal resolution, because the class of retrievable targets is characterized (in terms of spatial frequencies) by a low-pass behavior along the abscissa and a band-pass behavior along the depth. In particular, there is an irresolvable ambiguity specifically linked to this fact, and there is a question about whether two superposed spots should refer to two distinct targets or to the top and the bottom of a unique structure; this

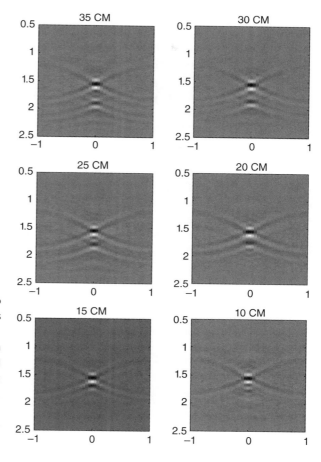

Figure 15.31. Reconstruction by Kirchhoff migration of two circular superposed targets (the upper one buried at 155 cm) progressively closer to each other. The title of each panel refers to the distance between the centers of the two targets. The axes are in meters. The data are noisy with SNR = 60 dB, evaluated on the total field.

is actually deduced (when possible) from the context of the case history at hand and not really from the involved physics–mathematics.

15.4 EXERCISES ON THE NUMBER OF TRIAL UNKNOWNS

In the two previous sections, we have shown examples on the achievable resolution and (related to this) on the needed spatial and frequency steps, which means examples on the number of data to be gathered. Clearly, in order to discretize the problem for a numerical inversion, we also need some criterion to choose the number and kind of unknowns formally looked for. We say formally looked for because, within a regularized SVD-based approach, the kind of unknowns really looked for are the expansion coefficients of the object function along the singular functions in the space of the unknowns (see Chapter 14). On the other hand, we don't know the singular functions in a closed form, and so we cannot choose them "directly" to represent the object function. Consequently, we have to adopt a trial basis through which we can represent in an approximate but adequate way the singular functions of the linear scattering operator. In particular, the basis expansion should not be too coarse because otherwise it would spuriously limit or alter the capabilities of the linear scattering operator with respect to the class of theoretically retrievable targets and should not be uselessly refined because, beyond a certain limit, this would increase the computational burden without improving the resolution. The kinds of adoptable trial bases functions are infinite, of course, but whatever the choice, a quantification of a convenient number of coefficients that are worth looking for can be based on DT, as will be shown in a while. In particular, in this chapter we will make use of bilateral Fourier harmonics along the horizontal direction and step functions along the depth. This specifically means that, indexing the basis functions after two indexes (namely m, ranging from 1 and $2M + 1$, for the horizontal variability, and n, ranging from 1 to N, for the vertical variability) the generic basis function is expressed as

$$\varphi_{m,n}(x,z) = \varphi_{hm}(x)\varphi_{vn}(z) \tag{15.5}$$

with the horizontal expansion given by

$$\phi_{hm}(x) = \exp\left(j\frac{\pi}{a}(m-M-1)x\right)\Pi\left(\frac{x}{a}\right) \tag{15.6}$$

and the vertical expansion given by

$$\phi_{vn}(z) = \Pi\left(\frac{Nz-0.5nb}{b}\right) \tag{15.7}$$

In Eqs. (15.6) and (15.7), a and b are the horizontal and vertical size of the investigation domain, respectively, according to Figure 4.1, and the Π step function is defined as usual as

$$\Pi(w) = \begin{cases} 1 & \text{if } |w| \leq \frac{1}{2} \\ 0 & \text{if } |w| > \frac{1}{2} \end{cases} \tag{15.8}$$

Under this choice, the comprehensive number of unknowns formally looked for is given by $N(2M+1)$ and the problem at hand is just to address a convenient choice of N and M. With regard to the horizontal variability, following the same reasoning exposed with reference to Figure 9.11, we can approximate the horizontal extent (HE) of the retrievable spectral set as

$$HE = 4k_{sc}\sin(\theta_{e\,max}) \tag{15.9}$$

where k_{sc} is the central wavenumber in the soil and the spectral set is symmetric with respect to the ς-axis; that is, it extends along the η-axis from $-2k_{sc}\sin(\theta_{e\,max})$ to $2k_{sc}\sin(\theta_{e\,max})$. Therefore, it is reasonable to choose M such that the maximum-order Fourier harmonic, given by $\phi_{h(2M+1)}(x) = \exp\left(j\frac{\pi}{a}(2M+1-M-1)x\right) = \exp\left(j\frac{\pi M}{a}x\right)$, has the same spatial period as $\exp(jHEx) = \exp(j2k_{sc}\sin(\theta_{e\,max})\,x)$, which yields the equation

$$\frac{M\pi}{a} = 2k_{sc}\sin(\theta_{e\,max}) = \frac{4\pi}{\lambda_{sc}}\sin(\theta_{e\,max}) \Leftrightarrow M = \frac{4a}{\lambda_{sc}}\sin(\theta_{e\,max}) \qquad (15.10)$$

In general, the result of Eq. (15.10) is not an integer number, and therefore now M should be meant as the first integer number equal to or greater than $\frac{4a}{\lambda_{sc}}\sin(\theta_{e\,max})$. Equation (15.10) shows that the number of horizontal harmonics also depends on the sine of the effective maximum view angle, which means that we might progressively relax the number of Fourier harmonics at deeper levels. This is coherent with the loss of resolution versus the depth, of course. At any rate, we will not do that in this book, and we will conservatively evaluate $\sin(\theta_{e\,max})$ with respect to the central point at the top of the investigation domain, according to Figure 15.12. Clearly, some redundancy is accepted in this way. In the cases considered throughout this chapter (frequency band ranging from 200 and 710 MHz, $\varepsilon_{sr} = 5$, $a = 2$ m), this evaluation yields $M = 25$.

With regard to the number of vertical steps, we propose an evaluation slightly looser, because the DT spectral set is intrinsically matched with complex Fourier harmonic functions. At any rate, again approximating the spectral set with a rectangle as shown in Figure 9.11, the vertical extent (VE) of the spectral set is given by

$$VE = 2k_{s\,max} - 2k_{s\,max} = \frac{4\pi\sqrt{\varepsilon_{sr}}}{c_0}(f_{max} - f_{min}) = \frac{4\pi\sqrt{\varepsilon_{sr}}B}{c_0} \qquad (15.11)$$

and it seems reasonable to guarantee that the length of the vertical step function is equal to one-half of the period of the spatial harmonic function $\exp(jVEz)$. This yields

$$\frac{b}{N} = \frac{1}{2}\frac{2\pi}{VE} = \frac{c_0}{4\sqrt{\varepsilon_{sr}}B} \Leftrightarrow N = \frac{4bB\sqrt{\varepsilon_{sr}}}{c_0} \qquad (15.12)$$

Again, N is to be meant as the first integer number equal to or greater than $4bB\sqrt{\varepsilon_{sr}}/c_0$. In the cases dealt with before in this chapter ($b = 2$ m, $B = 510$ MHz, $\varepsilon_{sr} = 5$), this drives to $N = 31$.

In all the inversion results shown in this chapter, we have had $M = 25$ and $N = 45$, for a comprehensive number of trial unknowns equal to $(2 \times 25 + 1) \times 45 = 2295$, which is therefore a reasonable (even if redundant) number of trial unknowns. In Figure 15.32, the behavior of the singular values for three choices of the trial unknowns are shown. Figure 15.32 confirms that the proposed evaluation is slightly redundant but reasonable.

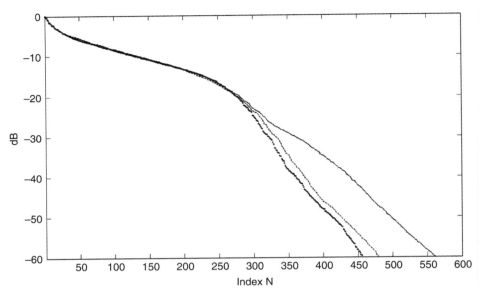

Figure 15.32. Singular values looking for 31 horizontal Fourier harmonics and 31 vertical step functions (dotted line), 51 horizontal Fourier harmonics and 45 vertical step functions (dashed line), and 63 horizontal Fourier harmonics and 63 vertical step functions (solid line).

In particular, up to a threshold of −20 dB, the three curves of singular values are substantially superposed, and beyond a threshold of −40 dB we should make twice larger the number of unknowns in order to increase the number of upper threshold of about 20%. However, in previous examples we have also seen that the use of a lower regularization threshold can improve the result only if the data are sufficiently clean, which in turn is related also to the depth of the targets of interest. To give a final rule of thumb, in the choice of the number of trial unknowns, some redundancy with respect to a DT-based evaluation is in general well-advised, both because the DT calculations are derived under several approximations and assumptions and because a larger number of basis functions will provide a better approximation of the actual singular function of the linear scattering operator. However, it is pointless (further than computationally burdening and potentially even harmful) to increase of an order of magnitude the number of trial unknowns with respect to a DT evaluation.

15.5   EXERCISES ON SPECTRAL AND SPATIAL CONTENTS

In Chapter 14 (Section 14.4) we have defined the spectral content and the spatial content of the upper threshold singular functions. In this section, we show a few examples about the possibilities that these quantities offer in order to characterize (in a nonrigorous but effective way) the class of the retrievable targets. In particular, this implicitly completes the numerical analysis of the previous sections, where the nonredundancy of the data has been related to some saturation of the behavior of the singular values. Indeed, the saturation of the behavior of the singular values is not a sufficient condition in order to show that the class of retrievable profiles is saturated too. In fact, the class of the retrievable profiles is the span of the singular function upper threshold; and even if the number of singular values higher than a fixed threshold is saturated, theoretically this does not mean that the retrievable targets are definitively (namely, in a way not modifiable through the gathering of further data) characterized. In particular, the span of the upper threshold singular functions might hypothetically still change when increasing the number of data. The spectral and the spatial content of the singular function are qualitative but immediate instruments to show that indeed not only the number but also the class of retrievable targets tends to become stabilized when the data sampling, either in space or in frequency (or in time), is slightly more intense than the DT values. In Figure 15.33 we show the spectral content [as defined in Eq. (14.38)] of the singular function relative to the singular values depicted in Figure 15.25. In particular, Figure 15.33 shows the spatial frequencies that we have some hope to retrieve making use of 5, 9, 18, and 35 time frequencies, respectively. The line describes the bound of the retrievable spectral set according to DT (see Chapter 9). The DT set has been calculated neglecting the conductivity and accounting from the maximum view (geometrical) angle, evaluated with respect to the central upper point of the investigation domain, as shown in Figure 15.12. Figure 15.33 makes us appreciate that there is a noticeable agreement between the DT spectral set and maximum achievable support of the spectral content

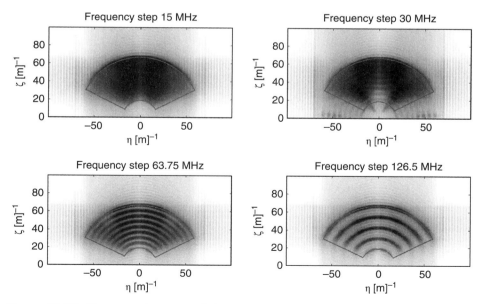

Figure 15.33. The spectral content of the upper threshold singular functions (threshold = -20 dB) at variance of the frequency step. The frequency step is after the title of each panel.

of the singular functions. However, when the frequency step is too large, this maximum achievable support is not completely filled. To make narrower the frequency step is equivalent to fill progressively better the available spectral set. In this regard, in Figure 15.33 it is evident that we can even distinguish the contributions of each involved frequency that is associated with a well-localized "spectral arch." Some effect of the frequency step on the reconstruction can be appreciated from Figure 15.34.

The examples of Figure 15.34 represent a variation on the theme of the first case presented in Figure 15.27. In particular, the first (upper left-hand side) panel is the same in the two cases. Then, in Figure 15.34, the frequency step has been progressively enlarged in the subsequent panels to 30, 63.75, and 126.5 MHz.

From Figure 15.34, we can appreciate that the main effect of the spectral holes is the introduction of false targets, the majority of which are actually related to aliasing effects

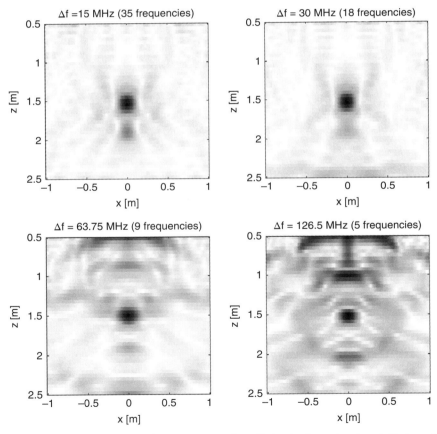

Figure 15.34. Reconstruction of two superposed small circular (ray = 1 cm) metallic targets at distance 35 cm from each other at variance of the frequency step within the comprehensive frequency band 200–710 MHz. The title of each panel indicates the frequency step. SVD threshold = –20 dB.

that replicate (in a distorted way) the actual targets. We can also appreciate that it is not wrong to take data with some (nonexcessive, i.e., not an order of magnitude) redundancy with respect to the DT previsions [which in the case at hand would suggest a frequency step of 33 MHz according to Eq. (9.45)]: The expected improvement is marginal but clearly visible from the comparison of the two upper panels in Figure 15.34. Finally, Figures 15.33 and 15.34 also show that the class of the retrievable targets can be different even at a parity (or almost at a parity) of singular value behavior. In particular, from the two upper panels of Figure 15.33 we appreciate that, even if the upper threshold singular values are almost the same (281 making use of 35 frequencies and 275 making use of 18 frequencies, according to Figure 15.25), actually the span of the upper threshold singular function is not the same, and that with 35 frequencies allows a better coverage of the lower central part of the theoretically available spectral set.

Let us pass to an example regarding the spatial content of the singular functions. To do that, now we propose an example with higher losses in the soil than those considered up to now. In particular, let us consider a soil with electrical conductivity $\sigma = 0.05$ S/m (the value considered before was $\sigma = 0.01$ S/m). The other parameters keep unchanged with respect to the previous case relative to Figure 15.33. In Figure 15.35 we show the spatial content, defined in Eq. (14.39), normalized to its maximum value, at variance of the regularization threshold for the singular values. The gray tones make us appreciate that the bottom of the investigation domains is much clearer than the top if the threshold is fixed at -10 or at -20 dB. In other words, the span of the singular functions upper threshold in these two cases does not portray adequately the bottom of the investigation

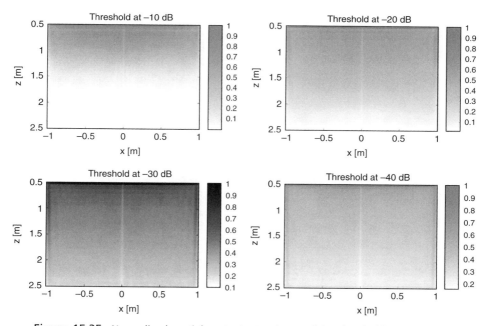

Figure 15.35. Normalized spatial content at variance of the threshold: $\sigma = 0.05$ S/m.

domain. This is made still more evident if we threshold the normalized spatial content at 0.1—that is, at one order of size with respect to its maximum value. We mean this quantity as a function whose value is 1 in the points where the normalized spatial content is equal to or greater than 0.1 and 0 elsewhere. The relative image is shown in Figure 15.36, and we can appreciate that with threshold equal to −20 dB (or even more −10 dB), the bottom of the investigation domain is not adequately "covered."

To show the effect of the lack of spatial coverage, let us propose the reconstruction of two metallic circular targets at the depth of 2.4 m (with reference to their centers: the ray is 1 cm). The distance between the two targets is 80 cm. The data are achieved, as before, by means of a two-dimensional FDTD simulation performed with GPRMAX, and a white Gaussian noise has been superposed to the data so that we have SNR = 60 dB with respect to the total field data. From Figure 15.37, we can appreciate that in this case we cannot regularize at −20 dB (or even more at −10 dB) the singular values if we want to see these two targets; we need more singular functions in order to image the bottom of the investigation domain. In particular, from Figure 15.37 we also appreciate that the threshold at −30 dB is certainly the best one among the four presented trial values. In fact, it allows the imaging of the two targets without artifacts, which instead appears when the threshold is further on decreased to −40 dB, in which cases the same image of the two targets is distorted. In other words, Figure 15.37 documents, in the case at hand, the progressive passage from an overregularization (threshold at −10 and −20 dB) to a reasonable regularization (−30 dB) to an underregularization (−40 dB). In previous cases

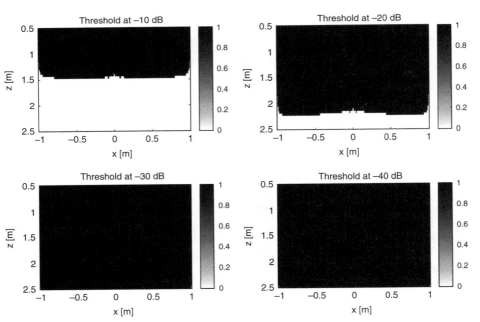

Figure 15.36. The normalized spatial contents of Figure 15.35 thresholded at one-tenth of their maximum value: $\sigma = 0.05$ S/m.

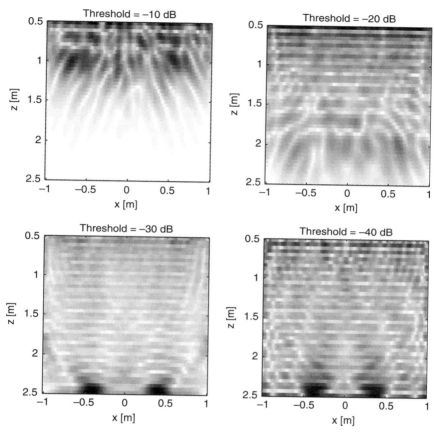

Figure 15.37. Reconstruction of two metallic targets (ray = 1 cm) at the depth of 2.5 cm at variance of the regularization threshold. The distance between the two targets is 80 cm.

(see Figures 15.16–15.19), we have seen that, at parity of signal-to-noise ratio, the best threshold might be lower or higher than that exploited here. This confirms that the signal-to-noise ratio is not a sufficient parameter to establish any optimal degree of regularization, because it depends on the losses and on the position of the targets of interest too. Before closing this section, we find it interesting also to show the data relative to the reconstruction of Figure 15.37. In particular, in Figure 15.38 the noisy data after zero timing and muting of the interface are shown, as they appear (upper panel) and after a filtering in the band 200–710 MHz (that is, the band of the source and the band exploited for the inversion) in the lower panel. From Figure 15.38, we can appreciate that the two targets are well-visible from the data. In other words, the data speak and tell us up to what depth we can reasonably try to focus the targets (even if under some progressive degradation of the resolution).

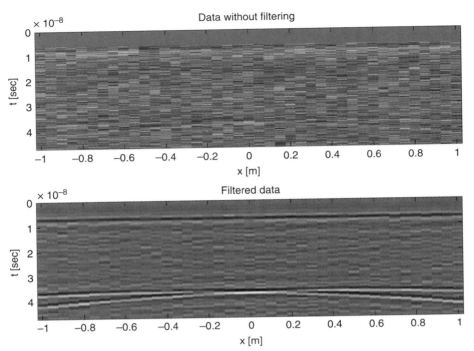

Figure 15.38. The noisy data (SNR = 60 dB, calculated on the total field data) exploited for the reconstructions of Figure 15.37, after zero timing and interface muting. **Upper panel:** The data as they appear. **Lower panel:** The data after filtering in the band of the source.

15.6 EXERCISES ON THE EFFECT OF THE HEIGHT OF THE OBSERVATION LINE

In the previous sections the data were at the air–soil interface. In this section we show some examples regarding the effect of the height of the observation domain. In particular, here we consider a soil with relative dielectric permittivity $\varepsilon_{rs} = 5$ and electric conductivity $\sigma = 0.01$ S/m. The investigation domain is sized 1.5×1.5 m^2 and ranges in depth from 1 cm to 151 cm. This investigation domain is discretized with 31 Fourier harmonic functions along the horizontal direction and 31 step functions along the depth. The frequency band ranges from 200 to 600 MHz with frequency step 25 MHz. The spatial step of the data is 2.5 cm, within a multimonostatic configuration where the offset between source and observation point is equal to zero. The observation line is 1.5 m long and is superposed to the investigation domain. Figure 15.39 shows the behavior of the singular values for $h = 0$ cm, $h = 10$ cm, $h = 20$ cm, $h = 30$ cm, $h = 40$ cm, and $h = 50$ cm. The figure shows that, as expected, the curve of the singular values becomes lower and lower when increasing the height of the observation line. Actually, in its initial part, the curve for $h = 0$ cm is lower than some among the other ones; this is due to some

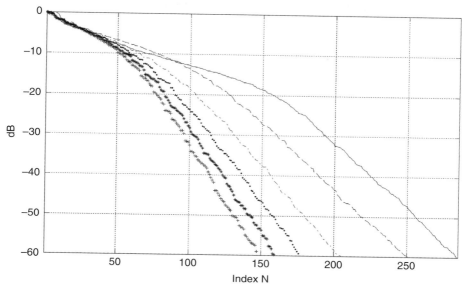

Figure 15.39. Singular value behavior at variance of the height h of the observation line. Solid line (upper curve): $h = 0$ cm. Dashed line: $h = 10$ cm. Dashes and dots: $h = 20$ cm. Dots: $h = 30$ cm. Asterisks: $h = 40$ cm. Crosses (lower curve): $h = 50$ cm.

redundancy of the data (Persico, 2006). However, by adopting a reasonable level of regularization (let's say with threshold chosen within the range between -15 and -40 dB), the behavior of the singular values tells us that the height of the observation line makes the retrievable amount of information decrease.

In order to show the relationship between this loss of information and the class of retrievable targets, let us remind that the height of the observation line is related to a decrease in the effective maximum view angle, as shown in Section 9.10. The progressive reduction of the maximum view angle while increasing the height of the observation line is shown in Figure 15.40.

In Figure 15.40 the minimum depth of the investigation domain (d_{min}) has been voluntarily exaggerated with respect to the data of the example of Figure 15.39 (where $d_{min} = 1$ cm and the investigation domain is sized 1.5×1.5 m^2). However, this is only for a graphical evidence: Indeed, for any fixed height and extent of the observation line, the maximum view angle is dictated by Snell's law on the refraction, and therefore the maximum view angle does not vanish while decreasing d_{min}. Instead, of course the maximum view angle vanishes for $h \rightarrow +\infty$, because we are considering an observation line of finite length. In particular, we have found numerically the effective maximum view angles solving Eq. (2.16) for the refraction point at the air–soil interface. According to Figure 15.40, the reference buried point is the upper central point of the investigation domain, so that, with reference to the symbols in Eqs. (2.16) and (2.17), in the case at hand we have $x = 1.5$ m, $x_0 = 0$ m, $c = c_0/\sqrt{\varepsilon_{sr}} = (3 \times 10^8)/\sqrt{5}$ m/s $\approx 1.342 \times 10^8$ m/s, $ct_{01}/2 = d_{min} = 0.01$ m. After solving for the refraction point x_1, the maximum effective

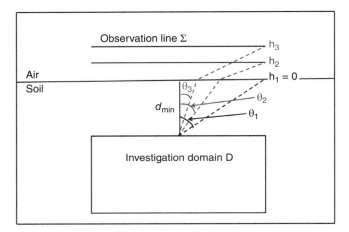

Figure 15.40. Pictorial of the reduction of the maximum view angle when increasing the height of the observation line. For color detail, please see color plate section.

TABLE 15.1 Numerical Results at Variance of the Height h

h (m)	$\theta_{e\,\text{max}}$ (degrees)	HR_h (cm)
0.0	89.24	16.8
0.1	34.6	29.5
0.2	32.8	31
0.3	31	32.6
0.4	28.37	35.3
0.5	25.24	39.3

angle is given by $\theta_{e\,\text{max}} = \arcsin\left(x_1 / \sqrt{x_1^2 + d_{\text{min}}^2}\right)$ and the expected horizontal resolution, according to Eqs. (9.62) and (9.63), is given by $HR_h = \lambda_{sc}/2\sin(\theta_{e\,\text{max}}) = \lambda_{sc}\sqrt{x_1^2 + d_{\text{min}}^2}/2x_1$. In the case at hand, the central frequency is equal to $f_c = 400$ MHz, and therefore the central internal wavelength is equal to $\lambda_{sc} = c/f_c = 33.54$ cm. The numerical results at variance of the height h are presented in Table 15.1.

In Figure 15.41, we show the spectral content of the singular functions at variance of the height, with reference to a threshold equal to −20 dB. Together with the spectral content, we show the DT retrievable spectral set, accounting for the maximum view angle at variance of the height as shown in Table 15.1. As can be seen, the agreement between the DT contour of the retrievable set (red lines) and the edges of the support of the spectral content is very good for $h \geq 20$ cm. At $h = 10$ cm this matching is still good, even if the spectral content extends slightly beyond the lateral bounds. This discrepancy is probably due to near-field effects, which as said are not accounted in DT. Instead, for $h = 0$ cm the support of the spectral content is substantially smaller than the DT bounds. This is because the spectral weight of the DT algebraic relationship is equal to zero at the extremes of the visible interval, as already outlined in Section 9.2.

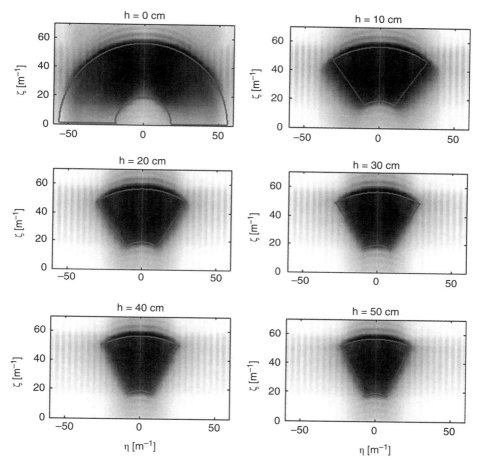

Figure 15.41. Spectral content of the singular functions (threshold of the singular value at −20 dB), together with the DT bound of the spectral retrievable set (red lines) at variance of the height h of the observation line. For color detail, please see color plate section.

Figure 15.41 shows that the degradation of the reconstruction expected for higher observation lines is mainly related to a decreasing of the maximum effective view angle, which essentially (see Chapter 9) makes us expect a degradation of the horizontal resolution, as explicitly calculated in Table 15.1 according to Eqs. (9.62) and (9.63). In order to show this on a reconstruction, we have simulated the data relative to two small circular metallic targets with GPRMAX. The centers are at the depth of 3 cm and the rays are 2 cm. Thus, the targets are superficial and the horizontal resolution limits in Table 15.1 can be tested (otherwise a further degradation due to the depth of the buried targets should be accounted for). The data are noisy with SNR = 60 dB with respect to the total field. The reconstruction is regularized by thresholding the singular values at −20 dB. The distance between the centers of the two targets is 30 cm. Since the targets are quite shallow,

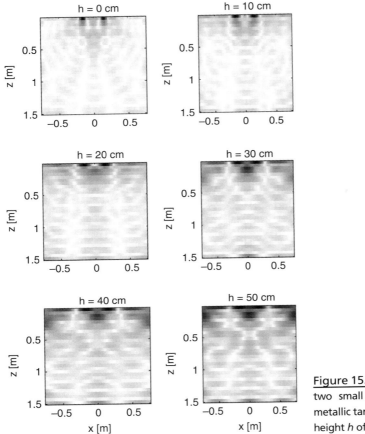

Figure 15.42. Reconstruction of two small and shallow circular metallic targets at variance of the height h of the observation line.

the relative diffraction curves were partially embedded in the echo of the airsoil interface. Therefore, background removal has been applied on the data. From Figure 15.42, we see the progressive degradation of the horizontal resolution. In particular, the two targets are well-distinguished for $h = \leq 20$ cm, are hardly distinguished for $h = 30, 40$ cm (and a meaningful artifact centered below the two targets appears too), and are substantially fused together for $h = 50$ cm (at this height the artifacts are still more relevant). These results are in good agreement with the DT horizontal resolution previsions of Table 15.1.

15.7 EXERCISES ON THE EFFECT OF THE EXTENT OF THE INVESTIGATION DOMAIN

Up to now, we have discussed two linear focusing methods, one essentially based on the migration and one essentially based on the linear inversion. The expected characteristics of the reconstruction in terms of retrievable spatial harmonic components are not much different, because in both cases the physical "point" driving the resolution is the filtering behavior of the linear scattering operator. These are more explicitly foreseeable under a DT model but they are essentially confirmed by the numerical SVD of the linear operator, even if this one accounts for losses, near-field effects and truncation of the observation line without introducing further approximations. The trade-off usually reported between migration and inversion is that the first one is a faster but mathematically less refined focusing method with respect to the second one. This is true, but there is an important corollary that customarily is not stated: The stronger computational burden required for an inversion compels us to restrict the investigation domain that (depending on the single case and on the available computational power) is usually extended up to a few central wavelengths within an inversion approach, whereas it can be extended up to hundreds of wavelengths within a migration-based approach. Now, in common GPR practices, a B-scan is often quite long in terms of wavelength (hundreds or even thousands). Therefore, in such a case, it is not possible to apply an inversion approach to the whole set of data. What has been done, up to now, is to join side by side several inversion results (Pettinelli et al., 2009; Persico, Soldovieri, and Utsi, 2010), according to the scheme of Figure 15.43, where the B-scan line has been divided into several observation lines Σ and the underlying portion of soil of interest has been divided into several corresponding investigation domains D.

As depicted in Figure 15.43, it is well-advised that, for each subportion of the B-scan, the subobservation line is of the same length of the underlying sub-investigation domain and is perfectly superposed to it. This is because, for each inversion, the adopted model (see Section 4.1) assumes that the targets are present only within the current investigation domain. This is not true in the real world, and therefore an observation line quite longer than the underlying investigation domain would enhance the effects of targets external to the investigation domain on the achieved image. However, even choosing observation lines superposed to the investigation domains, the targets within each investigation domain can influence the reconstruction in adjacent investigation domains, and the targets in the adjacent investigation domains can in turn affect the reconstruction in the current investigation domain. The possible situations are pictorially shown in Figure 15.44.

In particular, with reference to Figure 15.44, a target T_1 is expected to be optimally reconstructed, because it is centered with respect to the sub-observation line superposed to it, so that it is seen under a presumably adequate maximum effective view angle.

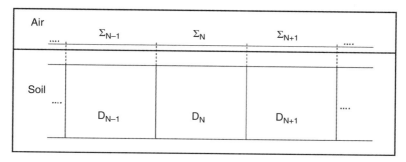

Figure 15.43. Subdivision of a long B-scan into several observation and investigation domains.

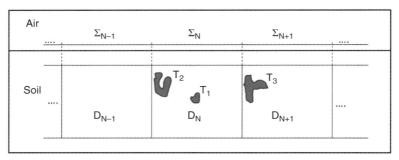

Figure 15.44. Pictorial of the possible situations: with respect to the current investigation and investigation domains D_n and Σ_n, the targets might be centered as T_1, peripheral as T_2 or external as T_3.

Instead a target like T_2 is seen under an asymmetric maximum view angle, because the sub-observation line is not centered on it. This makes us expect that the reconstruction of T_2 will be somehow poorer than that of T_1. Moreover, the external target T_3 is certainly "seen" by the GPR system when it reaches the right-hand bound of the sub-observation line Σ_N, and therefore it is a potential source of artifacts within the investigation domain D_N. For the same reason, T_2 is a potential source artifact within the investigation domain D_{N-1} and T_3 is a peripheral target within the investigation domain D_{N+1}, so that its reconstruction is subject to the effects of an asymmetric effective maximum view angle. At this point, it is clear that the presence of possibly many peripheral targets and the production of artifacts due to close external targets are the two sides of the same coin and are a consequence of the fact that we have partitioned the original whole observation line and the underlying whole investigation domain of interest. Figure 15.44 comprises also apparently more complicated cases. For example, a target that crosses the lateral bound between two investigation domains can be seen as a couple of adjacent targets, one of which is peripheral and another one external to the current investigation domain. Similarly, a long target that crosses the whole investigation domain on both sides can be seen as five adjacent targets: two external (one on the right-hand side and another one on the left-hand side), two internal but peripheral (one on the right-hand side and another one on the left-hand side), and one internal and centered. In order to show some effects of the subdivision, we propose an example constructed as follows: The relative dielectric permittivity of the soil is equal to $\varepsilon_{rs} = 5$, and its electric conductivity is equal to $\sigma = 0.001$ S/m. The offset between source and observation point is equal to zero, the spatial step of the data is 2.5 cm, and a B-scan 4.5 m long is considered. With these parameters, we have 181 observation points all together. For the inversion results, the B-scan at hand has been divided into three adjacent observation domains, each of which is 1.5 m long. In this way, we have 61 measurement points for each investigation domain, and the last point of the first (second) observation domain is also "re-exploited" as the first point of the second (third) observation domain. For each observation domain, the corresponding investigation domain starts from the interface and ends at the depth of 1.5 m. The investigation domain has been discretized by means

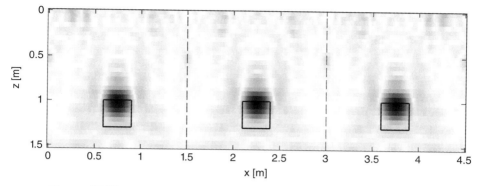

Figure 15.45. Reconstruction of three centered cavities at the depth of 1 m.

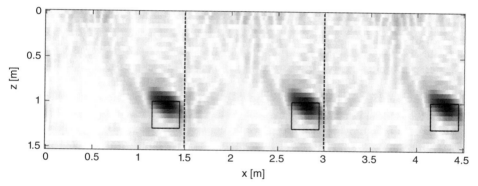

Figure 15.46. Reconstruction of three cavities of Figure 15.45 shifted 55 cm toward the right-hand side.

of 31 bilateral Fourier harmonic functions along the horizontal direction and 31 step functions along the depth. The inversion has been performed in the frequency domain, and the exploited band for the inversion has been 200–600 MHz sampled with frequency step 25 MHz. The SVD has been regularized by thresholding the singular values at −20 dB with respect to the first (maximum) one. The data have been simulated with GPRMAX, making use of a Ricker source with central frequency at 300 MHz. We have checked that the band at −3 dB of such a source is about equal to the interval 200–600 MHz. In Figure 15.45 we show the reconstruction of three centered square cavities, each of which is 30×30 cm^2, with the depth of the top at 1 m, and in Figure 15.46 we show the reconstruction that we achieve if the cavities are shifted 55 cm toward the right-hand side.

It is evident that the reconstruction in Figure 15.46 is worse than that in Figure 15.45, because the targets are peripheral. In particular the three spots relative to the reconstructions appear as spuriously "pulled" toward the centers of the investigation domains. This example shows that, if possible, the subdivision of the investigation domain should account of the most relevant targets in order to make them, if possible, centered.

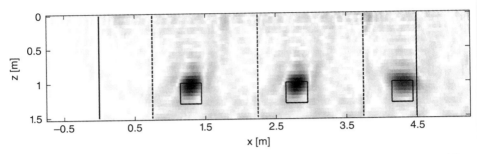

Figure 15.47. Reconstruction of three cavities of Figure 15.46 achieved after a zero padding that makes the targets centered.

This can be achieved in some cases with a shift of the subdivision, which in turn can require some zero padding of the B-scan at its beginning and/or at its end. Incidentally, some zero padding is customarily needed in any case if we want to invert the entire gathered B-scan. In fact, the length of the physically gathered observation line in general is not an integer multiple of the chosen length of the subportion selected for the inversions. In particular, with some zero padding on the same data of Figure 15.46, we have performed the same inversion on four investigation domains instead of three, and we have achieved the result of Figure 15.47. In Figure 15.47 the vertical solid lines indicate the start and end point of the real B-scan. As can be seen, in Figure 15.47 the same targets of Figure 15.46 are made centered with respect to the subportions of the observation line exploited for each inversion, and this improves the reconstruction. In particular, some limited view angle effect is perceived only with regard to the third cavity on the right-hand side, because this cavity is placed close to the end of the gathered observation line. In other words, it is a "really peripheral" target and therefore the lack of a symmetric and adequate maximum view angle is an intrinsic (physical) unavoidable phenomenon in this case.

For completeness, in Figure 15.48 the result of a migration-based processing is shown too. In particular, the data have been imported in Reflexw and have been migrated in the time domain (Schneider, 1978). The exploited propagation velocity has been that corresponding to the real part of the complex permittivity, which is licit because the soil is low-lossy. The number of traces considered for the migration has been 181—that is, the total and not a subset, because the diffraction curves visible from the data appeared as large as the comprehensive length of the B-scan. From Figure 15.48 we can see some residual effect of the diffraction curves and the image appears somehow noisier than that in Figure 15.47. Moreover, as in Figure 15.47, also in Figure 15.48 we appreciate some limited view angle effect only with regard to the cavity on the right-hand side (which appears slightly pulled toward the left-hand side).

We find also interesting to present some examples about the problem of artful anomalies due to targets external to the observation domain. In the previous example this problem was formally present too, but not meaningful because the three targets were scattering objects of the same shape, size, and nature. Instead, let us now consider a situation where both strong and weak scattering targets are present. In particular, let us consider,

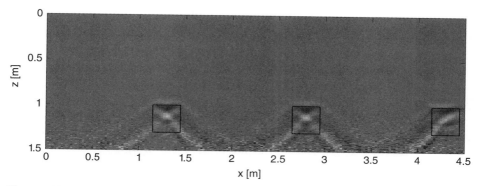

<u>Figure 15.48.</u> Reconstruction of three cavities of Figure 15.47 achieved from Kirchhoff migration.

at a parity of measurement and scenario parameters, three new targets with the same geometry (namely size, position, and orientation) as that of the targets above. However, now the left-hand side target is a perfect electric conductor, the central target is a weak scattering object with relative permittivity equal to 6 and with the same conductivity of the surrounding soil, and the right-hand target is a void.

This time we show first of all the data (Figure 15.49) after zero timing and muting of the interface contribution. This is done because we are considering a scenario with weak and strong scattering targets put together. In particular, it is natural that a strong scattering object can partially or even totally cover the visibility of a weaker scattering targets, and therefore we want to show preliminarily that the weak target is visible in the raw data and therefore it should be visible also after a correct focusing processing. The data of Figure 15.49 make us appreciate that, even if it is meaningfully weaker than the other two, the diffraction curve of the weak target—that is, the central one among the three diffraction curves—is clearly visible. In Figure 15.50 the result with the investigation domain sequence of Figure 15.46 is shown. As can be seen, the central weak scattering target has been substantially erased, because it is completely masked by the false target produced by the tail of the diffraction curve of the perfect electric conductor within the central investigation domain. In Figure 15.51, instead the result of a Kirchhoff migration similar to that of Figure 15.48 is shown. Indeed, the image of Figure 15.51 has been enhanced by means of some gain variable versus the depth and some saturation, which is a common "trick" to enhance the achieved image. However, the shape of the retrieved target and the comparison with the data provide a trustful result. Incidentally, we have tried similar and also further or alternative manipulations also on the image of Figure 15.50, but we didn't achieve any reliable image of the central target.

Let us now show the reconstruction of the same underground scenario according to the sequence of four investigation domains considered in Figure 15.47. The result is shown in Figure 15.52, and the image of the central target has been slightly enhanced by means of some darkening of the image. Alternatively, one might normalize each of the achieved adjacent reconstructions to its own maximum level. In this way, the image of Figure 15.53 is achieved.

Figure 15.49. GPR data with three targets. From left to right: A perfect electric conductor, a weak scattering target, and a void.

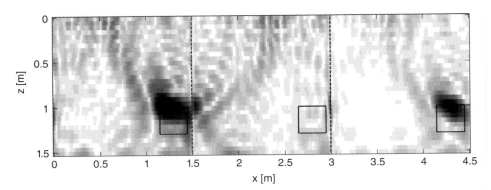

Figure 15.50. The reconstruction of three targets. From left to right: A perfect electric conductor, a weak scattering target, and a cavity.

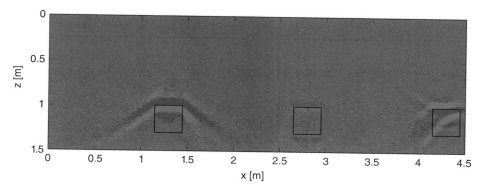

Figure 15.51. The migration result achieved for three targets. From left to right: A perfect electric conductor, a weak scattering target, and a cavity.

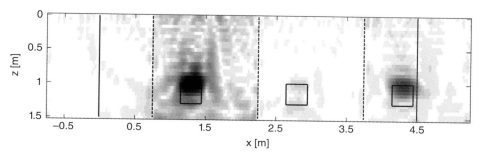

Figure 15.52. The reconstruction of the three targets of Figure 15.50, making use of four-investigation domain.

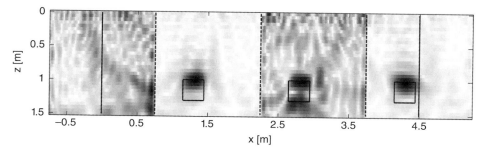

Figure 15.53. The reconstruction of the three targets of Figure 15.50, making use of the four-investigation domain and normalizing each reconstruction to its own level.

Any choice has its pros and cons, and clearly in this example we have an intrinsic physical problem due to the different amount of energy scattered by the targets at hand. However, from Figure 15.50 it is clear that the adopted investigation domain sequence makes much worse the effect of this problem.

Summarizing, we work out a first conclusion that, when possible, the choice of the investigation domain should be done such that the targets of interest are all relatively centered with respect to the investigation domains.

However, such a sequence is not identifiable in all cases, because it can happen that the distribution of the targets is too "random" and/or the buried targets are too long and/or too many to allow us to identify the desired sequence of adjacent investigation domains. In these cases, a possible solution is the use of a multiple sequence of investigation domains. In particular, let us explain this concept starting with a double sequence of investigation domains. The double sequence is illustrated in Figure 15.54. Of course, some further trivial zero-padding operations are needed at the beginning and at the end of the B-scan in order to identify the two complete sequences.

It is clear that, adopting a double sequence of investigation domain, only the central part of each investigation domain is retained, thereby mitigating the problems due to the maximum view angle. On the other hand, each piece of reconstruction is as large as one-half of the width of the investigation domain (labeled "a" in Figure 15.54, analogously to Figure 4.1) instead of its entire width. Actually, it is unavoidable to see a sort of seam point between any two adjacent sub-investigation domains, due to the different average values of the image relative to two adjacent sub-reconstructions. Consequently, when adopting a double sequence, we will see on the reconstruction about twice the seam points with respect to the case with a single sequence. Therefore, the choice should be data-driven. With regard to the computational burden, there is some additional time required for the double sequence but it is marginal. In fact, most of the required RAM and computational time is due to the calculation of the matrix and of its SVD, and this is done once and for all with regard to one single sub-investigation (and sub-observation) domain. The idea of a multiple series of investigation domains is a straightforward extension of the double sequence. In particular, Figure 15.55 shows how a triple sequence might be chosen. In this case the retained part of the reconstruction is still more centered, but the number of seams is about threefold with respect to those needed for a single sequence of investigation domains. However, on the other hand, the difference between the average

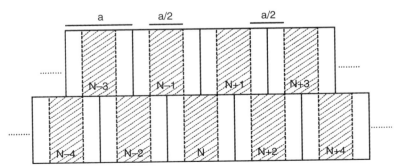

Figure 15.54. Double sequence of investigation domains. The second sequence is shifted with respect to the first one-half of an investigation domain. For each investigation domain the central half of the focused area is retained. Then the retained pieces of reconstruction are joined side by side after the indicated order.

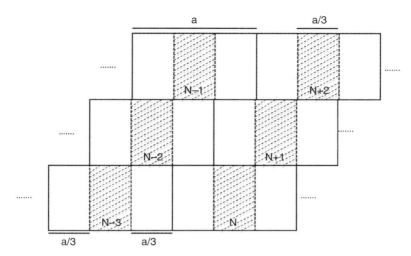

Figure 15.55. Triple sequence of investigation domains. The second sequence is shifted with respect to the first one-third of an investigation domain, and the third sequence is shifted with respect to the second one-third of an investigation domain. For each investigation domain the central one-third of the focused area is retained. Then the retained pieces of reconstruction are joined side by side after the indicated order.

value of the reconstruction between two adjacent domains is expected to be reduced because the amount of the shift between two adjacent sets of data is reduced. Finally, there will also be a further marginal increase of the required computational time with respect to the double sequence.

At this point, we provide an example of reconstruction achieved from a double investigation domain. In this example, the targets are nine square cavities sized 30×30 cm^2 with the depth of the top equal to 1 m. The abscissas of the centers of the cavities are at 0.20 m, 0.75 m, 1.30 m, 1.70 m, 2.25 m, 2.80 m, 3.20 m, 3.75 m, and 4.30 m, respectively. The scenario and measurement parameters are the same as those given in the previous examples. In this case, the investigation domains are too "crowded" to allow us to identify any well-advised single sequence. In particular, making use of the single sequence with three investigation domains (as in Figure 15.46), we achieve the result of Figure 15.56, and making use of the single sequence of four investigation domains (as in Figure 15.47) we achieve the result of Figure 15.57. In both cases, limited view angle effects are quite evident with regard to some of the peripheral targets. In particular, in Figure 15.57 some of the cavities cross the boundary between two adjacent investigation domains, and this affects in a particularly negative fashion their focusing. In Figure 15.58, instead, the result achieved from the double sequence (with seven investigation domains) is shown. The seam-points have not been indicated in this case, in order not to make the figure confusing; this also shows that their locations are intrinsically evident. The focusing of all the targets is meaningfully better by making use of a double sequence of investigation domains. Actually, some artifacts appear too, but these are due to some unavoidable interference between the targets, which

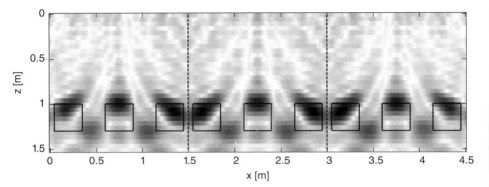

Figure 15.56. The reconstruction of nine cavities with three investigation domains.

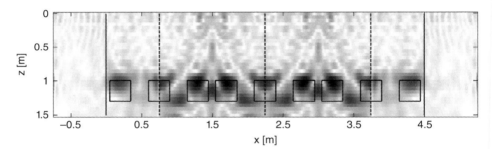

Figure 15.57. The reconstruction of nine cavities with four investigation domains.

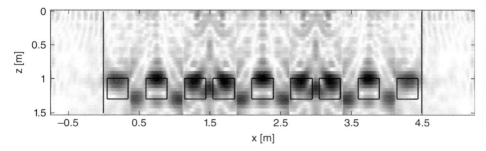

Figure 15.58. The reconstruction of nine cavities with a double sequence of investigation domains.

are quite close to each other. Such artifacts are present (and actually appear even worse) if the reconstruction is achieved from a Kirchhoff migration, as shown in Figure 15.59. As in some previous cases, the image of Figure 15.59 has been enhanced with some gain variable versus the depth.

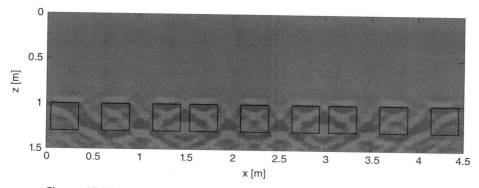

Figure 15.59. The reconstruction of nine cavities with a Kirchhoff migration.

For the interested reader, the effects of the observation and investigation domain subdivision have been tested also versus experimental data (Persico and Sala, 2014).

15.8   EXERCISES ON THE EFFECTS OF THE BACKGROUND REMOVAL

In this section, some examples about the background removal are shown. Preliminarily, let us show how to discretize the linear scattering operator so that the operation of the background removal is included. Let us consider at the moment a single frequency. To do this, let us start considering the matrix relative to a common offset prospecting and let us focus on a single-frequency case with K measurement points and H trial basis functions. It is taken for granted that the observation line is of the same length of the underlying investigation domain and is perfectly superposed to it. Let us label s the spatial step, so that the length of the investigation and observation domains is equal to $s(K-1)$. Let us label $\mathbf{M}$ the matrix of the system and $\mathbf{c}$ the column vector of the coefficients. Given any combination of coefficients, the product $\mathbf{Mc}$ is, by definition, the scattered field at the first order (with respect to the Born series, defined in Section 8.1) produced by the chosen combination of coefficients, which in turn represents some chosen buried target. Let us now consider, as chosen target, the first trial basis function. In this case the first element of the column vector of the coefficients is equal to 1, and the remaining $H-1$ are equal to 0. Consequently, the relative scattered field at the first order is given by

$$\mathbf{Mc} = \begin{pmatrix} M_{11} & M_{12} & ...M_{1H} \\ M_{21} & M_{22} & ...M_{2H} \\ ... & & \\ M_{K1} & M_{K2} & ...M_{KH} \end{pmatrix} \begin{pmatrix} 1 \\ 0 \\ ... \\ 0 \end{pmatrix} = \begin{pmatrix} M_{11} \\ M_{21} \\ ... \\ M_{K1} \end{pmatrix} \qquad (15.13)$$

Equation (15.13) shows that the first column of the matrix represents the scattered field relative to the first trial basis function. In the same way, the generic hth column of the matrix represents the scattered field relative to the hth trial basis function. At this point, in order to perform a background removal on $2N+1$ traces, we need N additional traces before the original scan and N additional traces after it. This means that we have to calculate the common offset matrix relative to $K+2N$ measurement points, and these $K+2N$ points have to be referred to an enlarged equivalent observation line of length $(K-1)s + 2Ns$, centered on the investigation domain. These $K+2N$ measurement points lead to a matrix (at single frequency) with $K+2N$ lines. Let us call M_1 this "augmented" matrix. Now, referring the background removal to the first-order samples of the scattered fields—that is, to the columns of the matrix—it is immediate that

$$M_b(k,h) = M_1(k+N,h) + \frac{1}{2N+1} \sum_{n=-N}^{N} M_1(k+n+N,h) \quad k=1\ldots K \qquad (15.14)$$

where M_b is the matrix relative to the scattering operator including the background removal. As can be seen, M_b has K rows as the original common offset matrix. With regard to the data, instead, the extra data needed for BKGR can be fictitious or actual if we are considering a long observation line partitioned into adjacent sub-lines for inversion purposes as discussed in the previous section. The best choice is case-dependent.

In a first example, we consider a stratified medium with a thin cavity between the two media as shown in Figure 15.60. The cavity is large (55 cm) and thick (2 cm). The depth of its top is 69 cm. The data have been simulated with GPRMAX, and the soil shows a relative dielectric permittivity equal to $\varepsilon_{sr} = 5$ and an electric conductivity equal to $\sigma = 0.01$ S/m. The underlying rock shows a relative dielectric permittivity equal to $\varepsilon_{rr} = 35$ and an electric conductivity equal to $\sigma_r = 0.01$ S/m The soil–rock interface is at the depth of 70 cm. The data are shown in Figure 15.61.

In particular, a central measurement line 1.5 m long has been considered for the common offset data, prolonged with an additional 25 cm on both sides for gathering

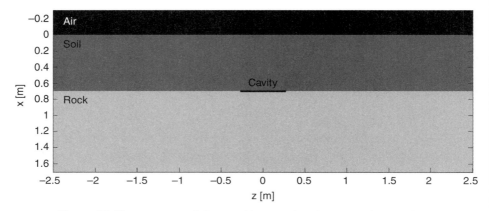

Figure 15.60. Geometry of the case history at hand in a stratified medium.

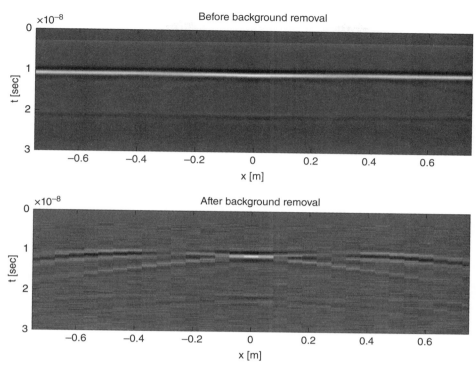

Figure 15.61. The data before (**upper panel**) and after (**lower panel**) a background removal on 11 traces in the layered medium of Figure 15.60.

the extra data needed for the background removal, which is performed on 11 traces (corresponding to a distance of 50 cm). The data have been gathered with spatial step 5 cm, and the source is a Ricker pulse with central frequency 300 MHz. Gaussian noise has been added to the data with signal-to-noise ratio equal to 60 dB, evaluated on the total field data. The data shown in Figure 15.61 have been filtered in the band 300–620 MHz, which is the band used for the inversion, about corresponding to the (–6 dB) band of the source. From Figure 15.61, we appreciate that the operation of background removal makes the target much more visible in the case at hand. This is obvious, because the target is embedded in the interface between two buried layers.

In Figure 15.62, the result of an inversion of these data without and with applying background removal on 11 traces is shown. In particular, the two inversion results of Figure 15.62 are achieved from two different data (namely the data without and with background removal) and the two different matrixes (namely the common offset matrix relative to an observation line 1.5 m long and the matrix including the background removal operation on the same observation line, which is achieved from a common offset matrix relative to an augmented common offset observation line, 2 m long). It can be seen that the target is visible only in the lower panel, because the upper one substantially

Without background removal

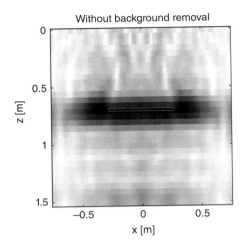

With background removal

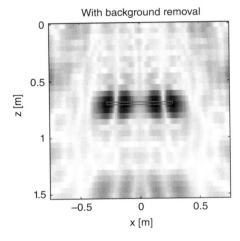

Figure   15.62. Reconstruction   without (**upper panel**) and with (**lower panel**) background removal. The target is the same as in Figure 15.60. The reconstruction is based on a linear inversion.

represents the interface between the soil and the underlying rock. The reconstruction in the lower panel, however, shows a band-pass version of the target along both the horizontal and the vertical directions, in agreement with the considerations discussed in Chapter 9. Let us note that, in the case at hand, the critical length as defined in Eq. (9.85) is just 55 cm long, so the length of the target is equal to the critical length. This is the reason why the background removal affects its reconstruction but does not "destroy" it completely. In Figure 15.63, the homologous migration-based results are shown.

Let us note that Figure 15.63 is based on the same procedure applied to two different sets of data.[2] In other words, since we have migrated the data with a commercial code,

---

[2] Unlike Figure 15.62, where the matrix (i.e., the model) combined with the two different sets of data was not the same.

Without background removal

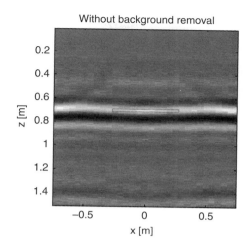

With background removal

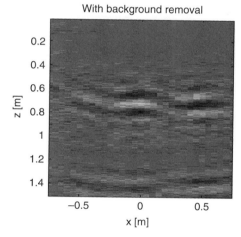

Figure 15.63. Reconstruction without (**upper panel**) and with (**lower panel**) background removal. The target is the same as in Figure 15.60. The reconstruction is based on Kirchhoff migration.

we didn't have the possibility to apply a migration algorithm conceived "ad hoc" for data undergoing background removal, and this can cause artifacts. In particular, in Figure 15.63 (lower panel) we have a spurious prolongation of the target, due to the "weight" of the cavity on the average trace at the time corresponding to its depth. This did not happen in the inversion result of Figure 15.62 (lower panel), where the filtering effect is accounted for in the model. The distortion of the spectrum of the target can be reduced adopting a background removal on all the (31) traces. In Figure 15.64 we compare the migration result achieved after background removal on all the traces (upper panel) and background removal on 11 traces (lower panel). It can be seen that the target is better retrieved with background removal on all the traces but, on the other hand, the background removal on all the traces leaves a "contrail" as long as the entire scan.

BKG removal on all the traces

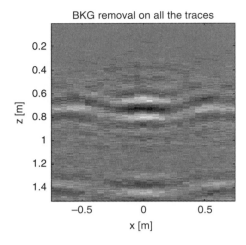

BKG removal on 11 traces

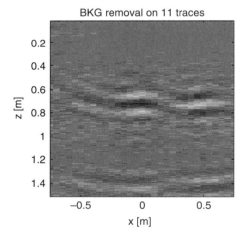

Figure 15.64. **Upper panel**: Migration performed with Reflexw after background removal on all the traces. **Lower panel**: Migration performed with Reflexw after background removal on 11 the traces. The scenario is that of Figure 15.60.

This happens because the weight of the target at its time depth is spread out over all the traces [the phenomenon is analogous to the belt shown in Figure 7.3 (panel b)].

At this point, let us consider the case analogous in a homogeneous (not layered) half-space. The geometry of the problem is shown in Figure 15.65. The values of the parameters are the same as in the previous example, with the only exception that the rock of Figure 15.60 is replaced with the same overlying soil. The data without and with background removal are shown in Figure 15.66.

The data shown in Figure 15.66 are noisy (SNR = 60 dB) and filtered in the band 300–620 MHz, analogously to those in Figure 15.60. It can be appreciated that the data this time appear clearer without background removal than with background removal. Indeed, in this case the background removal is not really needed at the depth level of the target, because the host medium is homogeneous. This is confirmed by the inversion results shown in Figure 15.67 without (upper panel) and with (lower panel) background

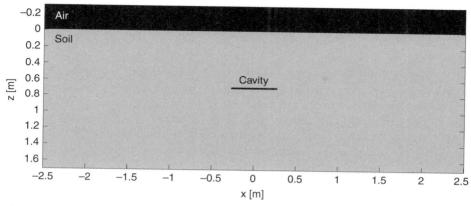

Figure 15.65. Geometry of a problem analogous to that represented in Figure 15.60, in the case of a homogeneous (not layered) soil.

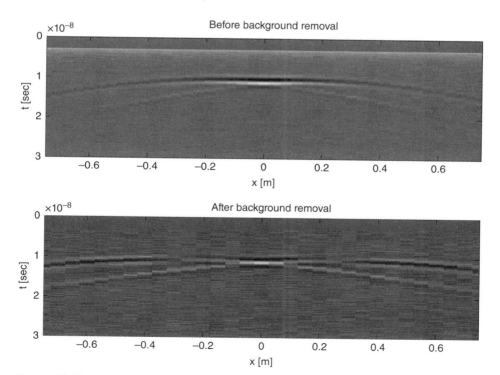

Figure 15.66. The data before (**upper panel**) and after (**lower panel**) a background removal on 11 traces in the homogeneous medium of Figure 15.65.

removal on 11 traces. The reconstruction of the upper panel this time is much better than that of the lower panel, because it doesn't suffer the filtering effect accounted for in the inversion algorithm, nor does any particular distortion occur in the data with the possible generation of horizontal artifacts.

In Figure 15.68, two migration results are shown, obtained from the commercial code Reflexw. The upper panel refers to a Kirchhoff migration applied without any background removal, whereas the lower panel refers to a Kirchhoff migration applied after background removal on 11 traces. As said, this is a commercial migration code and does not account for the fact that one has applied any background removal on the data. At any rate, the results of Figure 15.68 confirm the fact that in this case the best result is achieved without any background removal.

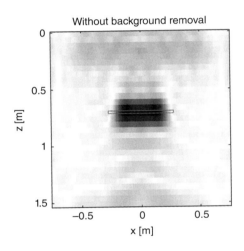

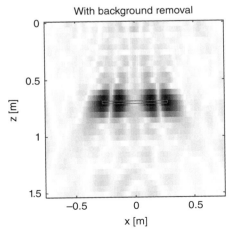

Figure 15.67. Reconstruction without **(upper panel)** and with **(lower panel)** of the target in Figure 15.65, based on a linear inversion.

Without background removal

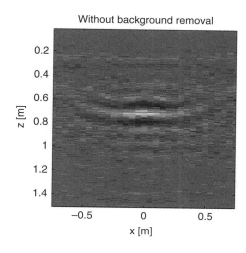

With background removal

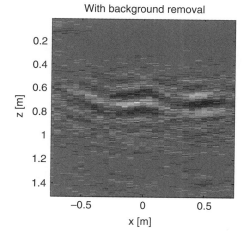

Figure 15.68. Reconstruction without **(upper panel)** and with **(lower panel)** background removal. The target is the same as in Figure 15.65. The reconstruction is based on Kirchhoff migration.

In Figure 15.69 we see that the background removal on all the traces drives to a better reconstruction of the target, but prolongs the contrail of the cavity along the entire scan, analogously to the case of Figure 15.64. In order to mitigate this problem, one might apply a filtering of the data after the migration, in order to erase the false horizontal belts possibly produced in the data by the background removal procedure. This erasing should be result-driven, because it should not erase the targets of interest too. In particular, we have applied such a filtering to the result of Figure 15.69 relative to background removal on all the traces. The filter was an ideal high-pass one with respect to the horizontal frequencies equal or greater than twice the inverse of the critical length. The filter has been implemented by a homemade MATLAB code, and the result is shown in Figure 15.70. As can be seen, the contrail is attenuated by the filtering without any meaningful distortion of the focused target.

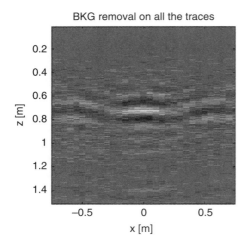

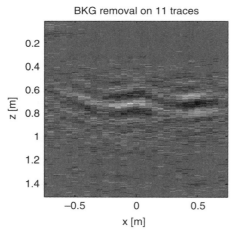

Figure 15.69. **Upper panel**: Migration performed with Reflexw after background removal on all the traces. **Lower panel**: Migration performed with Reflexw after background removal on 11 the traces. The scenario is that of Figure 15.65.

15.9   2D AND 3D MIGRATION EXAMPLES WITH A SINGLE SET AND TWO CROSSED SETS OF B-SCANS

Marcello Ciminale, Giovanni Leucci, Loredana Matera, and Raffaele Persico

In this section we aim to show both a 2D and a 3D migration algorithm applied to field data. Moreover, since it can be an aspect of interest too, we will show both results achieved from a set of B-scans parallel to each other and from two sets of crossed B-scans. In this last case, quite common in GPR prospecting, the final result can be considered as achieved as some combination of the two images achieved from the two sets of B-scans. The data were gathered in the archaeological park of Egnazia (Apulia, Southern Italy). The park is one of the largest in Apulia. However, its importance lies

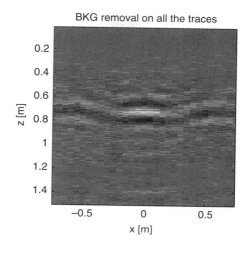

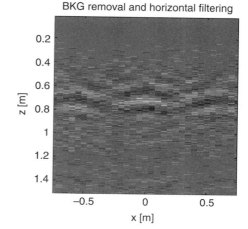

Figure 15.70. **Upper panel**: The same migration result of the upper panel in Figure 15.69. **Lower panel**: The migration result of the upper panel after high-pass horizontal filtering.

not only in its extension, but also in the continuous frequentation that occurred over time, which is well-documented in the archaeological remains. The first occupation concerning Egnazia dates back to the Bronze Age; and after a period of abandonment, the site was occupied again. In particular, very important are the periods of Messapian (from 8th to 3rd century B.C.) and Roman (starting from the year 267 B.C.) civilizations. Testimonies of the Messapian period are the mighty defensive walls and the necropolis, while the main remains of the Roman town are the civil basilica, the amphitheater, an arcaded square, and the well-preserved cryptoporticus. The plunder and destruction (545 A.D.) carried out by Totila, the Goths' king, could not impede a further rebirth of Egnazia, at least up to the 8th century A.D., when its definitive decline started.

The GPR investigated area is a square zone sized $12 \times 12 \, \text{m}^2$, located near an underground and accessible portico (the cryptoporticus) where a previous, more extended magnetic survey revealed, at a shallow depth, an articulated plan composed of a main rectangular building (15 m × 17 m) and several rooms connected to it (Caggiani et al., 2012).

The GPR data have been gathered in the framework of a Ph.D. research activity in collaboration between the University of Bari and the Institute for Archaeological and Monumental Heritage IBAM-CNR and these data have been used to test an innovative stepped-frequency GPR system (Persico et al., 2013a). This innovative system has been projected by IBAM-CNR, IDS, and University of Florence within the research project AITECH (http://www.aitechnet.com/home.html), financed by the Puglia Region. Here, the purpose is to show the results achieved from both a 2D and a 3D migration algorithm on this area, exploiting data gathered with this prototypal system. The system at hand, in particular, is equipped with three equivalent couples of antennas, achieved by means of suitable switches programmable versus the frequency, according to the scheme of Figure 15.71. In particular, the equivalent comprehensive covered band ranges from 50 MHz to 1 GHz and the system can gather up to three sets of data for each going through, one with the "short antennas" one with the "medium antennas," and one with the "long antennas." As is well known, the band and the central frequency of each equivalent couple of antennas tend to increase when the antennas become shorter.

With reference to the scheme of Figure 15.71, in this section we show results obtained from the "medium length" antennas, whose band (evaluated directly from the spectrum of the data) in the case at hand ranged between approximately 150 and 600 MHz.

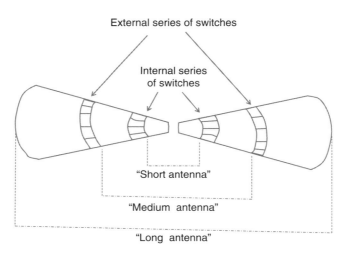

Figure 15.71. Schematic (nonquantitative) of the exploited reconfigurable GPR antennas. The system is equipped with "short antennas" if all the switches are open. It is equipped with "medium antennas" if the internal switches are closed and the external ones are open. It is equipped with "long antennas" if all the switches are closed.

The data have been gathered with an interline space (transect) of 50 cm, and two sets of orthogonal B-scans have been gathered. With regard to the spatial step of the data along any single B-scan, for technical reasons, the system is equipped with an odometer that records both the position at which each burst is radiated, but does not guarantees by itself a constant spatial step during the data acquisition. This does not cause any undersampling problem because the maximum recorded spatial step is in any case small enough to avoid aliasing. However, in order to achieve data with a constant spatial step, a homemade interpolation code has been implemented. In the case at hand, we have interpolated the data with a 2-cm spatial step along each B-scan. The data have first undergone a 2D processing, essentially based on the 2D Kirchhoff migration in the time domain (the complete sequence of steps has been zero timing, background removal, and numerical gain variable versus the depth, migration, and time slicing). The processing has been implemented twice, the first time making use of the commercial software Reflexw (Sandmayer, 2003) and the second time with the commercial software GPRSLICE (Goodman and Piro, 2013). In Figure 15.72, three depth slices are obtained from the software Reflexw. The chosen depth is 47 cm, because it seemed to be the most interesting one. Some division wall structures are well visible, and an oblique rectangular anomaly may be ascribable to a floor.

In Figure 15.73, the homologous of Figure 15.72 achieved from a 2D processing performed with GPRSLICE is shown.

The fundamental anomalies in the two cases are the same, but some important differences also appear between the Refexw and the GPRSLICE results. In particular, the slices obtained with GPRSLICE appear more "definite" in the case at hand.

Here, it is of interest to note that the use of crossed B-scan had in both cases a well-perceivable, even if not dramatic, effect. In particular, rather than "adding" features, the effect of the crossed B-scans seem to have cleaned the image. Let us now discuss the case of a 3D migration, performed again by means of GPRSLICE. In Figure 15.74 the results obtained from the x-directed B-scans, the y-directed B-scans, and both joined together are shown.

Unlike the 2D cases, in Figure 15.74 it seems that the use of two crossed sets of of B-scans improves the image in a meaningful fashion.

In the case at hand the 2D migration-based images appear better than the image obtained from a 3D migration-based image. However, this should not lead to the precipitous conclusion that 2D algorithms are better than 3D ones. Indeed, a nonnegligible point is the fact that the transect is much larger than one-fourth of the central wavelength, which for the 3D cases yields a meaningful undersampling. On the other hand, the data have been obtained with a GPR system able to gather data along a single B-scan for each going-through, and so it would have been experimentally prohibitive to gather data with an adequate interline step. Things are probably going to change in the future with a larger use of GPR systems equipped with arrays of antennas. In this case, it is possible to gather up to 14 or more parallel B-scans with an interline space of the order of 10 cm (Sala and Linford, 2010) with a unique "going-through." Under these conditions, the performance of a 3D migration algorithm versus field data can improve meaningfully.

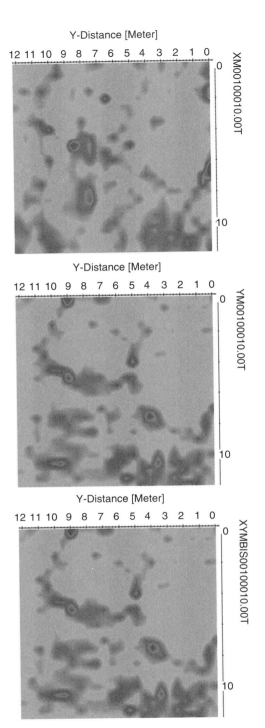

Figure 15.72. Depth slices at 47 cm obtained with Reflexw through a 2D migration-based processing. **Upper panel**: Result obtained from the B-scans parallel to only the *x*-axis. **Middle panel**: Result obtained from the B-scans parallel to only the *y*-axis. **Lower panel**: Result obtained from the B-scans parallel to both the *x*-axis and the *y*-axis. The axes are in meters. For color detail, please see color plate section.

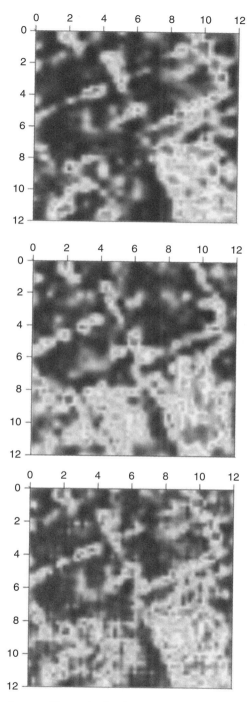

Figure 15.73. Depth slices at 47 cm obtained with GPRSLICE through 2D migration-based processing. **Upper panel**: Result obtained from the B-scans parallel to only the *x*-axis. **Middle panel**: Result obtained from the B-scans parallel to only the *y*-axis. **Lower panel**: Result obtained from the B-scans parallel to both the *x*-axis and the *y*-axis. The axes are in meters. For color detail, please see color plate section.

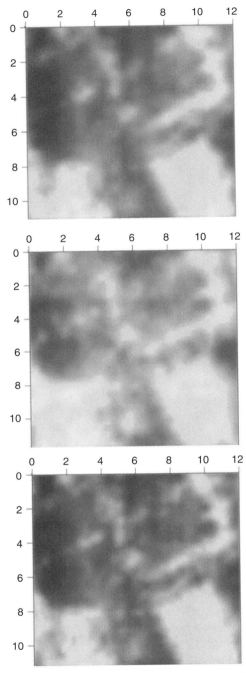

Figure 15.74. Depth slices at 47 cm obtained with GPRSLICE through a 3D migration-based processing. **Upper panel**: Result obtained from the B-scans parallel to only the *x*-axis. **Middle panel**: Result obtained from the B-scans parallel to only the *y*-axis. **Lower panel**: Result obtained from the B-scans parallel to both the *x*-axis and the *y*-axis. The axes are in meters. For color detail, please see color plate section.

15.10 2D AND 3D INVERSION EXAMPLES

Ilaria Catapano and Raffaele Persico

This section aims at investigating a full 3D linear inversion-based GPR data processing versus the commonly adopted pseudo 3D imaging, which is achieved by interpolating 2-D reconstructions obtained by processing independently the radargrams collected along each B-scan, as shown in the previous section. We will show both tests with synthetic and experimental data. In the first examples, synthetic data are considered. These data have been generated by means of the free software GPRMAX (the 3D version). In particular, let us now refer to the case of a square box, whose side is 2 m long, filled by dry sand having relative permittivity $\varepsilon_B = 4$ and conductivity $\sigma_B = 10$ m S/m, hosting one or more voids with different shape. The data have been acquired along five parallel 1.5-m-long measurement lines and moving the antenna system in seven different positions spaced 0.25 m apart. This spatial step is electrically large (as it will be clear in a while), but this example is introductive. The offset between the antennas is 0.25 m too. The simulated primary source is a y-oriented Hertzian dipole, according to the reference system introduced in Figure 11.1, that is fed with a Richer pulse signal having central frequency $f_c = 300$ MHz. The data have been simulated by assuming a time windows of 45 ns and discretizing the box with cubes whose edge is 0.02 m long. For the inversion, the investigation domain has been discretized in 3D cells with edge equal to 0.0625 m. The simulated data have been pre-processed by means of a background removal filtering procedure. Moreover, since the inversion procedure works in the frequency domain, a discrete Fourier transform has been applied to transform the filtered time domain data; and the meaningful bandwidth, which ranges between $f_{min} = 150$ MHz and $f_{max} = 450$ MHz, has been sampled with a frequency offset of 30 MHz (11 frequencies).

As a first example, let us consider the case of two voids both having size 0.30 m × 0.30 m × 0.5 m and centered in (−0.25 m, −0.25 m, 0.5 m) and (0.25 m, 0.25 m, 0.5 m), respectively (see Figure 15.75). The reconstruction obtained by applying the full 3D imaging approach are shown in Figure 15.76 panels a and b, while those provided by the interpolation of the 2D results are in Figure 15.77, panels a and b. The reconstruction figures show the amplitude of the reconstructed contrast function normalized to its maximum value.

By comparing these images, we see that, in the case at hand, the use of a full 3D imaging procedure allows an improvement of the localization and shape reconstruction of the upper side of the objects.

As a second case, we have considered an L-shaped cavity, made of two parallelepipeds having size 0.25 m × 0.80 m × 0.25 m and 0.55 m × 0.25 m × 0.25 m, respectively, hidden into the sandy box as shown in Figure 15.78, panels a and b. The reconstruction obtained with the full 3D imaging approach is shown in Figure 15.79, panels a and b, while that provided by the interpolation of the 2D results is in Figure 15.80, panels a and b. By comparing these figures, we can clearly appreciate the advantages offered by the use of a full 3D data processing procedure. As a matter of fact, it allows us to correctly estimate the shape of the target (more precisely, the shape of its upper face) as well as its size in the (x, y) plane.

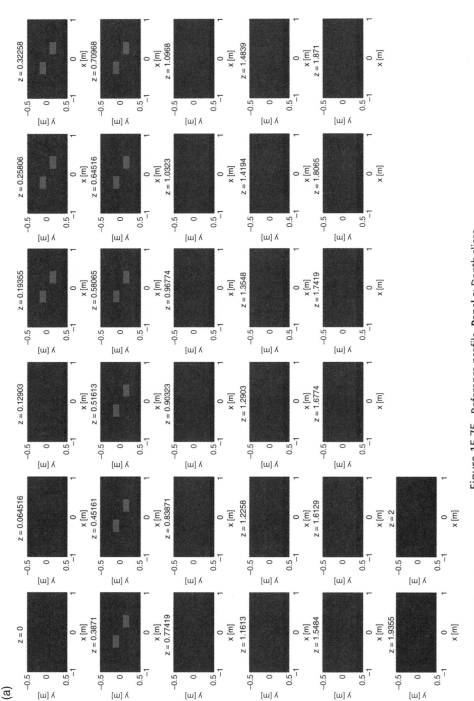

Figure 15.75. Reference profile. Panel a: Depth slices.

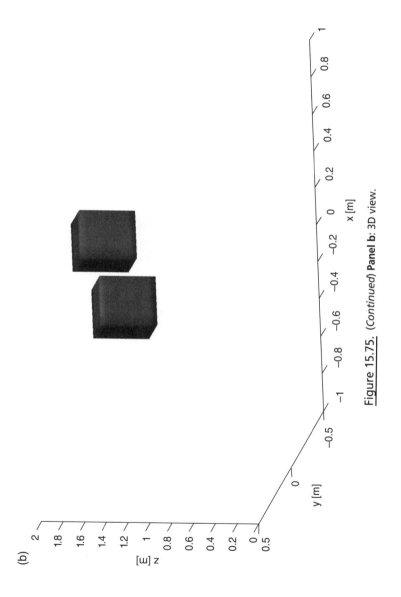

Figure 15.75. *(Continued)* **Panel b**: 3D view.

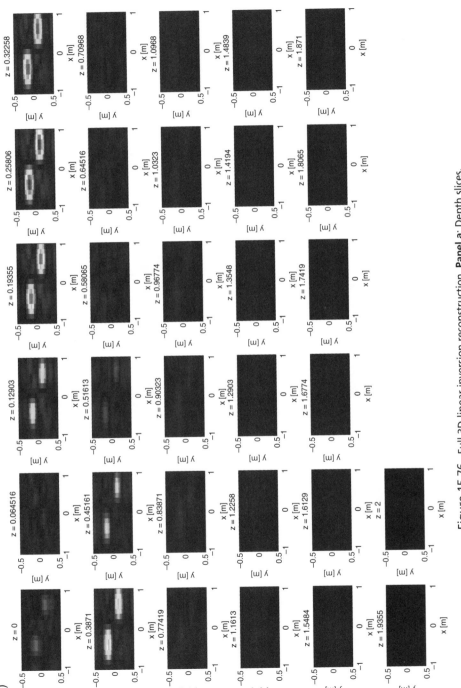

Figure 15.76. Full 3D linear inversion reconstruction. **Panel a:** Depth slices.

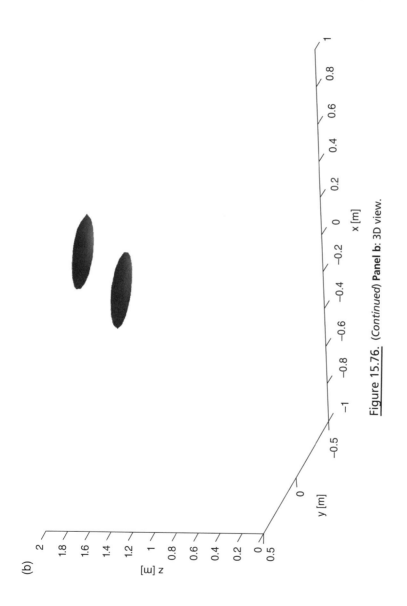

Figure 15.76. (*Continued*) **Panel b**: 3D view.

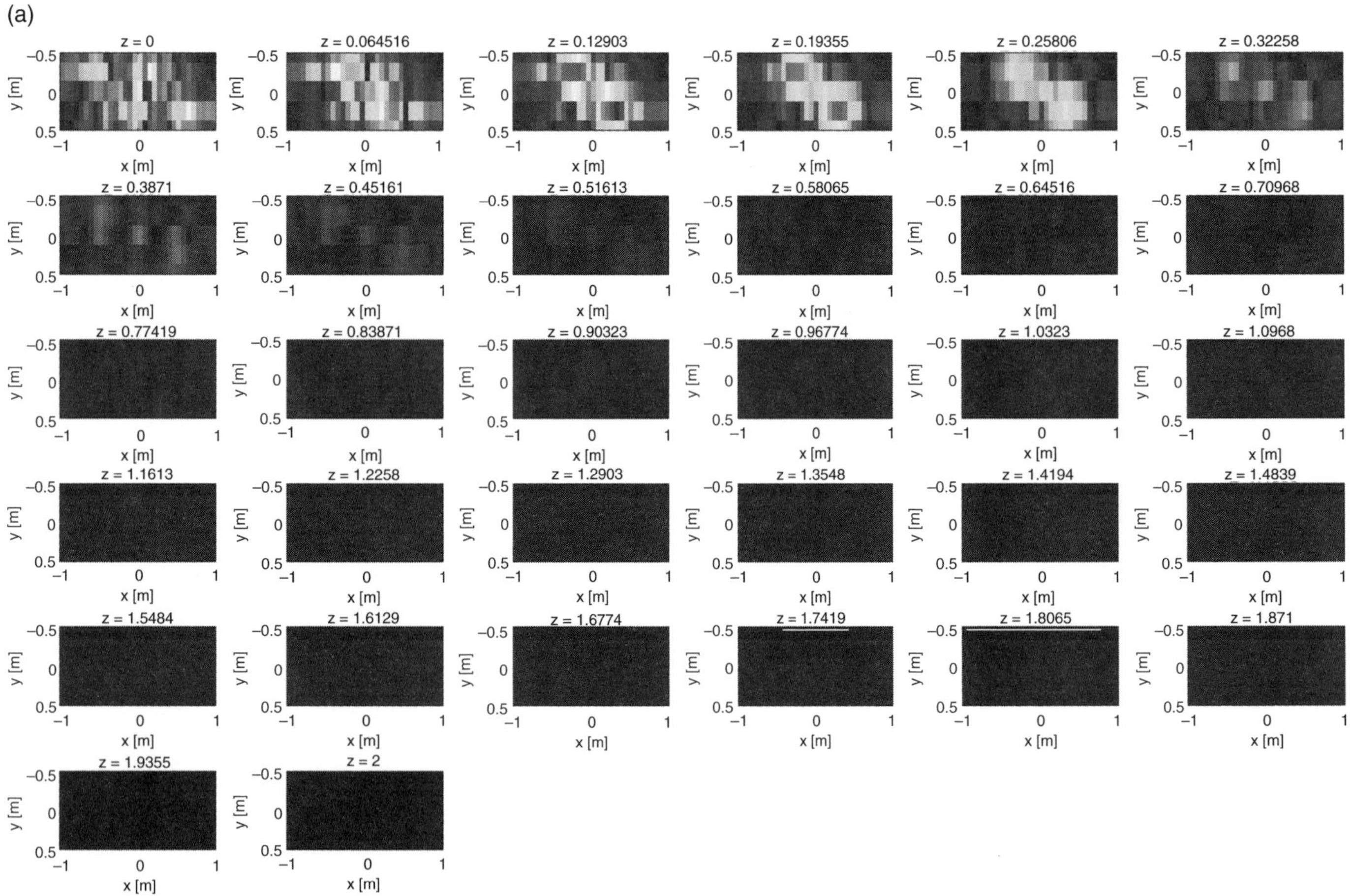

Figure 15.77. Tomographich reconstruction via 2D slice reconstruction method. **Panel a:** Depth slices.

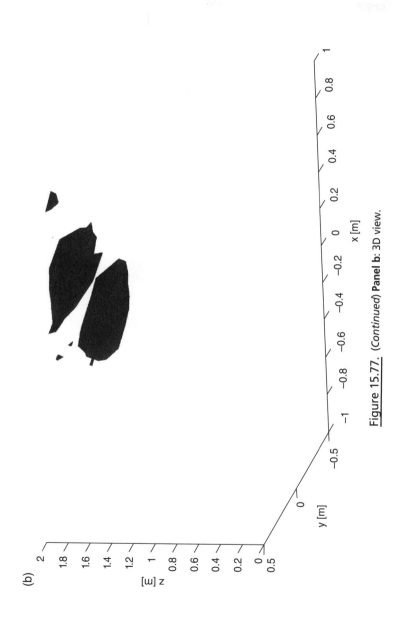

(b)

Figure 15.77. (*Continued*) **Panel b:** 3D view.

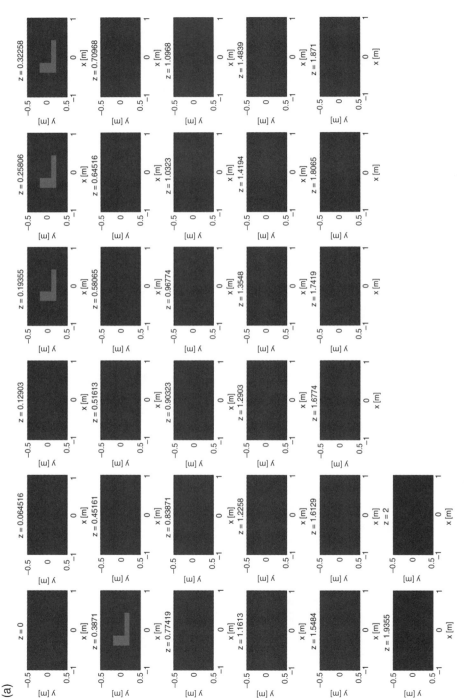

Figure 15.78. Reference profile. **Panel a:** Depth slices.

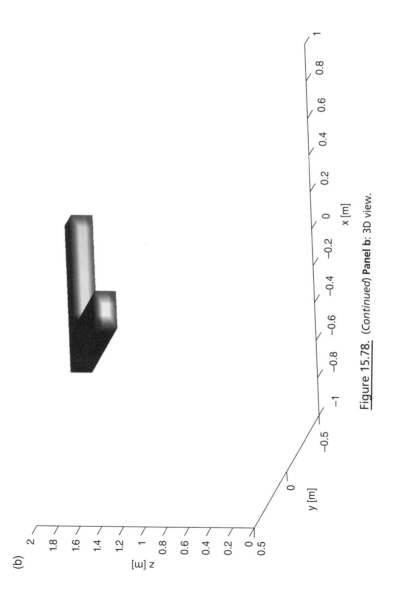

Figure 15.78. (*Continued*) **Panel b**: 3D view.

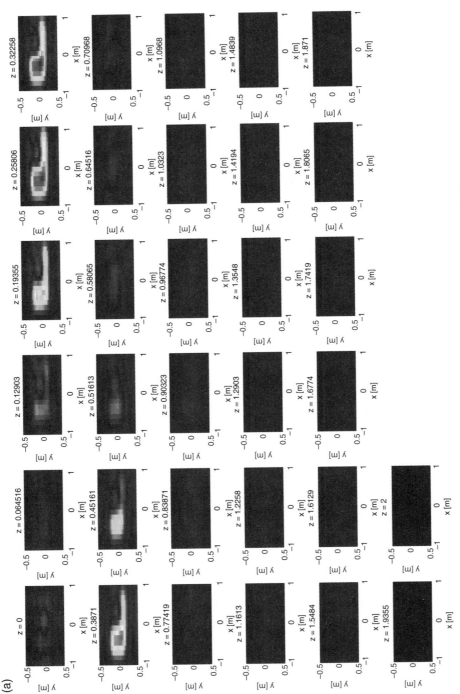

Figure 15.79. Full 3D linear inversion reconstruction. **Panel a:** Depth slices.

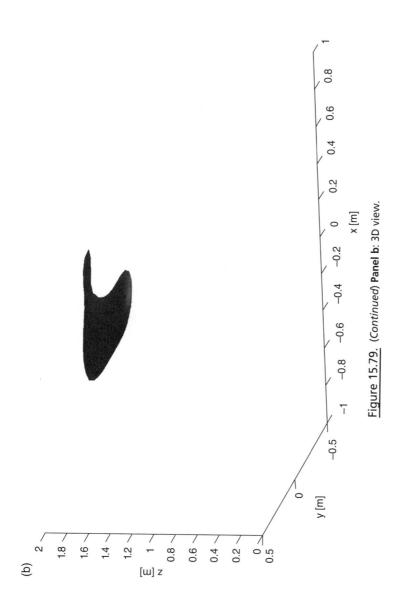

(b)

z [m]

y [m]

x [m]

Figure 15.79. (*Continued*) Panel **b**: 3D view.

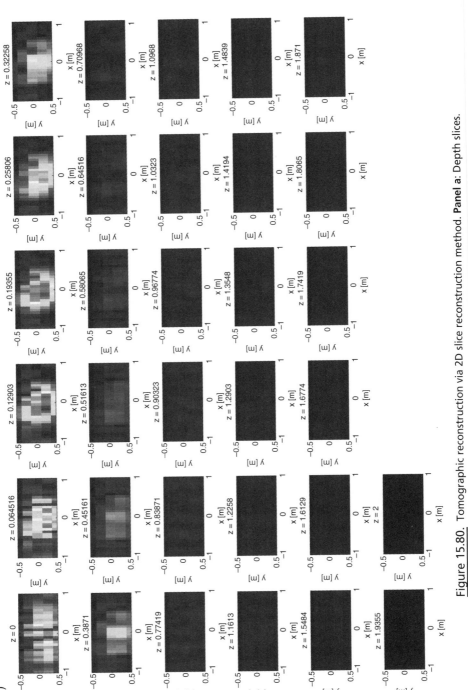

Figure 15.80. Tomographic reconstruction via 2D slice reconstruction method. **Panel a:** Depth slices.

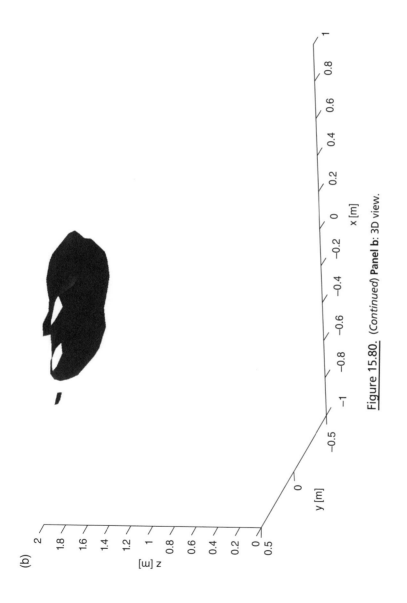

(b)

z [m]

y [m]

x [m]

Figure 15.80. *(Continued)* **Panel b**: 3D view.

Let us now consider an example concerning experimental data. The experiment has been carried out at the Electromagnetic Diagnostic Laboratory of IREA-CNR. The data have been gathered by using a GPR system, manufactured by IDS Corporation, working in the time domain and equipped with a single polarization antenna pair with central frequency $f_c = 2$ GHz. The antenna has been manually moved along the upper surface of a solid polystyrene box wherein the object to be imaged was concealed. The data have been collected over a rectangular measurement domain $\sum = 30 \times 30$ cm^2 at height $h = 0.001$ m above the air–polystyrene interface. The measurements were taken with a spatial discretization step equal to 0.05 m along both the x- and y- axis (49 points) in a time window of 32 ns, which was sampled in 512 points spaced 0.0625 ns from each other. The measured data had been pre-processed by means of a background removal filtering procedure in order to eliminate the direct coupling between the antennas and the signal due to

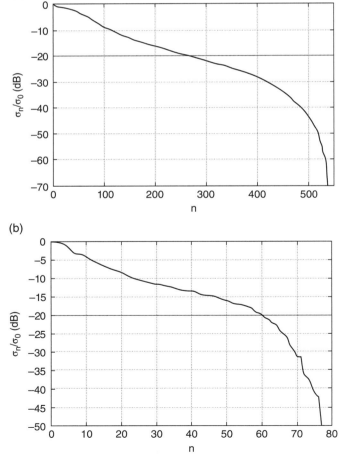

Figure 15.81. Singular values of the discretized operator: (a) Full 3D case; and (b) 2D case.

the air–polystyrene interface, and then they were transformed into the frequency domain. The frequency range [850, 1950] MHz with a step of 110 MHz (11 points) has been considered for the inversion. The relative permittivity of the box ε_B is assumed constant and equal to 2.7. With regard to the selection of the investigation region Ω to be used for the full 3D inversion, it is a cubic domain with edge 0.6 m and its discretization for the pixel basis expansion has been set to 0.02 m along x, y, and z directions.

The following example accounts for a U-shaped chipboard object made by three parallelepipeds, each of which are $0.2 \times 0.05 \times 0.02$ m^3 and located about 0.1 m apart from the air–polystyrene interface. The typical value of the relative dielectric permittivity of the chipboard is around 2, so the object is likely to act as a weak scattering target.

Such an example enables a comparison between the imaging capabilities of the full 3D inversion strategy and the 2D slice reconstruction one. In this respect, the amplitude

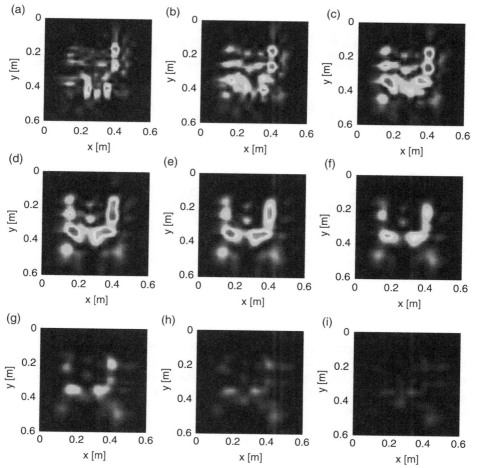

Figure 15.82. Full 3D tomographic reconstruction of the chipboard C. (a) $z = 0.02$ m, (b) $z = 0.04$ m, (c) $z = 0.06$ m, (d) $z = 0.08$ m, (e) $z = 0.1$ m, (f) $z = 0.12$ m, (g) z 0.14 m, (h) $z = 0.16$ m, (i) $z = 0.18$ m.

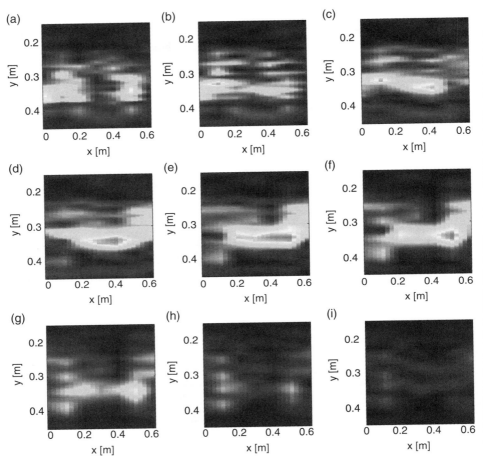

Figure 15.83. Tomographic reconstructions of the chipboard C via the 2D slice reconstruction method. (a) $z = 0.02$ m, (b) $z = 0.04$ m, (c) $z = 0.06$ m, (d) $z = 0.08$ m, (e) $z = 0.1$ m, (f) $z = 0.12$ m, (g) $z = 0.14$ m, (h) $z = 0.16$ m, (i) $z = 0.18$ m.

of the reconstructed contrast $\chi(\cdot)$ is shown over cut planes XY located at different depths with respect to the interface. According to Figure 15.81, panels a and b, which show the behavior of the singular values of the discretized version of the 3D and 2Dlinear operators, respectively, the provided reconstructions have been obtained by setting the threshold in a specific manner to neglect the singular values that are lower than -20 dB with respect to the first one.

Figure 15.82 shows the reconstructions attained with the full 3D linear imaging approach. As can be seen, the object can be clearly identified at a depth of about 0.1 m. Moreover, its size can be estimated from this reconstruction, and it closely resembles the actual one. The results of the 2D slice reconstruction algorithm are shown in Figure 15.83. From Figure 15.83, we can detect the presence of a buried target, but its actual shape cannot be identified at all.

APPENDICES

APPENDIX A

Raffaele Persico and Raffaele Solimene

The purpose of this appendix is to show that

$$
\int_{-\infty}^{+\infty} \frac{\exp\left(j\sqrt{\left(\frac{2\omega}{c}\right)^2 - \eta^2}\, z'\right)}{j\sqrt{\left(\frac{2\omega}{c}\right)^2 - \eta^2}} \exp(-j\eta(x-x'))\, d\eta = \int_{-\infty}^{+\infty} \frac{\exp\left(j\frac{2\omega}{c}\sqrt{(x-x')^2 + z'^2 + y^2}\right)}{\sqrt{(x-x')^2 + z'^2 + y^2}}\, dy
$$

$$(A.1)$$

To do that, let us start from the following equation:

$$
\nabla^2 G(x,y,z) + k^2 G(x,y,z) = -\delta(x-x_0)\delta(y-y_0)\delta(z-z_0) \tag{A.2}
$$

Equation (A.2) is, as is well known, the equation of the three-dimensional scalar Green's function in a homogeneous medium with wavenumber k.

Let us Fourier transform Eq. (A.2) according to the convention

$$
\hat{\hat{\hat{G}}}(k_1,k_2,k_3) = \int_{-\infty}^{+\infty}\int_{-\infty}^{+\infty}\int_{-\infty}^{+\infty} G(x,y,z)\exp(-jxk_1 - jyk_2 - jzk_3)\, dx\,dy\,dz \tag{A.3}
$$

In the transformed domain, Eq. (A.2) can be rewritten as

$$
\left(k^2 - k_1^2 - k_2^2 - k_3^2\right)\hat{\hat{\hat{G}}}(k_1,k_2,k_3) = -\exp\left(-j\vec{k}\cdot\vec{r}_0\right) \Rightarrow \hat{\hat{\hat{G}}}(k_1,k_2,k_3) = -\frac{\exp\left(-j\vec{k}\cdot\vec{r}_0\right)}{k^2 - \left(k_1^2 + k_2^2 + k_3^2\right)}
$$

$$(A.4)$$

where $\vec{k} = k_1 i_x + k_2 i_y + k_3 i_z$ and $\vec{r}_0 = x_0 i_x + y_0 i_y + z_0 i_z$. The multiplication dot stands for scalar product.

Introduction to Ground Penetrating Radar: Inverse Scattering and Data Processing,
First Edition. Raffaele Persico.
© 2014 The Institute of Electrical and Electronics Engineers, Inc. Published 2014 by John Wiley & Sons, Inc.

Considering the inverse Fourier transform of Eq. (A.4), we have

$$G(x,y,z) = -\frac{1}{8\pi^3} \int_{-\infty}^{+\infty}\int_{-\infty}^{+\infty}\int_{-\infty}^{+\infty} \frac{\exp\left(j\vec{k}\cdot(\vec{r}-\vec{r}_0)\right)}{k^2-\left(k_1^2+k_2^2+k_3^2\right)} dk_1 dk_2 dk_3$$

$$= -\frac{1}{8\pi^3} \int_{-\infty}^{+\infty}\int_{-\infty}^{+\infty}\int_{-\infty}^{+\infty} \frac{\exp\left(-j\vec{k}\cdot(\vec{r}-\vec{r}_0)\right)}{k^2-\left(k_1^2+k_2^2+k_3^2\right)} dk_1 dk_2 dk_3 \qquad (A.5)$$

where $\vec{r} = xi_x + yi_y + zi_z$ and, in Eq. (A.5), a change of integration variable from $\vec{k}$ to $-\vec{k}$ has been performed too.

Now, let us rewrite Eq. (A.5) as

$$G(x,y,z)$$

$$= -\frac{1}{4\pi^2} \times \int_{-\infty}^{+\infty}\int_{-\infty}^{+\infty} \exp(-jk_1(x-x_0)-jk_2(y-y_0))dk_1 dk_2 \frac{1}{2\pi} \int_{-\infty}^{+\infty} \frac{\exp(-jk_3(z-z_0))}{\left[k^2-\left(k_1^2+k_2^2\right)\right]-k_3^2} dk_3$$

$$(A.6)$$

The integral in dk_3 is calculable on the basis of the theory of the residuals and of Jordan's lemma. They are both well known, so they will taken for granted here. In particular, since the poles of the function are placed at $k_3 = \pm k_z = \pm\sqrt{k^2-\left(k_1^2+k_2^2\right)}$, the result of the integral is

$$G(x,y,z) = -\frac{j}{8\pi^2} \int_{-\infty}^{+\infty}\int_{-\infty}^{+\infty} \frac{\exp(-jk_1(x-x_0)-jk_2(y-y_0))\exp(-jk_z|z-z_0|)}{k_z} dk_1 dk_2 \quad (A.7)$$

At this point, Eq. (A.1) can be rewritten in spherical coordinative as

$$\nabla^2 G(r,\theta,\varphi) + k^2 G(r,\theta,\varphi) = -\delta(\|\vec{r}-\vec{r}_0\|) \qquad (A.8)$$

Due to the intrinsic symmetries of Eq. (A.8), it is licit to postulate a solution that depends only on the distance between the observation point and the source; thus, under the change of variable $\vec{r}_1 = \vec{r} - \vec{r}_0$, we have

$$\nabla^2 G(r_1) + k^2 G(r_1) = -\delta(r_1) \qquad (A.9)$$

Making use of the expression of the scalar Laplacian operator in polar coordinates in the case of no dependence from the angular coordinates, we rewrite Eq. (A.9) as (Collin, 1985)

$$\frac{d^2}{dr_1^2}G(r_1) + \frac{2}{r_1}\frac{d}{dr_1}G(r_1) + k^2 G(r_1) = -\delta(r_1) \qquad (A.10)$$

Let us now examine the associated homogeneous equation:

$$\frac{d^2}{dr_1^2}G(r_1) + \frac{2}{r_1}\frac{d}{dr_1}G(r_1) + k^2 G(r_1) = 0 \qquad (A.11)$$

The general solution is in this case is easily found after putting $G(r_1) = g(r_1)/r_1$ and substituting in Eq. (A.11). This yields to a harmonic equation in the unknown $g(r_1)$ and consequently we obtain

$$G(r_1) = C\frac{e^{-jkr_1}}{r_1} + D\frac{e^{+jkr_1}}{r_1} \qquad (A.12)$$

Since waves propagate away from the sources toward the infinite but not vice versa, there is only one physically meaningful term, so that necessarily we have $D = 0$ and Eq. (A.12) is rewritten as

$$G(r_1) = C\frac{e^{-jkr_1}}{r_1} \qquad (A.13)$$

Of course, Eq. (A.13) also represents the solution with an impulsive source [i.e., the solution of Eq. (A.10)] in any point except the origin, where the impulsive source makes the function singular. The integration of Eq. (A.9) about the source provides the value of the constant C. In fact, substituting the solution (A.13) into Eq. (A.9) and considering the volume integral within a small sphere centered about the origin, we obtain

$$C\iiint_V \nabla\cdot\nabla\frac{e^{-jkr_1}}{r_1}\,dV + C\iiint_V k^2\frac{e^{-jkr_1}}{r_1}\,dV = -\iiint_V \delta(r_1) = -1 \qquad (A.14)$$

The second integral term vanishes for $r_1 \to 0$ because the integrand diverges as the inverse of the ray but the volume element $\left(r_1^2 \sin(\theta)\,d\theta d\varphi\,dr_1\right)$ vanishes more rapidly. Using the Gauss divergence theorem, we obtain

$$\iiint_V \nabla\cdot\nabla\frac{e^{-jkr_1}}{r_1}\,dV = \iint_S \nabla\frac{e^{-jkr_1}}{r_1}\,dS = \iint_S \frac{d}{dr_1}\frac{e^{-jkr_1}}{r_1}\,dS = 4\pi r_1^2\frac{d}{dr_1}\frac{e^{-jkr_1}}{r_1} \qquad (A.15)$$

Substituting (A.15) into (A.14), and taking the limit of $r_1 \to 0$, we obtain

$$\lim_{r_1 \to 0} \left[C4\pi r_1^2 \frac{d}{dr_1} \frac{e^{-jkr_1}}{r_1} + C\iiint_V k^2 \overbrace{\frac{e^{-jkr_1}}{r_1}}^{\to 0} dV \right] = -1 \Rightarrow$$

$$\Rightarrow \lim_{r_1 \to 0} C \left[4\pi r_1^2 e^{-jkr_1} \left(-\frac{jk}{r_1} - \frac{1}{r_1^2} \right) \right] = -1 \tag{A.16}$$

which becomes

$$C = \frac{1}{4\pi} \tag{A.17}$$

So, the final solution of Eq. (A.13) is

$$G(x,y,z) = \frac{e^{-jkr_1}}{4\pi r_1} = \frac{e^{-jk\|\vec{r}-\vec{r}_0\|}}{4\pi\|\vec{r}-\vec{r}_0\|} = \frac{e^{-jk\sqrt{(x-x_0)^2+(y-y_0)^2+(z-z_0)^2}}}{4\pi\sqrt{(x-x_0)^2+(y-y_0)^2+(z-z_0)^2}} \tag{A.18}$$

The comparison between Eq. (A.7) and Eq. (A.18) provides the equality

$$\frac{e^{-jk\sqrt{(x-x_0)^2+(y-y_0)^2+(z-z_0)^2}}}{\sqrt{(x-x_0)^2+(y-y_0)^2+(z-z_0)^2}} = \frac{1}{2j\pi} \int_{-\infty}^{+\infty}\int_{-\infty}^{+\infty} \frac{\exp(-jk_1(x-x_0)-jk_2(y-y_0)-jk_z|z-z_0|)}{k_z} dk_1 dk_2 \tag{A.19}$$

Integrating Eq. (A.19) in the variable $y - y_0$ we obtain

$$\int_{-\infty}^{+\infty} \frac{e^{-jk\sqrt{(x-x_0)^2+y^2+(z-z_0)^2}}}{\sqrt{(x-x_0)^2+y^2+(z-z_0)^2}} dy$$

$$= \frac{1}{2j\pi} \int_{-\infty}^{+\infty}\int_{-\infty}^{+\infty}\int_{-\infty}^{+\infty} \frac{\exp(-jk_1(x-x_0)-jk_2y-jk_z|z-z_0|)}{k_z} dk_1 dk_2 dy$$

$$= \frac{1}{2j\pi} \int_{-\infty}^{+\infty}\int_{-\infty}^{+\infty} \frac{\exp(-jk_1(x-x_0)-jk_z|z-z_0|)}{k_z} dk_1 dk_2 \int_{-\infty}^{+\infty} \exp(-jk_2y)\,dy$$

$$= \frac{1}{j} \int_{-\infty}^{+\infty}\int_{-\infty}^{+\infty} \frac{\exp(-jk_1(x-x_0)-jk_z|z-z_0|)}{k_z} \delta(k_2)\,dk_1 dk_2$$

$$= \int_{-\infty}^{+\infty} \frac{\exp\left(-jk_1(x-x_0)-j\sqrt{k^2-k_1^2}|z-z_0|\right)}{j\sqrt{k^2-k_1^2}} dk_1 \tag{A.20}$$

Equation (A.20) holds for any value of $x - x_0$, k, and $z - z_0$. At this point, let us remind ourselves that z' in Eq. (A.1) is nonnegative because it is the depth coordinate of the generic buried point within the investigation domain, which is positive because of the downward verse of the z-axis (see Figure 4.1). Consequently, it is easily recognized that Eq. (A.20) is fully equivalent to Eq. (A.1), which completes the demonstration.

APPENDIX B

The purpose of this appendix is to show the expression of the half-derivative in the time domain.

The half-derivative is defined as the operation that in frequency domains is expressed by the multiplication of the spectrum times $\sqrt{j\omega}$, and it is a particular case of the fractional derivative. Therefore, given any function $f(t)$ whose spectrum is $F(\omega)$, by definition we have

$$
\begin{aligned}
\frac{\partial^{0.5} f}{\partial^{0.5} t} &\equiv \frac{1}{2\pi} \int_{-\infty}^{+\infty} \sqrt{j\omega} F(\omega) \exp(j\omega t)\, d\omega \\
&= \frac{1}{2\pi} \int_{-\infty}^{+\infty} \frac{1}{\sqrt{j\omega}} j\omega F(\omega) \exp(j\omega t)\, d\omega
\end{aligned}
\tag{B.1}
$$

This inverse transform (B.1) is equal to the convolution of the inverse transforms of the two functions $1/\sqrt{j\omega}$ and $j\omega F(\omega)$. Now, $1/\sqrt{j\omega}$ back-transforms into $H(t)/\sqrt{t}$, where H is the Heaviside function, equal to 1 for positive arguments and equal to 0 for negative arguments. Moreover as is well known, $j\omega F(\omega)$ back-transforms into $\partial f/\partial t$. This leads to the expression of the half-derivative in the time domain, which is

$$
\frac{\partial^{0.5} f}{\partial^{0.5} t} = \int_{-\infty}^{+\infty} \frac{H(\tau - t)}{\sqrt{\tau - t}} \frac{\partial f}{\partial t}(\tau)\, d\tau
\tag{B.2}
$$

Introduction to Ground Penetrating Radar: Inverse Scattering and Data Processing,
First Edition. Raffaele Persico.
© 2014 The Institute of Electrical and Electronics Engineers, Inc. Published 2014 by John Wiley & Sons, Inc.

The purpose of this appendix is to provide the solution for the potential vector in a homogeneous space related to a current element z-directed and placed in the origin of the geometrical reference system. Let us start from the Helmholtz equation (11.25), here repeated for commodity of reading:

$$\nabla^2 \mathbf{A} + k_0^2 \mathbf{A} = -\mu_0 \mathbf{J} \tag{C.1}$$

since

$$\nabla^2 \mathbf{A} = i_x \nabla^2 A_x + i_y \nabla^2 A_y + i_z \nabla^2 A_z$$

We can decompose (C.1) as

$$\begin{aligned}
\nabla^2 A_x + k_0^2 A_x &= -\mu_0 J_x, \\
\nabla^2 A_y + k_0^2 A_y &= -\mu_0 J_y, \\
\nabla^2 A_z + k_0^2 A_z &= -\mu_0 J_z
\end{aligned} \tag{C.2}$$

In order to solve any of Eqs. (C.2), let us preliminary study the solution of the following scalar differential equation:

$$\nabla^2 \phi(\mathbf{r}) + k_0^2 \phi(\mathbf{r}) = -\delta(\mathbf{r}) \tag{C.3}$$

Since the source is point-like, it is expected that the solution has spherical symmetry. Therefore, using the spherical representation of the curl operator, we obtain

$$\frac{d^2}{dr^2}\phi(r) + \frac{2}{r}\frac{d}{dr}\phi(r) + k_0^2 \phi(r) = -\delta(r) \tag{C.4}$$

where $r = \sqrt{x^2 + y^2 + z^2}$. Actually, this equation has been already examined in Appendix A [see Eq. (A.10)]. Therefore, without repeating the procedure, the solution is

Introduction to Ground Penetrating Radar: Inverse Scattering and Data Processing,
First Edition. Raffaele Persico.
© 2014 The Institute of Electrical and Electronics Engineers, Inc. Published 2014 by John Wiley & Sons, Inc.

$$\phi(\mathbf{r}) = \frac{e^{-jk_0 r}}{4\pi r} \qquad (C.5)$$

Now, Eqs. (C.2) are linear, and so we can retrieve their solution integrating the solution of the corresponding impulsive problem. The procedure has been widely dealt with and exploited in Chapter 4, and so we will not discuss it again. So, Eq. (C.5) provides a result to be integrated all over the elementary equivalent source points [see Section 4.2, and in particular Eqs. (4.41) and (4.42)]. This provides immediately the result of Eq. (11.28) with regard to the potential vector and the result of Eq. (11.29) with regard to the scalar potential function.

APPENDIX D

In this appendix the decomposition of a generic plane wave impinging at the air–soil interface (from the air side) along its TE and TM components is retrieved, with the purpose of applying this decomposition to the plane waves that compose the homogeneous Green's function.

The generic impinging plane wave has the form

$$\mathbf{E} = \exp(ju(x-x'))\exp(jv(y-y'))\exp(jk_{z10}(z-z'))\left(v_x(u,v)i_x + v_y(u,v)i_y + v_z(u,v)i_z\right)$$
(D.1)

It is immediate to see that the solenoidality of the plane wave imposes the condition[1]

$$v_z(u,v) = -\frac{uv_x(u,v) + vv_y(u,v)}{k_{z10}(u,v)}$$
(D.2)

Substituting Eq. (D.2) into Eq. (D.1), we have that the expression of the generic plane wave composing the Green's function is given by

$$\mathbf{E} = \exp(ju(x-x'))\exp(jv(y-y'))\exp(jk_{z10}(z-z'))\left(v_x(u,v)i_x + v_y(u,v)i_y - \frac{uv_x(u,v) + vv_y(u,v)}{k_{z10}(u,v)}i_z\right)$$
(D.3)

The air–soil interface is the plane of the equation $z = 0$. The TE component of the plane wave of Eq. (D.3) has an electrical field parallel to this interface, by definition. Thus, the electric field of the TE component does not have the z-component, and consequently we can write it as

$$\mathbf{E}_{TE} = \exp(ju(x-x'))\exp(jv(y-y'))\exp(jk_{z10}(z-z'))\left(v_{x1}(u,v)i_x + v_{y1}(u,v)i_y\right) \quad \text{(D.4)}$$

[1] As an exercise, the reader can verify that the columns of the plane wave spectrum of the homogeneous dyadic Green's function [Eq. (11.55)] satisfy Eq. (D.2).

Introduction to Ground Penetrating Radar: Inverse Scattering and Data Processing,
First Edition. Raffaele Persico.
© 2014 The Institute of Electrical and Electronics Engineers, Inc. Published 2014 by John Wiley & Sons, Inc.

The solenoidality of the plane wave D.4 imposes the condition

$$uv_{x1}(u,v) + vv_{y1}(u,v) = 0 \tag{D.5}$$

The TM component of the plane wave of Eq. (D.3) has to be equal to the difference between Eq. (D.3) and Eq. (D.4):

$$\mathbf{E}_{TM} = \exp(ju(x-x'))\exp(jv(y-y'))\exp(jk_{z10}(z-z'))$$
$$\times \left((v_x(u,v) - v_{x1}(u,v))i_x + (v_y(u,v) - v_{y1}(u,v))i_y - \frac{uv_x(u,v) + vv_y(u,v)}{k_{z10}(u,v)}i_z \right) \tag{D.6}$$

The magnetic field relative to this TM component is given by

$$\mathbf{H}_{TM} = -\frac{1}{j\omega}\nabla \times \mathbf{E}_{TM} = -\frac{1}{j\omega}\begin{vmatrix} i_x & i_y & i_z \\ \dfrac{\partial}{\partial x} & \dfrac{\partial}{\partial y} & \dfrac{\partial}{\partial z} \\ E_{TMx} & E_{TMy} & E_{TMz} \end{vmatrix} \tag{D.7}$$

In particular, the z-component of this quantity has to be equal to zero, by definition of TM field. Therefore, substituting the component of the electric field from Eq. (D.6), we retrieve the condition

$$u(v_y(u,v) - v_{y1}(u,v)) - v(v_x(u,v) - v_{x1}(u,v)) = 0 \tag{D.8}$$

Equations (D.5) and (D.8) provide a linear algebraic system that allows us to retrieve v_{x1} and v_{y1} versus v_x and v_y. The solution is

$$v_{x1} = \frac{v^2 v_x(u,v) - uv v_y(u,v)}{u^2 + v^2},$$
$$v_{y1} = \frac{u^2 v_y(u,v) - uv v_x(u,v)}{u^2 + v^2} \tag{D.9}$$

Substituting Eq. (D.9) into Eqs. (D.4) and (D.6), we retrieve the TE and TM components of the plane wave, which are

$$\mathbf{E}_{TE} = \exp(ju(x-x'))\exp(jv(y-y'))\exp(jk_{z10}(z-z'))$$
$$\times \left(\frac{v^2 v_x(u,v) - uv v_y(u,v)}{u^2 + v^2}i_x + \frac{u^2 v_y(u,v) - uv v_x(u,v)}{u^2 + v^2}i_y \right) \tag{D.10}$$

$$\mathbf{E}_{TM} = \exp(ju(x-x'))\exp(jv(y-y'))\exp(jk_{z10}(z-z'))$$

$$\times \left(\frac{u^2 v_x(u,v) + uv v_y(u,v)}{u^2 + v^2} i_x + \frac{v^2 v_y(u,v) + uv v_x(u,v)}{u^2 + v^2} i_y - \frac{u v_x(u,v) + v v_y(u,v)}{k_{z10}(u,v)} i_z \right)$$

$$(D.11)$$

Substituting Eqs. (D.10) and (D.11) into the columns of the plane wave spectrum of the homogeneous Green's function in air [Eq. (11.62)], we achieve Eqs. (11.66) and (11.67).

APPENDIX E

In this appendix the reflection and transmission coefficients are retrieved for a TE and a TM plane wave impinging on the air–soil interface.

Let us consider a generic TE impinging plane wave. Since phase factors are not of interest in this appendix, we can make use of a simple general expression of the following kind[2]:

$$\mathbf{E}_i = E_0 \exp(j(ux + vy + k_{z10}z))(v_{xi}i_x + v_{yi}i_y) \tag{E.1}$$

There is no z-component of the field because the wave is TE with respect to the plane $z = 0$. Moreover, the condition of solenoidality of the field imposes the constraint

$$uv_{xi} + vv_{yi} = 0 \Leftrightarrow v_{yi} = -\frac{u}{v}v_{xi} \tag{E.2}$$

Substituting into Eq. (E.1), we can write the generic TE impinging wave as

$$\mathbf{E}_i = E_0 v_{xi} \exp(j(ux + vy + k_{z10}z))\left(i_x - \frac{u}{v}i_y\right) \tag{E.3}$$

Coherently, the reflected and the refracted waves are expressed in general as

$$\mathbf{E}_r = E_0 v_{xr} \exp(j(u_1x + v_1y - k_{z101}z))\left(i_x - \frac{u_1}{v_1}i_y\right) \tag{E.4}$$

$$\mathbf{E}_t = E_0 v_{xt} \exp(j(u_2x + v_2y + k_{z1s2}z))\left(i_x - \frac{u_2}{v_2}i_y\right) \tag{E.5}$$

with

---

[2] The dependence along z is related to the dependence along x and y by means of the relationship $k_{z10} = \sqrt{k_0^2 - u^2 - v^2}$, which is a constraint due to the Maxwell's equations.

Introduction to Ground Penetrating Radar: Inverse Scattering and Data Processing,
First Edition. Raffaele Persico.
© 2014 The Institute of Electrical and Electronics Engineers, Inc. Published 2014 by John Wiley & Sons, Inc.

$$k_{z101} = \sqrt{k_0^2 - u_1^2 - v_1^2},$$

$$k_{z1s2} = \sqrt{k_s^2 - u_2^2 - v_2^2} \tag{E.6}$$

where the square roots are meant with nonpositive imaginary parts.

Equations (E.4) and (E.5) account for the Maxwell's equations, the different propagation media and the fact that the reflected waves back-propagate with respect to the interface, which arises the minus sign before k_{z101} in Eq. (E.4).

At the interface, the Maxwell's equations impose that the tangential components of both the electric and magnetic fields are equal at the air and soil sides. This has to hold for any point of the interface, which leads to the conditions

$$u_1 = u_2 = u,$$

$$v_1 = v_2 = v \tag{E.7}$$

Substituting Eq. (E.7) into Eqs. (E.4) and (E.5) we rewrite the reflected and the transmitted wave as follows:

$$\mathbf{E}_r = E_0 v_{xr} \exp(j(ux + vy - k_{z10}z)) \left(i_x - \frac{u}{v} i_y\right) \tag{E.8}$$

$$\mathbf{E}_t = E_0 v_{xt} \exp(j(ux + vy + k_{z1s}z)) \left(i_x - \frac{u}{v} i_y\right) \tag{E.9}$$

Equations (E.1), (E.8), and (E.9) show that the incident, reflected, and transmitted electric fields are parallel to each other, and so the preservation of the tangential component of the electric field before and after the interface reduces to the equation

$$v_{xi} + v_{xr} = v_{xt} \tag{E.10}$$

Let us now consider the magnetic field. This is retrieved from the incident electric field from the first Maxwell equations. In particular, distinguishing also for the magnetic field an incident, a reflected, and a transmitted wave, we have

$$\mathbf{H}_i = \frac{1}{-j\omega\mu_0} \nabla \times \mathbf{E}_i,$$

$$\mathbf{H}_r = \frac{1}{-j\omega\mu_0} \nabla \times \mathbf{E}_r, \tag{E.11}$$

$$\mathbf{H}_t = \frac{1}{-j\omega\mu_0} \nabla \times \mathbf{E}_t$$

In Eq. (E.11) we have accounted for the fact that the soil does not show magnetic properties.

By expanding the expression of the incident magnetic field, we achieve

$$\mathbf{H}_i = \frac{E_0 v_{xi}}{-j\omega\mu_0} \begin{vmatrix} i_x & i_y & i_z \\ \dfrac{\partial}{\partial x} & \dfrac{\partial}{\partial y} & \dfrac{\partial}{\partial z} \\ \exp(\cdot) & -\dfrac{u}{v}\exp(\cdot) & 0 \end{vmatrix} \qquad (E.12)$$

where for brevity $\exp(\cdot)$ stands for $\exp(j(ux + vy + k_{z10}z))$. Calculating the derivatives, we obtain

$$\mathbf{H}_i = \frac{E_0 v_{xi}}{-j\omega\mu_0}\left(\frac{juk_{z10}}{v}i_x + jk_{z10}i_y + \left(-\frac{ju^2}{v} - jv\right)i_z\right)\exp(j(ux + vy + k_{z10}z))$$

$$= -\frac{E_0 v_{xi}k_{z10}}{\omega\mu_0}\left(\frac{u}{v}i_x + i_y - \left(\frac{u^2 + v^2}{vk_{z10}}\right)i_z\right)\exp(j(ux + vy + k_{z10}z)) \qquad (E.13)$$

Following the same passages with the reflected and refracted waves, we obtain

$$\mathbf{H}_r = \frac{E_0 v_{xr}k_{z10}}{\omega\mu_0}\left(\frac{u}{v}i_x + i_y + \left(\frac{u^2 + v^2}{vk_{z10}}\right)i_z\right)\exp(j(ux + vy - k_{z10}z)) \qquad (E.14)$$

$$\mathbf{H}_t = -\frac{E_0 v_{xt}k_{z1s}}{\omega\mu_0}\left(\frac{u}{v}i_x + i_y - \left(\frac{u^2 + v^2}{vk_{z1s}}\right)i_z\right)\exp(j(ux + vy + k_{z1s}z)) \qquad (E.15)$$

From Eqs. (E.13)–(E.15), the preservation of the tangential component of the magnetic field at the interface imposes the condition:

$$v_{xi} - v_{xr} = v_{xt}\frac{k_{z1s}}{k_{z10}} \qquad (E.16)$$

Considering the scalar complex amplitude of the incident, reflected, and refracted wave at the interface (i.e., E_i, E_r and E_t), the reflection and transmission coefficients are by definition given by $R_{TE} = E_r/E_i = v_{xr}/v_{xi}$ and $T_{TE} = E_t/E_i = v_{xt}/v_{xi}$, respectively. Let us note that, even if only the x-component appears, R_{TE} and T_{TE} are the ratios between the complex amplitudes of the field, because the y-component is proportional to the x-component of the field [see Eq. (E.2)]. Equation (E.10) and (E.16) provide the algebraic system

$$1 + R_{TE} = T_{TE}, \quad 1 - R_{TE} = T_{TE}\frac{k_{z1s}}{k_{z10}} \qquad (E.17)$$

Solving the algebraic system, we obtain

$$R_{TE} = \frac{k_{z10} - k_{z1s}}{k_{z10} + k_{z1s}}, \quad T_{TE} = \frac{2k_{z10}}{k_{z10} + k_{z1s}} \quad \text{(E.18)}$$

Substituting into Eqs. (E.8) and (E.9), the reflected and transmitted waves can be then expressed as

$$\mathbf{E}_r(x,y,z) = E_0 R_{TE} v_{xi} \exp(j(ux + vy - k_{z10}z)) \left(i_x - \frac{u}{v} i_y \right) = R_{TE} \mathbf{E}_i(x,y,0) \exp(-jk_{z10}z)$$

$$\text{(E.19)}$$

$$\mathbf{E}_t(x,y,z) = E_0 T_{TE} v_{xi} \exp(j(ux + vy + k_{z1s}z)) \left(i_x - \frac{u}{v} i_y \right) = T_{TE} \mathbf{E}_i(x,y,0) \exp(jk_{z1s}z)$$

$$\text{(E.20)}$$

Let us now consider the case of a TM impinging plane wave. In this case, by definition, the incident magnetic field is expressed as

$$\mathbf{H}_i = H_0 \exp(j(ux + vy + k_{z10}z)) \left(v_{xi} i_x + v_{yi} i_y \right) \quad \text{(E.21)}$$

Considering the solenoidality of the magnetic field, as well as the condition of continuity at any interface point of the tangential component of the magnetic field, dually to the TE case we can rewrite the incident, reflected, and transmitted magnetic fields as

$$\mathbf{H}_i = H_0 v_{xi} \exp(j(ux + vy + k_{z10}z)) \left(i_x - \frac{u}{v} i_y \right) \quad \text{(E.22)}$$

$$\mathbf{H}_r = H_0 v_{xr} \exp(j(ux + vy - k_{z10}z)) \left(i_x - \frac{u}{v} i_y \right) \quad \text{(E.23)}$$

$$\mathbf{H}_t = H_0 v_{xt} \exp(j(ux + vy + k_{z1s}z)) \left(i_x - \frac{u}{v} i_y \right) \quad \text{(E.24)}$$

The electric field is retrieved versus the magnetic one from the (homogeneous, i.e. without sources) Maxwell's equations, so that we have

$$\mathbf{E}_i = \frac{1}{j\omega\varepsilon_0} \nabla \times \mathbf{H}_i,$$

$$\mathbf{E}_r = \frac{1}{j\omega\varepsilon_0} \nabla \times \mathbf{H}_r, \quad \text{(E.25)}$$

$$\mathbf{E}_t = \frac{1}{j\omega\varepsilon_s} \nabla \times \mathbf{H}_t$$

Developing the spatial derivatives, Eq. (E.25) can be rewritten as

$$\mathbf{E}_i = \frac{H_0 v_{xi} k_{z10}}{\omega \varepsilon_0} \exp(j(ux + vy + k_{z10}z))\left(\frac{u}{v}i_x + i_y - \frac{u^2 + v^2}{k_{z10}v}i_z\right) \tag{E.26}$$

$$\mathbf{E}_r = -\frac{H_0 v_{xr} k_{z10}}{\omega \varepsilon_0} \exp(j(ux + vy - k_{z10}z))\left(\frac{u}{v}i_x + i_y + \frac{u^2 + v^2}{k_{z10}v}i_z\right) \tag{E.27}$$

$$\mathbf{E}_t = \frac{H_0 v_{xt} k_{z1s}}{\omega \varepsilon_s} \exp(j(ux + vy + k_{z1s}z))\left(\frac{u}{v}i_x + i_y - \frac{u^2 + v^2}{k_{z1s}v}i_z\right) \tag{E.28}$$

From Eqs. (E.22)–(E.24) and from Eqs. (E.26)–(E.28) we retrieve the conditions on the amplitudes of the waves in order to verify the continuity of the tangential components of the fields at the interface:

$$
\left.\begin{aligned}
v_{xi} + v_{xr} &= v_{xt} \\
\frac{v_{xi}k_{z10}}{\varepsilon_0} - \frac{v_{xr}k_{z10}}{\varepsilon_0} &= \frac{v_{xt}k_{z1s}}{\varepsilon_s}
\end{aligned}\right\} \Leftrightarrow
\begin{cases}
1 + \dfrac{v_{xr}}{v_{xi}} = \dfrac{v_{xt}}{v_{xi}}, \\[2mm]
1 - \dfrac{v_{xr}}{v_{xi}} = \dfrac{v_{xt}k_{z1s}\varepsilon_0}{v_{xi}k_{z10}\varepsilon_s}
\end{cases}
\tag{E.29}
$$

Now, let us remind that the reflection coefficient is defined the ratio between the tangential components of the reflected and incident electric fields, whereas the transmission coefficient is defined as the ratio between the tangential component of the transmitted and incident electric fields, so that we have

$$
\left.\begin{aligned}
R_{TM} &\equiv -\frac{v_{xr}}{v_{xi}} \\[2mm]
T_{TM} &\equiv \frac{k_{z1s}\varepsilon_0 v_{xt}}{k_{z10}\varepsilon_s v_{xi}}
\end{aligned}\right\} \Leftrightarrow
\begin{cases}
\dfrac{v_{xr}}{v_{xi}} \equiv -R_{TM}, \\[2mm]
\dfrac{v_{xt}}{v_{xi}} = \dfrac{k_{z10}\varepsilon_s}{k_{z1s}\varepsilon_0}T_{TM}
\end{cases}
\tag{E.30}
$$

Substituting Eqs. (E.30) in Eqs. (E.29), we achieve

$$1 - R_{TM} = T_{TM}\frac{k_{z10}\varepsilon_s}{k_{z1s}\varepsilon_0}, \quad 1 + R_{TM} = T_{TM} \tag{E.31}$$

which eventually provide

$$R_{TM} = \frac{k_{z1s}\varepsilon_0 - k_{z10}\varepsilon_s}{k_{z1s}\varepsilon_0 + k_{z10}\varepsilon_s}, \quad T_{TM} = \frac{2k_{z1s}\varepsilon_0}{k_{z1s}\varepsilon_0 + k_{z10}\varepsilon_s} \tag{E.32}$$

Finally, substituting Eqs. (E.32) into Eqs. (E.27) and (E.28), the reflected and transmitted waves can be rewritten as

$$
\mathbf{E}_r(x,y,z) = \frac{H_0 v_{xi} R_{TM} k_{z10}}{\omega \varepsilon_0} \exp(j(ux + vy - k_{z10}z)) \left(\frac{u}{v} i_x + i_y + \frac{u^2 + v^2}{k_{z10}v} i_z \right)
$$

$$
= \mathbf{E}_{ixy}(x,y,0) R_{TM} \exp(-jk_{z10}z) - \mathbf{E}_{iz}(x,y,0) R_{TM} \exp(-jk_{z10}z)
$$

(E.33)

$$
\mathbf{E}_t(x,y,z) = \frac{H_0 v_{xi} T_{TM} k_{z10}}{\omega \varepsilon_0} \exp(j(ux + vy + k_{z1s}z)) \left(\frac{u}{v} i_x + i_y - \frac{u^2 + v^2}{k_{z1s}v} i_z \right)
$$

$$
= \mathbf{E}_{ixy}(x,y,0) T_{TM} \exp(jk_{z1s}z) + \mathbf{E}_{iz}(x,y,0) T_{TM} \frac{k_{z10}}{k_{z1s}} \exp(jk_{z1s}z)
$$

(E.34)

where $\mathbf{E}_{ixy}$ is the (vector) component of the incident field along the plane xy and $\mathbf{E}_{iz}$ is the (vector) component of the incident field along the z-axis.

The calculated Fresnel coefficients are "from air to soil." If the plane waves impinge from underground, it is straightforward to recognize that the Fresnel coefficients are obtained by exchanging the permittivity and the wavenumber in the soil with those in air and viceversa. In particular, this means that the soil–air reflection coefficients are the opposite of the air–soil reflection coefficients.

APPENDIX F

Raffaele Persico and Raffaele Solimene

In this appendix we provide the solution in terms of Bessel's functions of the integral in polar coordinatives in Eq. (12.27), namely,

$$f(\rho_1, 2k_{sc}\sin(\theta_{e\max})) = \int_0^{2k_{sc}\sin(\theta_{e\max})} \rho \, d\rho \int_0^{2\pi} \exp(j\rho\rho_1\cos(\varphi)) \, d\varphi \tag{F.1}$$

where [see Eq. (12.26)]

$$\rho_1 = \sqrt{(x'-x_0)^2 + (y'-y_0)^2} \tag{F.2}$$

To do this, let us start reporting one of the integral expressions for Bessel's functions of the first kind and integer order (Abramowitz and Stegun, 1972), namely,

$$J_n(x) = \frac{j^{-n}}{\pi} \int_0^\pi \exp(jx\cos(\varphi))\cos(n\varphi) \, d\varphi = \frac{j^{-n}}{\pi} \int_0^\pi \exp(-jx\cos(\varphi))\cos(n\varphi) \, d\varphi \tag{F.3}$$

The sign of the imaginary exponential is indifferent because the real part of the integrand is even with regard to the central point $\varphi = \pi/2$ and the imaginary part is odd with respect to the same central point.

Particularizing to the two cases $n = 0, 1$, we have

$$J_0(x) = \frac{1}{\pi} \int_0^\pi \exp(jx\cos(\varphi)) \, d\varphi = \frac{1}{\pi} \int_0^\pi \exp(-jx\cos(\varphi)) \, d\varphi \tag{F.4}$$

$$J_1(x) = \frac{1}{j\pi} \int_0^\pi \exp(jx\cos(\varphi))\cos(\varphi) \, d\varphi = \frac{1}{j\pi} \int_0^\pi \exp(-jx\cos(\varphi))\cos(\varphi) \, d\varphi \tag{F.5}$$

Introduction to Ground Penetrating Radar: Inverse Scattering and Data Processing,
First Edition. Raffaele Persico.
© 2014 The Institute of Electrical and Electronics Engineers, Inc. Published 2014 by John Wiley & Sons, Inc.

From Eq. (F.3), it is also easy to recognize that Bessel's functions of integer order are real for real values of their argument (which is the only case that we will consider). Moreover, from Eqs. (F.3) and (F.4), it is easily recognized that J_0 is an even function and J_1 is an odd function and that $J_0(0) = 1$ and $J_1(0) = 0$ (as due because of the oddness).

That said, by integrating Eq. (F.3) for parts, we obtain

$$J_n(x) = \frac{j^{-n}}{\pi} \int_0^\pi \exp(jx\cos(\varphi)) \cos(n\varphi) \, d\varphi$$

$$= \frac{j^{-n}}{n\pi} [\exp(jx\cos(\varphi)) \sin(n\varphi)]_0^\pi - \frac{j^{-n}}{n\pi} \int_0^\pi \exp(jx\cos(\varphi))(-jx\sin(\varphi)) \sin(n\varphi) \, d\varphi$$

$$= \frac{xj^{-n+1}}{n\pi} \int_0^\pi \exp(jx\cos(\varphi)) \sin(n\varphi) \sin(\varphi) \, d\varphi \Rightarrow$$

$$\Rightarrow \frac{j^{-n+1}}{\pi} \int_0^\pi \exp(jx\cos(\varphi)) \sin(n\varphi) \sin(\varphi) \, d\varphi = \frac{nJ_n(x)}{x} \qquad (F.6)$$

Let us now consider the derivative of Bessel's function of integer order

$$\frac{dJ_n(x)}{dx} = \frac{j^{-n}}{\pi} \int_0^\pi j\cos(\varphi) \exp(jx\cos(\varphi)) \cos(n\varphi) d\varphi$$

$$= \frac{j^{-(n-1)}}{\pi} \int_0^\pi \exp(-jx\cos(\varphi)) \cos(n\varphi) \cos(\varphi) d\varphi$$

$$= \frac{j^{-(n-1)}}{\pi} \int_0^\pi \exp(-jx\cos(\varphi))(\cos(n\varphi) \cos(\varphi)$$

$$+ \sin(n\varphi) \sin(\varphi) - \sin(n\varphi) \sin(\varphi)) \, d\varphi$$

$$= \frac{j^{-(n-1)}}{\pi} \int_0^\pi \exp(-jx\cos(\varphi)) \cos((n-1)\varphi) \, d\varphi$$

$$- \frac{j^{-(n-1)}}{\pi} \int_0^\pi \exp(-jx\cos(\varphi)) \sin(n\varphi) \sin(\varphi) \, d\varphi$$

$$= J_{n-1}(x) - \frac{j^{-n+1}}{\pi} \int_0^\pi \exp(-jx\cos(\varphi)) \sin(n\varphi) \sin(\varphi) \, d\varphi \qquad (F.7)$$

Substituting Eq. (F.6) into Eq. (F.7), we obtain

$$\frac{dJ_n(x)}{dx} = J_{n-1}(x) - \frac{nJ_n(x)}{x} \qquad (F.8)$$

Particularizing to the case $n = 1$, we obtain

$$\frac{dJ_1(x)}{dx} = J_0(x) - \frac{J_1(x)}{x} \tag{F.9}$$

As a consequence of Eq. (F.9), we have

$$\frac{d}{dx}(xJ_1(x)) = J_1(x) + x\frac{dJ_1(x)}{dx} = J_1(x) + x\left(J_0(x) - \frac{J_1(x)}{x}\right) = xJ_0(x) \tag{F.10}$$

Thus, we have worked out that the function $xJ_1(x)$ is a primitive of the function of $xJ_0(x)$.

At this point, coming back to Eq. (F.1), through some straightforward calculation steps, we obtain

$$f(\rho_1, 2k_{sc}\sin(\theta_{emax})) = \int_0^{2k_{sc}\sin(\theta_{emax})} \rho\,d\rho \int_0^{2\pi} \exp(j\rho\rho_1\cos(\varphi))\,d\varphi$$

$$= \int_0^{2k_{sc}\sin(\theta_{emax})} \rho\,d\rho \left[\int_0^{\pi} \exp(j\rho\rho_1\cos(\varphi))\,d\varphi + \int_{\pi}^{2\pi} \exp(j\rho\rho_1\cos(\varphi))d\varphi\right]$$

$$= \int_0^{2k_{sc}\sin(\theta_{emax})} \rho\,d\rho \left[\int_0^{\pi} \exp(j\rho\rho_1\cos(\varphi))\,d\varphi + \int_0^{\pi} \exp(j\rho\rho_1\cos(\varphi'-\pi))d\varphi'\right]$$

$$= \int_0^{2k_{sc}\sin(\theta_{emax})} \rho\,d\rho \left[\int_0^{\pi} \exp(j\rho\rho_1\cos(\varphi))\,d\varphi + \int_0^{\pi} \exp(-j\rho\rho_1\cos(\varphi'))d\varphi'\right]$$

$$= 2\pi \int_0^{2k_{sc}\sin(\theta_{emax})} \rho J_0(\rho\rho_1)d\rho = \frac{2\pi}{\rho_1^2} \int_0^{2k_{sc}\rho_1\sin(\theta_{emax})} \alpha J_0(\alpha)\,d\alpha$$

$$= \frac{2\pi}{\rho_1^2}[\alpha J_1(\alpha)]_0^{2k_{sc}\rho_1\sin(\theta_{emax})} = \frac{2\pi}{\rho_1^2}2k_{sc}\rho_1\sin(\theta_{emax})J_1(2k_{sc}\rho_1\sin(\theta_{emax}))$$

$$= \frac{4\pi k_{sc}\sin(\theta_{emax})J_1(2k_{sc}\rho_1\sin(\theta_{emax}))}{\rho_1} = 8\pi k_{sc}^2\sin^2(\theta_{emax})\frac{J_1(2k_{sc}\rho_1\sin(\theta_{emax}))}{2k_{sc}\rho_1\sin(\theta_{emax})}$$

$$= 8\pi k_{sc}^2\sin^2(\theta_{emax})\frac{J_1\left(2k_{sc}\sin(\theta_{emax})\sqrt{(x'-x_0)^2 + (y'-y_0)^2}\right)}{2k_{sc}\sin(\theta_{emax})\sqrt{(x'-x_0)^2 + (y'-y_0)^2}} \tag{F.11}$$

Substituting into Eq. (F.11) into Eq. (12.27), we obtain Eq. (12.28).

APPENDIX G:
ANSWERS TO QUESTIONS

Chapter 2

1. No, GPR data essentially can help to retrieve the propagation velocity and (with some more difficulty) the losses in the soil. This is sufficient in most cases, but (for example) we cannot distinguish whether the soil has magnetic properties from the mere GPR data.

2. Yes they do, because the electric conductivity influences the imaginary part of the equivalent permittivity. The propagation velocity of the waves is inversely proportional to the real part of the square root of the equivalent complex permittivity, and thus it is rigorously influenced by both the permittivity and the conductivity of the soil (further than by possible magnetic properties of the soil). However, in the relatively common case of low lossy soil, a first-order approximation of the square root of the complex permittivity can be exploited, and in this case the influence of the conductivity of the soil on the propagation velocity is usually negligible with respect to the effect of the permittivity (the relative calculations are straightforward and are left as an exercise).

3. In general we can't, even theoretically. For example, if we know that the buried target is circular and we apply the diffraction equation (2.7) to any two points, we have two equations and four unknowns (the ray, the propagation velocity, the minimum return time, and the abscissa of the minimum return time). Of course we can read the couple (x_0, t_0) from the data too, but this is to say that "any two points" of the curve in any case are not enough. In the case of a point-like target, instead, two points are sufficient because there is no need to have or achieve information about the size and the shape of the target. However, this is not the general case and, even in this case, to evaluate the propagation velocity from two points it is not well-advised, because a more extended matching between model and data can average better all the sources of uncertainty.

4. Essentially because in CMP the reflection comes from the same point: This makes the diffraction curve less dependent on the size and shape of the buried target from which it comes. In particular, the hypothesis of point-like reflector can be relaxed.

5. A TDR probe forces a reflection from a point at known depth without a meaningful alteration of the consistence of the soil, which is impossible to achieve with a cooperative target buried on purpose and looked for with a GPR. Moreover, the wave received with a TDR probe is a guided wave, and this provides the possibility to retrieve not only the propagation velocity of the wave (related to the product between the dielectric permittivity and the magnetic permeability of the soil), but also the impedance of the transmission line (related to the ratio between the dielectric

Introduction to Ground Penetrating Radar: Inverse Scattering and Data Processing,
First Edition. Raffaele Persico.
© 2014 The Institute of Electrical and Electronics Engineers, Inc. Published 2014 by John Wiley & Sons, Inc.

permittivity and the magnetic permeability of the soil). Incidentally, GPR data are instead not able to provide the wave impedance, essentially because the datum is proportional to the electric field and does not provide information on the magnetic field in the observation point.

6. A TDR probe knocked in the soil provides a local measure, usually up to a depth not greater than 50 cm. A GPR system can prospect more extended areas and can provide data based on deeper targets, which means to average the propagation velocity on a more extended range of depths.

Chapter 3

1. No, it isn't. Actually, 75 MHz is a frequency step equal to one-half of the frequency step at which the nonambiguous depth is equal to 50 cm, which would guarantee against any spurious replica of targets embedded in the first 50 cm and would guarantee also against the Hermitian image of any target embedded in the first 50 cm. In other words, the frequency step of 75 MHz corresponds to a nonambiguous depth of 100 cm in the case at hand. However, in a dry sandy (in particular, not clay) soil the penetration of the wave is expected quite good, up to a depth depending on the exploited antennas (in particular their central frequency) but likely to be quite greater than 100 cm. So, for example a target with its top buried at 101 cm might provide a strong replica with its top at the depth of 1 cm. Even without a specific deeper knowledge about the penetration of the radiation in the case at hand, it is certainly better advised to hypothesize the possibility of a maximum penetration of 5 m (even more if low-frequency antennas were used, but in this case the targets of interest would probably be deeper than 50 cm), which leads to a frequency step of 15 MHz if the Hermitian images can be neglected, 7.5 MHz otherwise.

2. The required time step is given by the Nyquist criterion [Eq. (3.39)]. As said, lacking more specific information, we can assume that the band and the central frequency of the antennas are quantities of the same order, which leads to $B = 2$ GHz and consequently $\Delta t = 0.5$ ns. Let us be reminded that the effect of the soil might change the band of the antennas with respect to its nominal one. However, generally this effect is some decrease of the central frequency and not an increase of the band. That said, the bottom scale in time corresponding to a maximum reached depth of 50 cm is given by $T = 2p/c = 2p\sqrt{\varepsilon_{sr}}/c_0 = 3.3$ ns. This would lead to seven time samples. However, as it is shown in Chapter 9, in order to focus in the best way the buried targets up to 50 cm, it is well-advised that the bottom scale time be longer than that needed to reach the depth of 50 cm along a vertical path. Actually, this would make us "see" a target at 50 cm only when the GPR system is over it, whereas it is better to also have some possibility of lateral view. Moreover, in order to amortize possible uncertainties that we can have about the band of the antennas and the characteristics of the soil, and in order to better average (and so filter out) the noise, it is also well-advised to make narrow the time step. It is not possible to provide exact numbers for these choices, but let us suppose that we want to guarantee that the GPR can still receive an echo from a target at the depth of 50 cm when it is 2 m beyond this target. In this case, the maximum "communicative" distance between the target and the GPR can be estimated as $\sqrt{2^2 + 0.5^2}$ m $= 2.062$ m, and the corresponding time bottom scale is about 27 ns. Narrowing the time step to 0.25 ns, this leads to 109 samples.

3. The effect of the aliasing for a stepped frequency GPR system is the production of replicas of the echoes from the targets in time domain; practically, this means the production of spurious targets along the depth, which add to the actual ones. These replicas can be in some cases recognized

from the result because they are somehow similar to some actual target that gets "copied," but they can be eliminated only repeating the measure with a correct frequency step.

4. The effect of the aliasing for a pulsed system is to provide replicas of the spectra of the echoes from the targets in frequency domain. This generates a distortion of the targets and possibly artifacts, but it is not immediate to quantify it from the results. In other words, the aliasing in frequency domain is more easily recognized from the results with respect to that in time domain.

5. No, we can't. In particular, with a stepped frequency system we gather the data directly in the frequency domain. Thus, we obtain the time domain data to be truncated by back Fourier transforming the gathered stepped frequency data. However, these data are already sampled in frequency domain, and so they should already be aliasing-free.

6. No, we can't. In this case we obtain the data in the frequency domain by forward Fourier transforming the gathered pulsed GPR data. Also in this case, the sampling comes intrinsically before the Fourier transformation, and so we can't think of erasing the aliasing effects by filtering them.

Chapter 4

1. No. The scattering equations essentially link the targets with the scattered field in the observation point. Any antenna system in the observation point measures a voltage approximately proportional to the total field in the observation point. It is a task of the human operator to extract the scattered field data from the total field data.

2. No. The incident field in the subsoil will be not the same as what we would have in a homogeneous medium, even if the energy radiated toward the air is strongly dissipated. In particular, the air–soil interface is necessarily a direction of null of the radiation pattern. Notwithstanding, a model based on a homogeneous propagation medium can be exploited in GPR data processing providing useful results in many cases.

3. The reason of this dissymmetry is the fact that the primary source is an electrical filamentary current, or in any case a source that radiates an electric field parallel to the axis of invariance. In order to have a formulation where the dielectric permittivities appear explicitly instead of the magnetic permeabilities, we should solve the scattering problem for a filamentary magnetic current, and the datum should be the magnetic scattered field instead of the electric one.

4. No. In this way we would implement a long wire antenna where the current behaves as a wave propagating along the wire. In particular, in the case of a sinusoidal imposed voltage, we would have a sinusoidal current along the wire but the phase of this sine would change from point to point.

5. No. The reciprocity theorem assures us (under wide hypotheses) that the behavior in reception of an antenna is linked to its behavior in transmission mode, because the effective lengths are equal for the two modes. In particular, it is impossible that an antenna behaves as a 3D structure in transmission mode and as a 2D structure in reception mode.

6. No. In particular, for energy conservation reasons, in a lossless medium, the geometrical spreading of the far field is proportional to the inverse of the distance from the source in 3D and is instead proportional to the square root of the inverse of the distance from the source in 2D. Notwithstanding, in GPR prospecting, usually the phase behavior is sufficiently similar and the distances involved are not so large to make the different geometrical spreading cause dramatic effects. Therefore, a 2D model in many cases provides a useful representation of a 3D buried scenario.

Chapter 5

1. No, it is solenoidal in the absence of a net accumulation of charges. However, within a 2D model we have a current flux without having accumulation of charges, and this makes the electric field solenoidal.

2. No. What is solenoidal in any case is the magnetic induction. Within the kind of propagation media considered in this text, this is equivalent to saying that the product between the magnetic field and the magnetic permeability is a solenoidal quantity, but not necessarily the magnetic field alone.

3. No. The incident field is the field in the background medium without targets, and it depends only on the sources and on the characteristics of the soil.

4. Yes, because both quantities are involved in the determination of the internal field. Actually, the magnetic part and the dielectric part of the contrast interact as two targets placed in the same volume.

5. Yes, because the total field is the sum of the incident and scattered field.

6. Yes. In particular, if the source is a filamentary magnetic (instead of electric) current, then the magnetic field is parallel to the axis of invariance and the electric field is orthogonal to it. The same happens if the source is constituted by a continuous series of parallel Hertzian dipoles directed along any direction orthogonal to the axis of invariance. This situation is usually labeled as the vector 2D case (the 2D geometry dealt with in this text, instead, is more precisely labeled as the 2D scalar geometry).

Chapter 6

1. No, both the statements are not true because in general the Born series is not guaranteed to converge.

2. Yes, because the scattered field is a linear quantity with respect to the incident field, which in turn is a linear quantity with respect to the density current that generated it.

3. No, because the antennas are not structures that impose a density current but essentially structures where a voltage (real or equivalent) can be imposed. Now, especially if close to each other, the two antennas interact with each other (i.e., are coupled), and the current densities that develop along their arms are not the same that we would have on each of them in absence of the other one.

4. No, because the two targets interact with each other, which leads to the nonlinearity of the relationship between contrast and scattered field.

5. Yes, because the amount of energy that causes the interactions between the targets is decreased if they are placed far from each other.

6. Yes, because in this case the amount of energy that causes the interactions between the targets is decreased by the losses, which transform part of the electromagnetic energy into heat.

Chapter 7

1. No, because in any case there is the propagation time of the signal inside the GPR, which can be amortized by means of zero timing that is easily performed only in the time domain.

2. Yes: This can be done in a hardware way (e.g., by means of a differential configuration) or in a software way (e.g., by means of a background removal).

3. No, and actually a negative effect is expected in this case. In fact, in the limit for a very narrow spatial step, the N averaged traces become too close to each other (in terms of the central wavelength). In these conditions, the field does not have a sufficient "space" to vary meaningfully within the considered set of traces. Consequently, the central trace and the average trace become the same, and their difference vanishes.

4. No. We have to gather the maximum extractable information from the scattered field, and this makes it in any case well-advised to keep the spatial step not too large. A quantification in this sense is worked out in Chapter 9.

5. No, it provides an approximation of the scattered field. The only way to achieve the scattered field from the total one is to subtract the incident field, which should be measured or calculated some way. This is difficult to be performed in a reliable way in the case of the GPR prospecting.

6. Yes, because the incident field in the case of homogeneous soil (of course with a flat air soil interface) is the same for any observation point.

Chapter 8

1. When the target is embedded in the masonry, the impinging radiation crosses it more times, due to the presence of the far interface of the masonry that leads to a partial reflection of the waves impinging on it. This makes the mutual interactions between different parts of the targets change, and so it is highly probable that the ratio between the norms of the internal incident field and the internal total field is not the same in the two cases. This shows that the degree on nonlinearity of the scattering also depends on the kind of background medium. It is implicit that a masonry is expected to worsen the nonlinearity with respect to a homogeneous soil ideally composed of the same material.

Chapter 9

1. The visible interval depends on the frequency, the dielectric permittivity of the soil, the magnetic permeability of the soil and on the maximum effective view angle.

2. Formally they don't, because the retrievable spectral set is related to the visible intervals achieved at all the available frequencies and therefore depends on the frequency band, the dielectric and magnetic permeability of the soil and the maximum effective view angle. However, the spatial and frequency step are essential in order to "sample" the retrievable spectral set properly.

3. Formally they don't, but a proper time step is indispensable in order to calculate an un-aliased version of the spectrum, and an adequate time bottom scale is indispensable to include and focus properly the targets of interest.

4. No, because we have identified and quantified the Nyquist rate for the spatial and frequency (and also time) step. Below the Nyquist rate, there is no theoretical improvement still available, because the sampled function can be perfectly reconstructed from its samples.

5. The resolution improves when the observation line is longer, because the maximum view angle increases. However, this improvement is not progressive because at a certain point we reach the maximum effective view angle, which no longer increases even if the maximum geometrical view angle continues to increase.

6. No, because an excessive directivity makes the antennas look only under them and hinders the lateral view of the targets, which is instead essential in order to allow a processing that improves the resolution.

7. No, because the vertical resolution is essentially influenced by the available frequency band, as said.

8. Yes, this can be done muting the data in time domain before and after the time-depth range of interest. This truncation of the signal in time domain has to be done before the frequency resampling, and it allows us to relax the frequency step. This one, in fact, can be found on the basis of the Nyquist the criterion in the frequency domain relative to a "narrower" function in the time domain. The price paid for this is the renounce to focus targets before or beyond the chosen depth range.

9. Yes, this can be done filtering the data in the frequency domain before and after the band of interest. This truncation of the signal in the frequency domain has to be done before the time resampling, and it allows us to relax the time step. This one can be found on the basis of the Nyquist criterion in the time domain relative to a narrower function in frequency domain. The price paid for this is to renounce to retrieve part of the band of the signal, which leads to a loss of resolution.

10. An enlargement of the band can of course improve the reconstruction because the resolution can improve in this case. However, a progressive increasing of the band can also cause problems, both computational and related to the fact that the Born approximation, underlying the diffraction tomography deductions proposed in this chapter, eventually becomes poor because the targets become large in terms of wavelength. Moreover, a larger band can be more sensible to interferences caused by other sources. Notwithstanding, from a practical point of view, it is usually better to have at disposal a larger band because, after gathering the data, commercial codes for the processing usually make it possible to filter them, thereby making the results less detailed but more robust. The computational extra-price paid for this is usually not dramatic.

11. No, because the increasing is negligible beyond a certain limit. In particular, the number of averaged traces can be indefinitely increased by a progressive artificial prolongation of the B-scan, achieved by adding the average trace on both sides. In particular, the number of averaged traces in this way can go even beyond the same actual number of traces of the B-scan. However, the substantial effect is a progressive narrowing of the spatial filtering constituted by the complement to one of the Dirichlet sine (see Section 9.14), which involves progressively marginal effects.

12. Not necessarily. As said, the information we need in GPR prospecting is not merely quantitative but above all qualitative: We are not merely interested in the potentialities of the involved mathematical operator, but in the reliability and usefulness of the interpretation of the results that we can provide. For example, if a larger band provides a too noisy signal, it is better to use a narrower band. If some not perfectly horizontal layer is present, a background removal on a smaller number of traces can make the datum more readable, even if the available information is formally larger with a background removal on more traces (because the spatial filtering effect is reduced). Eventually, the quality of the result is optimal when we have a useful "combination" between the data and the exploited mathematical model (not between some mathematical model and itself).

Chapter 10

1. No. Migrations are linear algorithms derived under the Born approximation and do not account for the intrinsic nonlinearity of the scattering phenomenon.

2. No. They are derived under the Born approximation, but formally require the further conditions that the soil is lossless and the targets are not shallow.

3. It depends on whether the spectral weight is accounted for or not. In fact, the refraction at the air–soil interface is accounted for by means of the spectral weight.

4. Yes, because we can usually choose the number of traces involved in the migration algorithm (at least in the case of the Kirchhoff migration algorithm). This indirectly means to account for the maximum effective view angle, which is influenced, among other things, also by the radiation pattern of the antennas. Of course, this does not mean to account for the radiation pattern in a rigorous way.

5. No, because it depends on whether we reliably know the involved parameters—in particular, the effective length of the antennas.

6. The target at hand is a strong scattering object quite large with respect to the central wavelength in the soil (equal to 20 cm in the case at hand), and therefore a linear BA model cannot provide a precise model of the scattering in the case at hand, and so we cannot hope to retrieve the exact shape of such a big target with a standard migration algorithm, unless some a priori extra information is available or alternatively some nonlinear and/or iterative approach can be tried. That said, if the room is empty, the propagation velocity in it is 3 times larger than that in the surrounding soil. Since the migration has been supposed to account for the propagation velocity of the waves in the soil, the room is imaged compressed in the depth direction, and after time–depth conversion its thickness will appear to be of the order of 1 m instead of 3. More correctly, the distance between the top and the bottom of the reconstructed main anomaly appears of the order of 1 m instead of 3. Instead, if the room is filled with fresh water (whose relative dielectric permittivity is well known to be of the order of 81), then the propagation velocity in the room is 3 times smaller than that in the surrounding soil, and the room is imaged elongated. In particular, its top and bottom are imaged at the distance of 9 m (instead of 3) from each other.

Chapter 11

1. Yes it is. The proof might be performed as a straightforward extension of that exposed in Chapter 6 with regard to the 2D inverse scattering. In particular, a contrast varying as a sinusoid along all the three directions x, y, and z can be exploited. Also in this case, this would demonstrate that the problem is ill-posed but would not provide a full characterization of the nonretrievable targets.

2. No. In general the electric field (either incident, scattered, or total) has three components. However, at least for wire antennas parallel to the air–soil interface, the component of the incident field parallel to the transmitting antenna is usually the most important one.

3. The gauge of Lorenz applied in Chapter 5 was referred to the potential (Fitzgerald) vector relative to the electric scattered field in the case of secondary magnetic sources. In the 3D case dealt with in Chapter 10, the gauge of Lorentz is referred to a potential vector relative to the magnetic field.

4. In the scalar 2D case dealt with, we have only TM polarized waves—that is, waves where the electric field is directed along the axis of invariance and the magnetic field is orthogonal to it. So, in that case there is no reason to distinguish a TE from a TM reflection or transmission coefficient.

5. No, it isn't. In particular, the GPR datum is the same whatever (2D or 3D) processing we apply on it. In particular, the GPR gathers a scalar quantity in any observation point, given by the retrieved voltage, which is proportional to the scalar product of the electric field and of the effective length of the antenna.

Chapter 12

1. No, the ideal visible interval translates into the ideal visible circle, given by the rotation of the (symmetric) visible interval around the origin.

2. Rigorously not, essentially because the antenna pattern is not symmetrical around the broadcast (depth) direction. However, this dissymmetry can be neglected if an order of size for the resolution and the needed spatial step along x and y are looked for. In this case, the 3D retrievable spectral set is given by the 2D retrievable spectral set rotated around its symmetry axis.

3. No, the horizontal resolution is essentially the same, with a theoretical worsening of 22% in the 3D case [see Eqs. (12.30) and (9.44)].

4. Fortunately it is not indispensable, because the target might be relatively large and this makes somehow smoother the spatial behavior of the scattered field. In particular, as said, it is customarily quite easy to gather data at the Nyquist rate along the direction of the B-scan, but it is critical to do the same with respect to the transect. Therefore, as a matter of fact, along the transect direction we have in most cases a spatial undersampling. To gather B-scans along two orthogonal directions partially mitigates the effects of this undersampling, but not rigorously and not completely.

Chapter 13

1. No, and the physical reason has been explicitly shown in Section 12.3.

2. Yes, the calculations would be analogous to those exposed [see, in particular, Eq. (12.9)], but with different unit vectors, which would lead to the multiplication of two different columns of the matrixes of Eq. (12.9). The final result would be a formulation with the same retrievable spectral set but with a different spectral weight.

3. No, because in the case at hand two orthogonal B-scans would gather data in the same points but with the antennas oriented along two orthogonal directions: in particular, the scattered field also depends on the relative angle between the incident field and the buried target, that of course would change in the two cases.

4. Yes, but the array should be constituted by two couples of orthogonal dipoles, so that two orthogonal polarizations would be radiated and received at each observation point with a unique going-through.

Chapter 14

1. Yes, an SVD algorithm accounts for losses of the soil and near field effects. Moreover, the truncation of the investigation domain is accounted for without introducing further approximations.

2. No, it depends (among other things) on the reliability of our knowledge about the involved parameters, such as e.g., the losses and the characteristics of the antennas.

3. No, first of all there is a problem of RAM availability. In particular, commercial routine for the calculation of the SVD can fail if the RAM requirements are too high. Second, the computational burden implicates the necessity of the partition of the B-scan into several observation–inversion domains. This can cause some problems, described in detail in Chapter 15 (see in particular Section 15.7).

4. No, the SVD ideally provides a least square solution of the problem, which coincides with the exact one if and only if the problem admits an exact solution. With overdetermined linear problems, this happens very rarely.

5. No. The regularization parameter that we have at our disposal when performing a migration is the number of involved traces, which is related to the maximum view angle that we can exploit. This does not drive directly to a check of the investigated depth.

6. Yes, we can perform this "check" thanks to the spatial content of the upper-threshold singular functions.

REFERENCES

M. Abramowitz and I. Stegun (1972). *Handbook of Mathematical Functions with Formulas, Graphs, and Mathematical Tables*, Dover Publications, New York.

M. N. Afsar, J. R. Birch, R. N. Clarke, and G. W. Chantry (1986). The measurement of the properties of materials, *Proceedings of IEEE* **74**(1), 183–199.

G. Alberti, L. Ciofaniello, M. Della Noce, S. Esposito, G. Galiero, R. Persico, M. Sacchettino, and S. Vetrella (2002). A stepped frequency GPR system for underground prospecting, *Annals of Geophysics* **45**(2), 375–391.

S. A. Arcone, V. B. Spikes, and G. S. Hamilton (2005). Stratigraphic variation within polar firn caused by differential accumulation and ice flow: Interpretation of a 400 MHz short-pulse radar profile from West Antarctica, J. Glaciology **51**, 407–422.

C. A. Balanis (1989). *Advanced Engineering Electromagnetics*, John Wiley & Sons, New York.

M. Bertero and P. Boccacci (1998). *Introduction to Inverse Problems in Imaging*, Institute of Physics Publishing, Bristol and Philadelphia.

U. Böniger and J. Tronicke (2010). Symmetry based 3D GPR feature enhancement and extraction, in *Proceedings of the XIII International Conference on Ground Penetrating*, Lecce, Italy, June 21–25.

M. Born and E. Wolf (1999). Principles of Optics, 7th (expanded) edition, Cambridge University Press, New York.

M. C. Caggiani, M. Ciminale, D. Gallo, M. Noviello, and F. Salvemini (2012). Online non destructive archaeology: The Archaeological Park of Egnazia (Southern Italy) study case, *Journal of Archaeological Science* **39**, 67–75.

S. Caorsi, G. L. Gragnani, S. Medicina, M. Pastorino, and G. Zunino (1991). Microwave imaging method using a simulated annealing approach, *IEEE Microwave and Guided Wave Letters*, **MGWL-1** (11), 331–333.

E. Cardarelli, C. Marrone, and L. Orlando (2003). Evaluation of tunnel stability using integrated geophysical methods, *Journal of Applied Geophysics* **52**(2–3), 93–102.

A. Cataldo, E. De Benedetto, and G. Cannazza (2011). *Broadband Reflectometry for Enhanced Diagnostics and Monitoring Applications*, Springer-Verlag, New York.

I. Catapano, L. Crocco, R. Persico, M. Pieraccini, and F. Soldovieri (2006). Linear and non-linear microwave tomography approaches for subsurface prospecting: Validation on real data, *IEEE Transactions. on Antennas and Wireless Propagation Letters* **5**, 49–53.

Introduction to Ground Penetrating Radar: Inverse Scattering and Data Processing,
First Edition. Raffaele Persico.
© 2014 The Institute of Electrical and Electronics Engineers, Inc. Published 2014 by John Wiley & Sons, Inc.

W. C. Chew (1995). *Waves and Fields in Inhomogeneous Media*, Institute of Electrical and Electronics Engineers, Piscataway, NJ.

R. J. Chignell (2004). The radio licensing of GPR systems in Europe, *Proceedings of 10th International Conference on Ground Penetrating Radar*, Delft, the Netherlands.

P. C. Clemmow (1996). *The Plane Wave Spectrum Representation of Electromagnetic Fields*, Wiley-IEEE Press, Now York.

R. E. Collin (1985). *Antennas and Radiowave Propagation*, McGraw-Hill, New York.

D. Colton and R. Kress (1992). *Inverse Acoustic and Electromagnetic Scattering Theory*, Springer Verlag, New York.

D. Colton, H. Haddar, and M. Piana (2003). The linear sampling method in inverse electromagnetic scattering theory, *Inverse Problems* **19**(6), S105–S137.

L. B. Conyers (2004). Ground Penetrating Radar for Archaeology, AltaMira Press, Lanham, MD.

J. C. Cook (1975). Radar transparencies of mine and tunnel rocks, *Geophysics* **40**, 865–885.

T. J. Cui and W. C. Chew (2002). Diffraction Tomographic algorithm for the detection of three-dimensional objects buried in a lossy half space, *IEEE Trans. on Antennas and Propagation* **50**(1), 42–49, January 2002.

D. J. Daniels (2004). *Ground Penetrating Radar*, 2nd edition, IEE Press, Luxembourg.

J. L. Davis and A. P. Annan (1989). Ground-penetrating radar for high resolution mapping of soil and rock stratigraphy, *Geophysical Prospecting* **37**(5), 531–551.

R. Deming and A. J. Devaney (1996). A filtered backpropagation algorithm for GPR, *Journal of Environmental & Engineering Geophysics* **0**(2).

A. J. Devaney (1981). Inverse-scattering theory within the Rytov approximation, *Optics Letters* **6**(8), 374–376.

M. D' Urso, I. Catapano, L. Crocco, and T. Isernia (2007). Effective solution of 3D scattering problems via series expansions: Applicability and a new hybrid scheme, *IEEE Transactions on Geoscience and Remote Sensing* **45**, 639–648.

S. Ebihara, M. Sato, and H. Niitsuma (2000). Super-resolution of coherent targets by a directional borehole radar, *IEEE Transactions on Geoscience and Remote Sensing* **38**(4), 1725–1732.

N. Engheta, C.H. Papas, and C. Elachi (1982). Radiation pattern of interfacial dipole antennas, *Radio Science* **17**(6), 1557–1566.

L. B. Felsen and N. Marcuvitz (1994). *Radiation and Scattering of Waves*, Wiley-IEEE, New York.

G. Franceschetti. *Electromagnetics: Theory, Techniques and Engineering Paradigms*, Plenum Press, New York.

J. Francke (2010). Applications of GPR in mineral resource evaluations, in *Proceedings of the XIII International Conference on Ground Penetrating*, Lecce, Italy.

S. P. Friedman (2005). Soil properties influencing apparent electrical conductivity: A review, *Computers and Electronics in Agriculture* **46**(1–3), 45–70.

J. Gazdag and P. Squazzero (1984). Migration of seismic data, *Proceedings of IEEE* **72**, 1302–1315.

G. Gentili and U. Spagnolini (2000). Electromagnetic inversion in monostatic ground penetrating radar: TEM horn calibration and application, *IEEE Transactions on Geoscience and Remote Sensing* **38**(4), 1936–1946.

A. Giannopoulos (2003). GprMax2D V 1.5 (Electromagnetic simulator for Ground Probing Radar, the software is available at www.gprmax.org).

G. H. Golub and C. F. van Loan (1996). *Matrix Computations*, 3rd edition, Johns Hopkins University Press Baltimore.

D. Goodman and S. Piro (2013). *GPR Remote Sensing in Archaeology*, Springer-Verlag, New York.

G. Grandjean, J. C. Gourry, and A. Bitri (2000). Evaluation of GPR techniques for civil-engineering applications: Study on a test site, *Journal of Applied Geophysics*, **45**(3), 141–156.

F. Grasso, G. Leucci, N. Masini, and R. Persico (2011). GPR Prospecting in Renaissance and Baroque Monuments in Lecce (Southern Italy), *Proceedings of VI International Workshop on Ground Penetrating Radar*, IWAGPR, Aachen, Germany.

M. Grodner (2001). Delineation of rockburst fractures with ground penetrating radar in the Witwatersrand Basin, South Africa, *International Journal of Applied Geophysics* **38**(6), 885–891.

J. Groenenboom and A. Yarovoy (2002). Data processing and imaging in GPR system dedicated for landmine detection, *Subsurface Sensing Technologies and Applications* **3**(4), 437–452.

L. Gurel and U. Oguz (2003). Optimization of the transmitter–receiver separation in the ground penetrating radar, *IEEE Transactions on Antennas and Propagation* **51**(3), 362–370.

J. Hadamard (1923). *Lectures on Cauchy's Problem in Linear Partial Differential Equations*, Yale University Press,. New Haven.

W. S. Hammon, G. A. McMechan, and X. Zeng (2000). Forensic GPR: Finite-difference simulations of responses from buried human remains, *Journal of Applied Geophysics* **45**(3), 171–186.

R. F. Harrington (1961). *Time-Harmonic Electromagnetic Fields*, McGraw-Hill, New York.

K. Heidary (2003). Ultra-wideband radiation and scattering from a thin filament, in *Proceedings. Antennas and Propagation Society International Symposium*, June, pp. 551–554.

J. R. Higgins, Sampling Theory in Fourier and Signal Analysis, Oxford University Press, New York, 1996.

J. Hugenschmidt, M. N. Partl, and H. de Witte (1998). GPR inspection of a mountain motorway in Switzerland, *Journal of Applied Geophysics* **40**, 95–104.

M. Idemen and I. Ackduman (1990). Two dimensional inverse scattering problems connected with bodies buried in a slab, *Inverse Problems* **6**, 749–766.

K. Iizuka, A. P. Freundorfer K. H. Wu, H. Mori, H. Ogura, and V-K. Nguyen (1984). Step frequency radar, *Journal of Applied Physics* **56**(9), 2572–2583.

J. D. Jackson (1998). *Classical Electrodynamics*, 3 edition, John Wiley Sons, New York.

H. Jol (2009). *Ground Penetrating Radar: Theory and Applications*, Elsevier, New York.

A. Kirsh (1996). An Introduction to the Mathematical Theory of Inverse Problems, Springer-Verlag, New York.

K. S. Kunz and R. J. Luebbers (1993). *The Finite Difference Time Domain Method for Electromagnetics*, CRC Press, Boca Raton, FL.

S. Lambot, E. C. Slob, I. van den Bosch, B. Stockbroeckx, and M. Vanclooster (2004). Modeling of ground-penetrating radar for accurate characterization of subsurface electric properties, *IEEE Transactions on Geoscience and Remote Sensing* **42**(11), 2555–2568.

H. J. Landau and H. O. Pollak (1961). Prolate spheroidal wave functions, Fourier analysis and uncertainty—II, *The Bell System Technical Journal* **40**(1), 65–84.

H. J. Landau and H. O. Pollak (1962). Prolate spheroidal wave functions, Fourier analysis and uncertainty—III, *The Bell System Technical Journal* **41**(4), 1295–1336.

G. Leone, R. Persico, and R. Solimene (2003). A quadratic model for electromagnetic subsurface prospecting, *AEÜ, International Journal of Electronics and Communications* **57**(1), 33–46.

D. Lesselier and B. Duchene (1996). Wavefield inversion of objects in stratified environments: From back-propagation schemes to full solutions, in R. Stone, ed., *Review of Radio Science 1993–1996*, pp. 235–268, Oxford University Press, New York.

A. A. Lestari, A. G. Yarovoy, and L. P. Ligthart (2004). Numerical and experimental analysis of circular-end wire bow-tie antennas lossy a ground, *IEEE Transactions on Antennas and Propagation* **52**, 26–35.

G. Leucci (2008). Ground penetrating radar: The electromagnetic signal attenuation and maximum penetration depth, Scholarly Research Exchange, Article ID 926091, doi:10.3814/2008/926091.

N. Levanon (1988). *Radar Principles*, John Wiley & Sons, New York.

W. Li, H. Zhou, and X. Wan (2012). Generalized Hough transform and ANN for subsurface cylindrical object location and parameters inversion from GPR data, in *Proceedings of 14th International Conference on Ground Penetrating Radar*, pp. 285–289, Shanghai, China, June.

A. Liseno, F. Tartaglione, and F. Soldovieri (2004). Shape reconstruction of 2d buried objects under a Kirchhoff approximation, *IEEE Geoscience and Remote Sensing Letters* **1**(2), 118–121.

L. Lo Monte, D. Erricolo, F. Soldovieri, and M. C. Wicks (2010). Radio frequency tomography for tunnel detection, *IEEE Transactions on Geoscience and Remote Sensing* **48**(3), 1128–1137.

N. Masini, R. Persico, E. Rizzo, A. Calia, M. T. Giannotta, G. Quarta, and A. Pagliuca (2010). Integrated techniques for analysis and monitoring of historical monuments: the case of San Giovanni al Sepolcro in Brindisi, southern Italy, *Near Surface Geophysics* **8**(5), 423–432.

P. Meincke (2001). Linear GPR inversion for lossy soil and a planar air–soil interface, *IEEE Transactions on Geoscience and Remote Sensing* **39**(12), 2713–2721.

P. Meincke and T. B. Hansen (2004). Plane wave characterization of antennas close to a plane interface, *IEEE Transactions on Geoscience and Remote Sensing* **42**(6), 1222–1232.

M. Moghaddam and W. C. Chew (1992). Nonlinear two-dimensional velocity profile inversion using time domain data, *IEEE Transactions on Geoscience and Remote Sensing* **30**(1), 147–156.

M. Nabighian (1987). *Electromagnetic Methods in Applied Geophysics—Theory*, Vol. **1**, Society of Exploration Geophysics, Tulsa, OK.

D. A. Noon (1996). Stepped-Frequency Radar Design and Signal Processing Enhances Ground Penetrating Radar Performance, Ph.D. thesis, Department of Electrical & Computer Engineering, University of Queensland, Australia.

K. M. O'Connor and C. H. Dowding (1999). *Geomeasurements by Pulsing TDR Cables and Probes*, CRC Press, Boca Raton, FL.

C. P. Oden, M. H. Powers, D. L. Wright, and G. R. Olhoeft (2007). Improving GPR Image resolution in lossy ground using dispersive migration, *IEEE Transactions on Geoscience and Remote Sensing* **45**(8), 2492–2500.

G. P. Otto and W. C. Chew (1994). Microwave inverse scattering—Local shape function imaging for improved resolution of strong scatterers, *IEEE on Microwave Theory and Techniques* **42**(1), 137–141.

M. Pastorino (2010). Microwave Imaging, John Wiley & Sons, Hoboken, N J.

R. Persico (2006). On the role of measurement configuration in contactless GPR data processing by means of linear inverse scattering, *IEEE Transactions on Antennas and Propagation* **54**(7), 2062–2071.

R. Persico and G. Prisco (2008). A reconfigurative approach for SF-GPR prospecting, *IEEE Transactions on Antennas and Propagation* **56**(8), 2673–2680.

R. Persico and J. Sala (2011). Some possibilities and problems related to 2D linear inverse scattering algorithms on large scale GPR data, in *Proceedings of 6th International Workshop on advanced Ground penetrating Radar* (*IWAGPR*), Aachen, Germany, June.

R. Persico and J. Sala (2014). Single and double sequence of investigation. domains in 2D linear inversions applied to GPR data, *IEEE Geosciences and Remote Sensing Letters*. DOI 10.1109/LGRS.2013.220008.

R. Persico and F. Soldovieri (2004). One-dimensional inverse scattering with a Born model in a three-layered medium, *Journal of Optical Society of America Part A* **21**(1), 35–45.

R. Persico and F. Soldovieri (2006). A microwave tomography approach for a differential configuration in GPR prospecting, *IEEE Transactions on Antennas and Propagation* **54**(11), 3541–3548.

R. Persico and F. Soldovieri (2008). Effects of the background removal in linear inverse scattering, *IEEE Transactions on Geoscience and Remote Sensing* **46**(4), 1104–1114.

R. Persico and F. Soldovieri (2010). Recent issues relevant to GPR prospecting, invited paper, *Proceedings of EMTS, 2010*, Berlin Germany.

R. Persico and F. Soldovieri (2011). Two dimensional linear inverse scattering for dielectric and magnetic anomalies, *Near Surface Geophysics* **9**(3), 287–295.

R. Persico, F. Soldovieri, and R. Pierri (2002). On the convergence properties of a quadratic approach to the inverse scattering problem, *Journal of Optical Society of America Part A*, **19**(12), 2424–2428.

R. Persico, R. Bernini, and F. Soldovieri (2005). On the configuration of the measurements in inverse scattering from buried objects under the distorted Born approximation, *IEEE Transactions on Antennas and Propagation* **53**(6), 1875–1886.

R. Persico, F. Soldovieri, and E. Utsi (2010). Microwave tomography for processing of GPR data at Ballachulish, *Journal of Geophysics and Engineering*, **7**(2), 164.

R. Persico N. Romano, and F. Soldovieri (2011). Design of a balun for a bow tie antenna in reconfigurable ground penetrating radar systems, *Progress In Electromagnetics Research C* **18**, 123–135.

R. Persico, G. Leucci, L. Matera, M. Ciminale, D. Dei, F. Parrini, and M. Pieraccini (2013a). Applications of a reconfigurable stepped frequency GPR in the Chapel of the Holy Spirit, Lecce (Italy), in *Proceedings of 7th International Workshop on Advanced Ground Penetrating Radar* (*IWAGPR*), Nantes (France), July.

R.Persico, G. Leucci, and F. Soldovieri (2013b). The effect of the height of the GPR antennas on the diffraction curve, in *Proceedings of 7th International Workshop on Advanced Ground Penetrating Radar* (*IWAGPR*), Nantes (France), July.

E. Pettinelli, A. Di Matteo, E. Mattei, L. Crocco, F. Soldovieri, D. J. Redman, and A. P. Annan (2009). GPR response from buried pipes: Measurement on field site and tomographic reconstructions, *IEEE Transactions on Geosciences and Remote Sensing* **47**(8), 2639–2645.

G. Picardi, J. J. Plaut, D. Biccari, OrnellaBombaci, D. Calabrese, M. Cartacci, A. Cicchetti, S. M. Clifford, P. Edenhofer, W. M. Farrell, C. Federico, A. Frigeri, D. A. Gurnett, T. Hagfors, E Heggy, A. Herique, R. L. Huff, A. B. Ivanov, W. T. K. Johnson, R. L. Jordan, D. L. Kirchner, W. Kofman, C. J. Leuschen, E. Nielsen, R. Orosei, E. Pettinelli, R. J. Phillips, D. Plettemeier, A. Safaeinili, R. Seu, E. R. Stofan, G. Vannaroni, T. R. Watters, and E. Zampolini (2005). Radar soundings of the subsurface of Mars, *Science* **310**, 1925–1928.

M. Pieraccini, L. Noferini, D. Mecatti, C. Atzeni, R. Persico, and F. Soldovieri (2006). Advanced processing techniques for step-frequency continuous-wave penetrating radar: The case study of "Palazzo Vecchio" walls (Firenze, Italy), *Research on Nondestructive Evaluation* **17**, 71–83.

W. H. Press, B. P. Flannery, S. A. Teukolsky, and W. T. Vetterling (1987). *Numerical Recipes*, Cambridge University Press, Cambridge, UK.

J. C. Ralston and D. W. Hainsworth (2000). Use of ground penetrating radar in underground coal mining, in *Proceedings of the VIII International Conference on Ground Penetrating*, pp. 731–736.

L. Robinson, W. B. Weirand, and L. Yung (1974). Location and recognition of discontinuities in dielectric media using synthetic RF pulses, *Proceedings of the IEEE* **62**(1), 36–44.

J. Sala and N. Linford (2010). Processing stepped frequency continuous wave GPR system to obtain maximum value from archaeological data sets, in *Proceedings of the XIII International Conference on Ground Penetrating*, Lecce, Italy, June 21–25.

L. Sambuelli, G. Bohm, P. Capizzi, E. Cardarelli, and P. Cosentino (2011). Comparison between GPR measurements and ultrasonic tomography with different inversion algorithms: an application to the base o fan ancient Egyptian sculpture, *Journal of Geophysics and Engineering* **8**(3), 106–116.

K. J. Sandmeier (2003). Reflexw 3.0 manual, Sandmeier Software, ZipserStrabe1 D-76227, Karlsruhe, Germany.

M. Sato and K. Takahashi (2009). Development of dual sensors and deployment in mine affected countries, in K. Furuta and J. Ishikawa, eds., *Anti-personnel Landmine Detection for Humanitarian Demining*, pp. 27–44, Springer, London.

W. A. Schneider (1978). Integral formulation for migration in two and three dimensions, *Geophysics* **43**(1), 49–76.

R. E. Sheriff (1980). Nomogram for Fresnel-zone calculation, *Geophysics* **45**(5), 968–972.

M. Slaney, A. C. Kak, and L. E. Larsen (1984). Limitations of imaging with first order diffraction tomography, *IEEE Trans. Microwave Theory and Techniques* **MTT-32**, 860–874.

D. Slepian and H. O. Pollak (1961). Prolate spheroidal wave functions, Fourier analysis and uncertainty—I, *The Bell System Technical Journal*, **40**(1), 43–63.

E. Slob, M. Sato, and G. Olhoeft (2010). Surface and borehole ground-penetrating-radar developments, *Geophysics*, **75**(5), 75A103-75A120.

D. G. Smith and H. M. Jol (1997). Radar structure of a Gilbert-type delta, Peyto Lake, Banff National Park, Canada, *Sedimentary Geology* **113**(3–4), 195–209.

F. Soldovieri, R. Persico, and G. Leone (2005a). Frequency diversity in a linear inversion algorithm for GPR prospecting *Subsurface Sensing Technology and Applications* **6**(1), 25–42.

F. Soldovieri, R. Persico, and G. Leone (2005b). Effect of source and receiver radiation characteristics in subsurface prospecting within the DBA, *Radio Science* **40**, RS3006.

F. Soldovieri, J. Hugenschmidt, R. Persico, and G. Leone (2007). A linear inverse scattering algorithm for realistic GPR applications, *Near Surface Geophysics* **5**(1), 29–42.

F. Soldovieri, G. Prisco, and R. Persico (2009). A strategy for the determination of the dielectric permittivity of a lossy soil exploiting GPR surface measurements and a cooperative target, *Journal of Applied Geophysics* **67**(4), 288–295.

M. Solla, C. Caamano, B. Riveiro, and H. Lorenzo (2011). GPR analysis of a masonry arch for structural assessment, in *Proceedings of sixth International Workshop on Advanced Ground Penetrating Radar IWAGPR*. Aachen, Germany.

A. Sommerfeld (1912). Die Greensche Funktion der Schwingungsgleichung, *Jber. Deutsch. Math. Verein.* **21**, 309–353.

G. F. Stickley, D. A. Noon, M. Cherniakov, and I. D. Longstaff (1999). Gated stepped-frequency ground penetrating radar, *Journal of Applied Geophysics* **43**(2), 259–269.

D. E. Stillman and G. R. Olhoeft (2004). GPR and magnetic minerals at Mars temperatures, in *Tenth International Conference on Ground Penetrating Radar*, pp. 735–738.

R. H. Stolt (1978). Migration by Fourier transform, *Geophysics* **43**(1), 23–48.

W. L. Stutzman and G. A. Thiele (1998). Antenna Theory and Design, 2nd edition, John Wiley & Sons, New York.

W. Tabbra, B. Duchene, Ch. Pichot, D. Lesselier, L. Chommeloux, and N. Joachimowicz (1988). Diffraction tomography: Contribution to the analysis of some applications in microwave and ultrasonics, *Inverse problems* **4**, 305–331.

C. T. Tai (1991). Generalized Vector and Dyadic Analysis, IEEE Press, Piscataway, NJ.

C. T. Tai (1994). *Dyadic Green Functions in Electromagnetic Theory*, 2nd edition, IEEE Press Series on Electromagnetic Waves, New York.

A. N. Tikhonov and V. Y. Arsenine (1977). *Solution of Ill-Posed Problems*, Winston, Washington, DC.

G. S. Timofei and M. Sato (2004). Comparative analysis of UWB deconvolution and feature-extraction algorithms for GPR landmine detection, in R. S. Harmon, J. T. Broach, and J.H. Holloway, Jr., eds., *Proceedings of the SPIE, Detection and Remediation Technologies for Mines and Minelike Targets IX*, Vol. 5415, pp. 1008–1018.

G. C. Topp, J. L. Davis, and A. P. Annan (1980). Electromagnetic determination of soil water content: Measurements in coaxial transmission lines, *Water Resources Research* **16**(3), 574–582.

C. Torres-Verdín and T. M. Habashy (2001). Rapid numerical simulation of axisymmetric single-well induction data using the extended Born approximation, *Radio Science* **36**(6), 1287–1306.

V. Utsi and E. Utsi (2004). Measurement of reinforcement bar depths and diameters in concrete, in E. Slob, A. Yarovoy, and Rhebergen J, eds., *Proceedings of the Tenth International Conference on Ground Penetrating Radar*, pp. 658–662, Delft, the Netherlands.

J. Van Bladel (2007). Electromagnetic Fields, 2nd edition, IEEE Press, Hoboken, NJ.

G. Villain, X. Dérobert, Z. M. Sbartaï, and J. P. Balayssac (2010). Evaluation of concrete water content and other durability indicators by electromagnetic measurements, in *Proceedings of XIII International Conference on Ground Penetrating Radar*, pp. 165–170, Lecce, Italy.

A. Witten, A. Schatzenberg, and A. Devaney (1996). Vector radar diffraction tomography maximum likelihood estimation, *Journal of Environmental & Engineering Geophysics* **1**(B), 91–104.

R. Yelf (2004). Where is the true zero time?, in *Proceedings of Tenth International Conference on Ground Penetrating Radar*, pp. 279–282.

O. Yilmaz (1987). Seismic Data Processing, B. Edwin Neitzel, ed., Society of Exploration Geophysicists, p. 339, Tulsa, OK.

INDEX

Introduction to Ground Penetrating Radar: Inverse Scattering and Data Processing,
First Edition. Raffaele Persico.
© 2014 The Institute of Electrical and Electronics Engineers, Inc. Published 2014 by John Wiley & Sons, Inc.